Green Energy and Technology

Climate change, environmental impact and the limited natural resources urge scientific research and novel technical solutions. The monograph series Green Energy and Technology serves as a publishing platform for scientific and technological approaches to "green"—i.e. environmentally friendly and sustainable—technologies. While a focus lies on energy and power supply, it also covers "green" solutions in industrial engineering and engineering design. Green Energy and Technology addresses researchers, advanced students, technical consultants as well as decision makers in industries and politics. Hence, the level of presentation spans from instructional to highly technical.

Indexed in Scopus.
Indexed in Ei Compendex.

Consuelo Nava · Aurora Angela Pisano ·
Giuseppe Mangano · Francesca Giglio
Editors

Climatic and Structural Safety in Multi-Hazard Regime of Cultural and Natural Heritage

Methodological Advances and Case Study Applications

Springer

Editors
Consuelo Nava
Dipartimento Architettura e Design
Università degli Studi Mediterranea di Reggio Calabria
Reggio Calabria, Italy

Aurora Angela Pisano
Dipartimento Architettura e Design
Università degli Studi Mediterranea di Reggio Calabria
Reggio Calabria, Italy

Giuseppe Mangano
Dipartimento Architettura e Design
Università degli Studi Mediterranea di Reggio Calabria
Reggio Calabria, Italy

Francesca Giglio
Dipartimento Architettura e Design
Università degli Studi Mediterranea di Reggio Calabria
Reggio Calabria, Italy

ISSN 1865-3529 ISSN 1865-3537 (electronic)
Green Energy and Technology
ISBN 978-3-032-11925-4 ISBN 978-3-032-11926-1 (eBook)
https://doi.org/10.1007/978-3-032-11926-1

This work was supported by Università degli Studi Mediterranea di Reggio Calabria.

This Springer imprint is published by the registered company Springer Nature Switzerland AG
The registered company address is: Gewerbestrasse 11, 6330 Cham, Switzerland

Preface

The protection of a Country's cultural heritage, both architectural and environmental, is a highly relevant political and scientific concern, given its profound implications for the social and cultural fabric of societies.

Safety assessment requires a comprehensive understanding of the various sources of risk (climatic, geomorphological, structural, etc.), demanding interdisciplinary approach to the problem. The topics involved are of considerable interest for the scientific community and constitute a knowledge base for any planning strategy, both for preventive actions (maintenance, building and site safety) and reactive interventions (emergency management, evacuation plans).

This book addresses the issue of climatic and structural safety in multi-hazard regime of cultural and natural heritage, drawing on the outcomes of a three-year technology transfer project, funded by the Italian National Recovery and Resilience Plan -PNRR T4Y PP 4.7.1. In particular, the project is entitled Open Platform *phigital space* (physical and digital) of the type *user-profiling* for the advanced and dynamic codesign of interventions on the built and ex novo.

The book is organized into two main parts collecting peer-reviewed papers from approximately twenty involved researchers and begins with a chapter detailing the methodologies adopted and the final goals of the project.

The first part of the book aims to deepen theoretical knowledge related to cultural heritage analysis, proposing advanced models for damage assessment on structures and their surrounding environments. These models also incorporate uncertainties related to material properties and consider different environmental stressors under three projected climate scenarios for the years 2030, 2050, and 2085.

The second part of the book features some chapters with a more applied focus. Thanks to some large-scale case studies, it proposes a digital platform to support all the meta-files relating to the security data collected. This platform, along with the creation of a Living Lab (physical platform), will provide a methodology for the transfer of digitalized data to different stakeholders: including institutions, professionals, and local communities.

The Editors sincerely thank all the senior and the young researches who have contributed by their outstanding contributions to the quality of this book.

October 2025

Consuelo Nava
Aurora Angela Pisano
Giuseppe Mangano
Francesca Giglio

Acknowledgements

Finanziato dall'Unione europea
NextGenerationEU

Italiadomani
PIANO NAZIONALE DI RIPRESA E RESILIENZA

MUR
Ministero dell'Università e della Ricerca

This volume is the result of the synergistic activities related to the following National Recovery and Resilience Plan (Mission 4, Component 2 Investment 1.4) Research Project, funded by NextGenerationEU: Tech4You "Technologies for climate change adaptation and quality of life improvement" – "Pilot Project 4.7.1-Open platform 'phygital space' (physical and digital) of the type user profiling for the advanced and dynamic codesign of interventions on the built and ex novo" (Spoke 4 - Technologies for resilient and accessible cultural and natural heritage - Goal 4.7- Protection and enhancement of natural and cultural heritage and identity of the inner territories) - CUP C33C22000290006 (ECS_00000009).

The authors sincerely thank the project that funded their research activities and the open access publication of this volume.

Contents

Theoretical and Methodological Approaches

The Exploitation Research, the Actions of the Laboratories, the Pilot

Consuelo Nava(✉)

ABITAlab, dAeD, Mediterranea University of Reggio Calabria, Reggio Calabria 89124, Italy
cnava@unirc.it

Abstract. The scientific document presents the PNRR project – T4Y PP 4.7.1 entitled "Open platform "phigital space" (physical and digital) of the type "user profiling" (Call for tender for the presentation of proposals for the Strengthening of research structures and creation of R&D "innovation ecosystems", set up of "territorial leaders in R&D"—to be funded under the National Recovery and Resilience Plan (NRRP), Mission 4, Component 2 Investment 1.4, funded from the European Union - NextGenerationEU. Proposal: Tech4You - Technologies for climate change adaptation and quality of life im-provement. Spoke 4 - Technologies for resilient and accessible cultural and natural herit-age: Goal 4.7: Protection and enhancement of natural and cultural heritage and identity of the inner territories.Pilot Project 4.7.1 - Open platform "phigital space" (physical and digi-tal) of the type "user profiling" for the advanced and dynamic codesign of interventions on the built and ex novo.) for the advanced and dynamic codesign of interventions on the built and ex novo, carried out by the scientists and researchers of the dAeD and DICEAM departments of the Università Mediterranea degli Studi of Reggio Calabria. The frontier themes of information management and open knowledge for structural and environmental safety on cultural heritage contribute to the processes of "radical innovation" of the mission – spoke 4. The "mechanisms of the just transition", to which respond through the adoption of KET's, address the material and immaterial (physical and digital) management of the "data-information-resources" system, towards the production of models, methods, prototypes realized in "dynamic" regimes, such as "uncertain parameters" and "limit states" of the analysis and "predictive models of advanced design" for the adaptability to climate scenarios. The pilot project realizes a system-demonstrator, with the open platform "phigital space" of the type "user profiling" for the advanced and dynamic codesign of interventions on the built and ex novo. The experimentation and validation on n.2 case studies in n.2 historical identity settlements in the most fragile inland areas of Calabria (depopulation). The applied experimental activity, illustrated by the contributions of Session II, therefore has a digital information level that realises the transfer of information on a "user profile digital platform" and a physical information level with the creation of the Living Lab (physical platform) for the transfer to users (institutional, professional, community). This scenario realises what is called "open knowledge" in the PP objective.

Keywords: Climate change scenario · Living Lab · Agile Method · Digital Platform · Exploitation · Interdisciplinarity · Safety · Phigital · Open Knowledge · Frontier Research

C. Nava et al. (Eds.): *Climatic and Structural Safety in Multi-Hazard Regime of Cultural and Natural Heritage*, GREEN, pp. 3–28, 2026.
https://doi.org/10.1007/978-3-032-11926-1_1

1 Introduction

1.1 Theoretical Background: The Exploitation Research [1]

The proposal Tech4You – Technologies for climate change adaptation and quality of life improvement, "(…) main concept is "climate affects you" and "you act on the climate" towards a sustainable and competitive Europe. Our R&I programme will enhance the scientific expertise of researchers and bring R&D results into use to build a resilient community and to strengthen SME competitiveness reducing the gap between research, market, and society. The effects of climate change require a multi-stakeholder involvement (research, businesses, society and government) in a joint action towards resilient communities. The scientific data show the global impact of climate change: Southern and central Europe are seeing more frequent heat waves, forest fires, floods and drought. (…) Tech4You recognises both the urgency and the last chance to take an action. Tech4You will inspire a new generation of researchers showing that using knowledge to develop innovative solutions is the response to the "last call" for a green transition towards a sustainable, inclusive and fair future. Tech4You will address climate change from 5 perspectives: natural risk, energy, food, culture, and healthy communities talking about the specific objectives (SO) described below:

(…)

SO4 – Adaptation of cultural heritage: Tech4You aims to mitigate the impact of extreme weather events on cultural heritage and to improve the accessibility for all leveraging on digital technologies (i.e.: robots, sensors, artificial intelligence, big data and augmented reality) (Spoke4); it wants to reinforce the local identity leveraging on the enhancement of the tangible and intangible cultural heritage and, on the other side, wants to enable the accessibility for all without any restriction for older people or disabled.

(…) The spokes 1-5 will be focused on the implementation of the research and innovation activities that are all addressed to bring the most promising technologies already developed by the research organizations to a higher TRL as close as possible to a marketable stage so that they could be quickly transferred to companies for being industrialized and commercialized. R&I activities will be strongly addressed to have pilot demonstration projects in which the involvement of the final users will be fundamental for the validation. The use of cascade funding will be also very important to expand the ecosystem by including a wide range of companies (particularly the many start-ups and spin-offs that are interested in the Tech4You topics).[2]

The T4Y proposal therefore presents itself as an innovative exploitation research project, aimed at improving the transfer of the most mature research carried out by teams of researchers, with social, economic and knowledge benefits for the regions involved (Calabria and Basilicata).

---

[1] This section of the paper draws on the studies by C. Nava, presented in the text 'Emerging Technologies for Regenerative Design'. 5 theoretical questions on innovation and sustainability in frontier research (2023, edited by Aracne), in order to scientifically place the PP4.7.1 research proposal within the context of innovation and frontier research.

[2] The text outlines some key points of the T4Y proposal taken from the document. *Annex 1 – Project proposal (Article 10, paragraph 3 and Article 12 of the Call)*

In this direction, the theoretical framework of pilot project 4.7.1 "Open platform "phigital space" (physical and digital) of the type "user profiling", which is maturing through the work of university laboratories and living labs in the field, can be traced back to frontier research, with its applied innovation aimed at advancing knowledge by "applying known solutions to investigated problems", a path facilitated by the interdisciplinary skills of the research team.

Below are some references to theoretical studies that the author has examined in order to define the framework of "frontier research" in the field of "innovative knowledge processes", in the case of "exploitation research" (Nava, 2023).

For a Definition and Classification of Frontier Research

Pasteur's quadrant (Stokes, 1997, 2011) refers to the relationship between the relevance of advances in knowledge and their immediate application (Fig. 1), identifying the different operating conditions of frontier research. Experiencing these paths means questioning the conventional theory-method-project process, not so much in its necessary cognitive distinction, but in its operating condition and ability to report on the relationship between "pure applied research" and "pure basic research", through the experimental phase of concept and inspiration, which also requires application. This phase becomes fundamental for all trajectories related to technology and technologies, referring to transformations in the built environment and innovation in processes and products, through new performative conditions of the project.

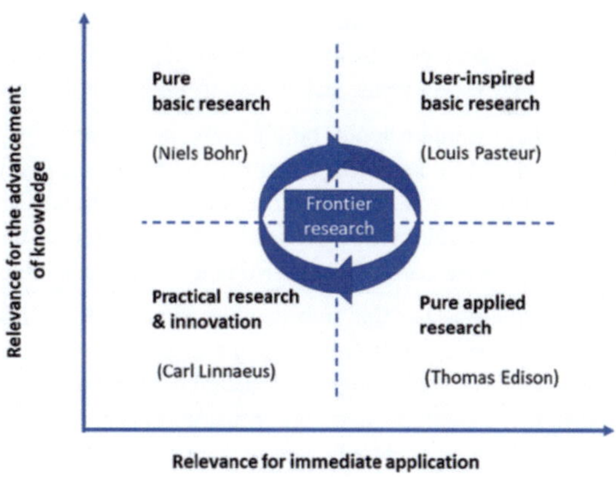

Fig. 1. Type of research in relation to the advancement of knowledge. Adapted from Stokes, (2011).

D. Holste, I. Roche, M. Hörlesberger, Marianne, D. Besagni, T. Scherngell, F. Claire, P. Cuxac, Pascal, E. Schiebel (2011), in their scientific paper "A Concept for Inferring "Frontier Research" in Research Project Proposals, investigate methods for recognising "frontier research" in research project proposals. This document discusses the issues sought in research project proposals submitted for peer review under the European Research Council (ERC) programme. The investigation has two aims: 1) to conceptualise and define, automatically extract and comparatively evaluate the attributes of frontier research in proposals; and 2) to construct and compare the results of a statistical model with the review decision, in order to gain further insights and reflect on the influence of frontier research in the peer review process. To this end, indicators (including scientific "novelty", "risk" and interdisciplinarity – see Table 1) were developed for all scientific disciplines in accordance with the European Council's strategic definition of frontier research, using textual information from the proposal and other bibliometric data from applicants for research funding.

Table 1. Relationship between the definition of frontier research and corresponding indicators. (Source: definition: EC (2005); indicator: own data)

Indicators	Taxonomy (of frontier research)
Novelty	"… is at the forefront of creating new knowledge and developing new insights. The people involved are responsible for fundamental discoveries and advances in theoretical and empirical understanding, occasionally achieving a breakthrough discovery that completely changes our knowledge of the world.".
Risk	"…is an inherently risky endeavour. In the newest and most exciting areas of research, the approach or trajectory that may prove the most fruitful for the development of the field is often unclear. Researchers must be bold and take risks. Indeed, only researchers are generally able to identify the most promising opportunities. The task of funding agencies is limited to supporting the best researchers with the most interesting ideas, rather than trying to identify priorities.".
Pasteuresqueness	"… may be interested in both gaining new knowledge about the world and generating knowledge that is potentially useful at the same time. Therefore, there is a much closer and more intimate link between science and the technology that derives from it, with few of the barriers that arise when basic research and applied research are conducted separately.".
Interdisciplinarity	'… pursues issues that transcend established disciplinary boundaries. It may involve multidisciplinary, interdisciplinary or transdisciplinary research that brings together researchers from different disciplines, with different theoretical and conceptual approaches, different techniques, methodologies and tools, and perhaps even different objectives and motivations.'

Kim's Matrix for Knowledge Innovation

For all experiences (project or research) involving disruptive innovation and therefore solutions based on emerging technologies, the "KIM matrix" is used to assess the transfer impact on external socio-technical contexts. This knowledge innovation matrix is divided into four quadrants: (1) Invention, (2) Improvement, (3) Adaptation and (4) Utilisation, based on two key dimensions: the maturity of the application domain (i.e.

opportunity, need, problem) and the maturity of knowledge (i.e. solution, artefact, theory) (see Figure 2).

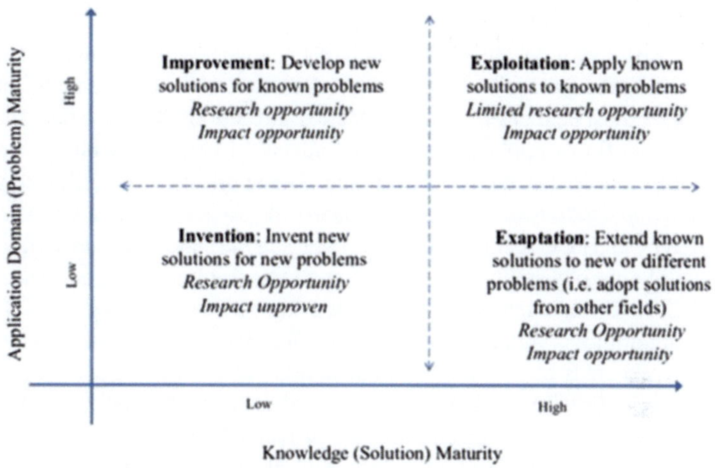

Fig. 2 Knowledge Innovation Matrix (KIM) with research opportunities and external impact results

Gregor S. and Hevner A. R. (2014), in “The Knowledge Innovation Matrix (KIM): A Clarifying Lens for Innovation”, propose different interpretations of the quadrants, breaking them down according to their various impacts on the sectors of innovation in which such experiences are practised.

They also demonstrate this evaluation strategy for the innovation sector in academia, where the focus is on rigorous knowledge production.

What is interesting about the KIM matrix, as it relates to this sector experience, are the definitions of the terms used in the quadrants, which are in some way capable of teaching us to recognise new values in frontier research, from the evaluation practice of innovative applications, as reported below:

Academic value proposition 1: in the *invention* quadrant:

(a) Research results will be significant, even if cases may be rare.
(b) Rigorous measurements of knowledge contributions are necessary to provide convincing evidence of new thinking.
(c) The impact on the real world is unproven and uncertain.

Academic value proposition 2: in both the *improvement and adaptation quadrants:*

(a) The research results will occur with relative frequency.
(b) Rigorous measurements of knowledge contributions are necessary to provide convincing evidence of new thinking.
(c) The impact on the real world can be substantial.

Academic value proposition 3: in the *utilisation* quadrant

(a) Research results will be negligible, both in terms of number of events and importance.

(b) The impact of the research will be evident in the form of articles on technology transfer in specialised journals.

In this scientific scenario, it is clear that the case of PP4.7.1, all research related to enabling and emerging technologies capable of expressing hypotheses, options, models and reference visions for design and research that challenges environmental and climate issues finds its space for innovation in the "adaptation and improvement" quadrants, where strategies must be measured against tactics and actions. This is a task and responsibility for the world of applied science and advanced research, which is capable of connecting different issues and tools available. The application of resilience strategies requires such joint and transdisciplinary action in order to pursue the "adaptive capacity" (also required by all IPCC reports) that is necessary in the state of vulnerability in which we operate, not only physically, in a time of great transition and neutrality scenarios.

We will discover that, once again, the relationship between enabling technologies and emerging technologies is very close and relevant when it becomes capable of transferring innovation to two areas of action of resilience, intercepting the regenerative project as an expression of radical and/or transformative resilience (Nava, Melis 2021), for which some reflections and scientific discussions are provided below.

The "phigital space" platform, of "open profiling" type, aims to reach the TRL7, as a demonstrator of a generative system of data-information-resources, prototyped in an operational environment. The architecture of the digital platform on digital protocols, together with the transfer of data-information and use testing on selected case studies, in addition to private users, provides useful DSS to the PA sector or operating entities, to size interventions on cultural heritage. The platform will be able to work with a system of modules able to draw from data-input of different format and configuration, able to report the most useful information to the construction of scenario searched and allow with an interface of applications and simulations, the return of models and protocols attached on the requests made. The level of skills aimed at information management and open knowledge for structural and environmental safety on cultural heritage (settlements and historic buildings) in transition scenarios, are related to the field of geotechnics, structures and analysis of their vulnerability, adaptive technologies identified in the advanced design of dynamic and parametric type on scenarios of physical and climatic context of evolutionary type. Through the contribution for external research activities, the project plans to implement skills and services in the field of information-design, IoT, ICT and computer programming. So the platform "phigital space" further innovates the informative character, with reference to transition scenarios and the opening with the production of protocols on models and prototypes of material and immaterial type, useful for the enhancement of cultural heritage.

1.2 The proposal of Pilot Project (PP 4.7.1) [3]

1.2.1 Open Platform "Phigital Space" (Physical and Digital) of the Type "User Profiling" for the Advanced and Dynamic Codesign of Interventions on the Built and ex novo

The pilot project achieves a system demonstrator with the open platform "phigital space" of the "user profiling" type for advanced and dynamic co-design of interventions in existing or new buildings. 2 case studies/prototypes will be tested and validated in 2 historical settlements in the hinterland of Calabria (Grecanica area).

Prototype 1 – Municipality of Palizzi Marina – coastal area/natural heritage with a test area in the former slaughterhouse area (municipal area).

Prototype 2 – Municipality of Bova – promontory and historic centre/artistic-monumental heritage with the testing area of Palazzo Mesiani (municipal area).

The applied experimental activity, therefore has a digital information level that realises the transfer of information on a "user profiling platform" and a physical information level with the creation of the Living Lab (physical platform) for the transfer to users (institutional, professional, community). This scenario realises what is called "open knowledge" in the PP objective.

The focus of the application activity is to operate on the heritage of the cases identified in order to create prototypes from which to draw data, resources and information for the purpose of "managing information on issues of structural and environmental safety" (together) of (cultural and natural) heritage, through open knowledge systems (of physical transfer and with digital technologies for multiple user profiles), under climate change, in the 3 scenarios 2030, 2050, 2085 (IPCC 2018-19), with declination to the possibility of identifying an existing "asset" and a new one, as a field of application. It is recalled that T4Y intends to profile the improvement of the "Quality of environmental and life".

1.2.2 The Executive Work Plan: 3 Milestones, 10 Actions

The project development period is set at 36 months (January 2023 – February 2026). The PP4.7.1 executive work plan is divided into three milestones that achieve the project objectives, progress on the technology readiness level (TRL4/TRL7) and propose three different levels of innovation, as described below.

Milestone 1

Formulation of work plan and detailed project strategy (TRL5) of the pilot project with two general objectives:

a) realization of a system-demonstrator, with the open platform "phigital space" of the type "user profiling" for the advanced and dynamic codesign of interventions on the built and ex novo, in three physical-climatic scenarios 2030, 2050, 2085 (IPPC data information)

[3] This section of the paper summarises, from paragraphs 1.2.1 to 1.2.4, the extract from the PP 4.7.1 proposal in the competitive phase and developed in the executive plan, drawn up by the author.

b) the experimentation and validation on n.2 case studies in n.2 historical identity settlements in the most fragile inland areas of Calabria

Construction of the architecture of the digital platform on the digital protocols (pilot 30%) and construction of the input dataset, on the design of the pilot plants with the choice of case studies and information derived from the activities of analysis and study, with the interpretation of the first laboratory activities and the processing of integrated technologies (TRL6)

Project innovation: The technology transfer will be realized through the inclusion in the open platform "phigital space" of data "open profile" to serve different users

Milestone 2

Application of methodologies and experimentation on case studies and in situ laboratory; definition of contextualized data and definition of the platform information level and user accessibility profiles (pilot 80%). Identification of prototypes of systems and components on the testing cases of historical urban aggregate and building technologies of interest of the testing activities referred to in the actions of the PP, inserted in the scenarios elaborated in the action and architecture of the information of the output dataset. (TRL 6)

Product innovation: Actions that study models, methods instruct protocols containing recommendations for the proper implementation of technology and simplified criteria for the design of interventions. Action contributes to the technological development of a new generation of locally resonant metamaterial structural components with inherent vibration reduction capabilities that are entirely innovative and 100% patentable. Action contributes to a new method of advanced parametric design on climatic and technological scenarios.

Milestone 3

Operation of the platform on case studies and in situ simulation with different user profiles (pilot 100%). Interpretation and verification of output data testing the methodology used on different actions and the effectiveness of integrated technologies. (TRL7)

Process innovation: Exportability of the Platform as a DSS useful also to the Public Administration sector (municipalities), professional to size interventions on cultural heritage (from case studies exportable models); patenting of research outputs (pilot project) under supervision of dedicated office at University.

10 Actions

The project starts with activities whose methodology and technology is validated in the laboratory (TRL4), with the use of processes, systems and instrumentation that map, catalog and characterize data, information, procedures, tests (actions 1, 2, 3,4) and with some cases of realization of models and prototypes of small-scale systems or full-scale component (actions 5, 6,7, 8, 9, 10) tested in the relevant environment (TRL 5, TRL6, TRL7). In detail:

- with actions 1,2, throught the inventory of building structural type and the capability to predict, by limit analysis theory, the collapse load and the failure mechanism of structural elements;

- with the action 3, through the use of RADP technology for the static and dynamic behavior of foundations;
- with action 4, through the use of dynamic analysis with uncertain parameters for structural bodies in brickworks;
- with action 5, through the development of tools for time-varying damage assessment on the structures of historic buildings, with models that consider the accumulation of damage;
- with actions 6, 7, through theoretical studies related to the dynamics of structural components in metamaterials with local resonance and devices with inerter;
- with actions 8,9, through methods and experimental projects on the themes of Advanced Sustainable Design (ASD) for Resilient Scenarios with bioclimatic approaches in urban and building design; through demonstrators, industrially validated and patented, on technologies applied to energy self-sufficient and circular buildings, of which prototypes have been made in scale with the techniques of additive manufacturing and industrial manufacturing.

The cognitive and experimental apparatus of the methodologies and technologies formulated and validated, as that related to the technologies and the realization of prototypes and systems is widely discussed and disseminated by the scientific products in national and international by the researchers of the team project and by research grants with the topics of pilot project. The expected result of the activities of departure, can only point to a system of interface between the products of the various actions and generate a new level of innovation of methodologies, technologies and the creation of prototypes in operational environments, assumable in a system "demonstrator" (action 10)

1.2.3 Description and Auantification of the Expected Results

The PP4.7.1 contributes to the value of T4Y research by establishing, during the proposal design phase, three levels of impact for achieving the objectives of transferring studies and applications to the territory and community involved.

The indicators refer to the extent of the effects that the research actions will have on the pilot project topics of interest to spoke 4.

The following are the operating systems for assessing impact: a. Economic Impact; b. Impact on Improving the Quality of Life; c. Technology Transfer.

a. **Economic Impact (e.i.)**

To e.i. in the project context. "The regeneration of a cultural asset can be a factor of sustainable development" (English Heritage, 2005). Reactivate assets, to the settlement and building, making them usable and safe from the structural and environmental point of view means to return "their value of use", already "economic value" recovered and further "incentivized".

To e.i in the project activities. The sector related to the fruition of cultural heritage has been affected by the economy of knowledge and smart growth (H2020 and EU), "Producing culture is therefore a frontier economic activity in the era of the knowledge society and global markets and is increasingly complex, because it must adapt to goods and services very different in content and technology, as well as absolutely heterogeneous users" (Santagata et al. 2014).

Value from cost saving. The availability platform useful to instruct interventions can reduce by 20% the economic cost of investigations in the cognitive phase of preliminary technical feasibility.

Indicators

(to e.i. in the project context)

Quality indicators: Knowledge of the area and of its major structural fragilities allows for targeted and calibrated public/private interventions; allows also the possibility of investment over time and local economy activities based on the new attractiveness of the assets.

Quantity indicators: % of new interventions on the assets subject to experimentation; effectiveness (if) of transferable project results (processes and products) for attractiveness of funds for territorial investments on heritage.

(to e.i in the project activities)

Quality indicators: construction of new competencies based on the value of knowledge; investments in R&D; profitability of the initial know-how and the know-how acquired.

Quantity indicators: Investments in service providers and supplies; economies of scale for project management on a platform of integrated and interdisciplinary internal competencies (co-financing)

b. **Impact on Improving the Quality of Life** (environmental well-being, social and cultural development) of citizens.

The Calabrian territory has suffered significant losses of historical and cultural heritage due to its multiple fragilities (geomorphological, structural, climatic). The inland areas of Calabria are the most "sensitive" but also the most "identity", the depopulation of villages, has increased the levels of fragility, decresed the care of heritage, together with the lack of primary services, strongly affecting the state of degradation and disruption of settlements and historic buildings. The protocols and models validated in the laboratory, on the geotechnical improvement of soils, on the integrated adaptation of the structures of historic buildings and the energy and environmental performance and durability of structures and artifacts in variable scenarios under different conditions, also restore ecological and social stability to the assets, aiming at new levels of reliability and comfort perceived by users and communities for their sense of belonging.

Indicators

The project aims to provide new levels of security and physical and digital accessibility to the spaces, in the different arrangements of use and over time. It reactivates the collective value of the settlement spaces and buildings, which for their guaranteed usability, increase the level of environmental and social well-being of the community and become real "attractors" for both possible promoters and users.

Quality indicators: level of usability and accessibility of assets; environmental and structural safety; flexible use of spaces; cultural interest in the open digital platform.

Quantity indicators: no. and duration of upgrading interventions on buildings and aggregates in the next 3 years; no. of months of accessibility and operation of upgraded spaces in a year; no. and types of activities open to use in a year; demographic increase / repopulation index; no. of visitors to the assets; no. of users (various profiles) of the open digital platform.

c. **Technology Transfer**

The total impact is defined by 3 levels of "design driven innovation" to technology transfer:

Project innovation: The technology transfer will be realized through the inclusion in the open platform "phigital space" of data "open profile" to serve different users (Action 10)

Product innovation: Actions 1-2-3-4-5 that study models, methods instruct protocols containing recommendations for the proper implementation of technology and simplified criteria for the design of interventions. Action 6-7 contributes to the technological development of a new generation of locally resonant metamaterial structural components with inherent vibration reduction capabilities that are entirely innovative and 100% patentable. Action 8 contributes to a new method of advanced parametric design on climatic and technological scenarios.

Process innovation: Exportability of the Platform as a DSS useful also to the PA sector, professional to size interventions on cultural heritage (from case studies exportable models) (Action 8,9)

1.2.4 Goal Output and Goal Results the PP4.7.1

The frontier themes of information management and open knowledge for structural and environmental safety on cultural heritage contribute to the process-es of "radical innovation" of the mission – spoke 4, goal 7. The "mechanisms of the just transition", to which respond through the adoption of KET's, address the material and immaterial (physical and digital) management of the "data-information-resources" system, towards the production of models, methods, prototypes realized in "dynamic" regimes, such as "uncertain parameters" and "limit states" of the analysis and "predictive models of advanced design" for the adaptability to climate scenarios.

The **goal outputs** aim to bring the following technologies and systems up to TRL5 @M12 and TRL7 @M36:

1) A database of structural typologies for the historical buildings in the in-land settlement of Calabria selected as case studies; (nr.1 database with no 2 modules for 3 scenarios)
2) Direct methods numerical analysis for the prediction of mechanisms and collapse load of historical buildings falling in fragile territories; (no 1 method for 2 cases to 3 scenarios)
3) Advanced solutions in the design and implementation of ground improvements in fragile areas of settlement of historic heritage buildings; (no 2 series for 3 scenarios)
4) Dynamic modeling of historic brickworks; (no 2 models for 3 scenarios)
5) Time-varying vulnerability assessment tools and advanced protocols for cataloging existing historic buildings on identity structural types; (no 2 protocols for 3 scenarios)

6) Digital active control devices, based on the concept of inerter, for the pro-tection and security of monumental heritage assets; (no.1 series for 3 scenarios)
7) Structural components in locally resonant metamaterials for structural and environmental safety of historic buildings; (no.2 series x 3 scenarios)
8) Atlas of predictive models and adaptive technologies on climate scenari-os and reliability (technological and material) of the artifacts of historical heritage buildings (no 3 scenarios for 2 cases);
9) Enabling Technologies and Additive Manufacturing (3D printing) of pro-totypes of systems and components on testing cases (at settlement and building scale).(no. 8 prototypes)
10) Digital Platform User profiling Architecture and data implementation, in-formation and models – Demonstrator

The following **results are expected** for the goal:

1. Promote the knowledge economy by means of the reactivation the use- value of heritage.

Contribute to the impact on the local economy of the cultural heritage identity, through innovative processes of reactivation, regeneration and securing of settle-ments and historic buildings, including in the economy of open knowledge and smart growth all the stakeholders of different profiles: researchers and profes-sionals, users-prosumers, companies, public municipalities.

2. Realize a dual-interface (phigital) demonstrator of an "open knowledge" physical platform (phisical-space)

Connect different skills, methodologies and validated technologies in environ-ments relevant to the topics of interest and the techno-scientific fields of investi-gation, according to the model of co-design, open to the exchange of information and elaborations, among researchers of the team, IR and external companies. Use the transition scenarios as the "common space" to refer to in their experiments and advanced design techniques, as the module of translation of the different results, in models and configurations of morphologies, technologies, architectures and structures of historical heritage assets (settlements and buildings).

3. Realize a dual-interface demonstrator (phigital) of a digital platform "us-er profiling" (digital-space)

Demonstrate, with an open digital platform of the type "user profile", the pos-sibility of drawing from data-input of different formats and configurations in or-der to report the most useful information to the construction of the scenario sought and, through an interface of applications and simulations, be able to re-turn a response to the requests made.

The implementation, populating and use-testing of data-information-models, extracted from the products of actions, allow the possibility of integrating digital outputs, in dedicated protocols to support design choices and intervention (evolved DSS and validation phase with case studies).

2 Methodology: Laboratories and Living Lab [4]

2.1 Ecophigital Map

The work carried out by the team in order to achieve the objectives expressed in Milestone 1, took place in terms of methodological and desk research work and with a first start in the field, already in the first semester, through all the activities competing with the final objectives of the PP: and with reference to the adopted methodology of the TRL and the impacts to be produced at the end of the project. In order to integrate all the activities and actions aimed at the production of the deliverables, pursuing the PP's own objectives, a cognitive-applicative map was formulated (Ecophigital MAP), to read the research process and control the organisational logic of the industrial research and experimental development, with which the team operationally produces "the exploitation research", according to the objectives required by the spoke and the HUB (Fig. 3).

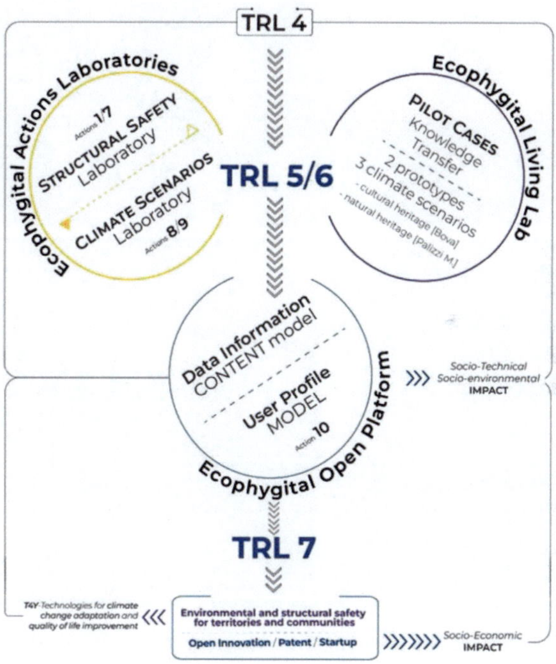

Fig. 3 . Ecophigital map for PP 4.7.1 (Diagram Method by C.Nava -Graphic Elaboration by G.Mangano (2024)

From the Ecophigital MAP, the three research dimensions/spaces of the PP emerge:

[4] The section of the paper that describes the methodology refers to the implementation of the programme involving the activities and deliverables achieved in Milestone I and part of Milestone II, with the contribution of researchers who worked in teams and achieved the objectives in university laboratories and by participating in the living lab set up for the various actions.

(1) *Ecophigital Actions Laboratories*: which expresses all the activities related to the actions, organised and carried out by 2 thematic/disciplinary Laboratories – SSL Structural Safety Laboratory (for actions 1/7) and – CSL Climate Scenarios Laboratory (for actions 8/9).

The activities carried out by the laboratories with the heads of the actions and the working group, are realised through the provision of instrumental resources (soft and equipment) which are already in the possession of the research groups and which are foreseen to be upgraded with the T4Y project's expenditure plan. Figure 4

Fig. 4. Meeting and dissemination activities in university laboratories

(2) *Ecophigital Living lab:* this refers to the transfer of the research actions and activities to the knowledge levels for the applied experimentation on the 2 pilot cases selected (Bova municipality and Palizzi Marina municipality – municipalities of the internal areas of the Grecanica Area of the Metropolitan City of Reggio Calabria, , studied on the 3 climate scenarios (2030, 2050, 2085), as prototypes to be used for the acquisition of the data and information that will populate the data-information of the digital user profiling platform (Figure 5) (Lucanto D, Nava C, Mangano G; (2024).

The pilot cases were chosen with reference to the profile expressed by Spoke 4 and as far as useful for the PP.

The activities of the on-site laboratories will be carried out according to the Living Lab approach, which will be physically activated in the municipality of Bova, with the start-up of the activities planned for December 2023.

A preparatory document (sect. 1) has been drawn up to justify the choice and articulate the activities to be carried out in an experimental way by the SSL and CSL laboratories.

Fig. 5 . On-site activities and Living Lab event with municipalities

TRL Method

From the starting TRL 4, the Ecophigital Actions Laboratories (1) and the Ecophigital Living lab (2), scale up to TRL5/6 levels, reporting on research workspaces capable of moving from model verification (reduced scale) in the rilevant environment (1) at university laboratories to demonstration (full scale) in the rilevant environment (2) in situ. The output of these activities will serve to produce data and models for Ecophigital Open Platform (content)

Impact Value

The activities carried out through the Ecophigital Actions Laboratories (1) and the Ecophigital Living lab (2), contribute to producing value in achieving the final levels of socio-technical and socio-environmental impact.

(3) *Ecophigital Open Platform:* it is realised by organising the data-information space, transferred from the results of the activities of the Ecophigital Actions Laboratories (1) and the Ecophigital Living lab (2), with which the archiving of the contents of

the Digital Platform is constructed; then the structure of the inter-face for user profiling operations.

On this tool, in order to guide and instruct all the actions of the project towards the construction of the aforementioned platform a first preliminary document has already been advanced, (sect.2).

TRL Method

Ecophigital Open Platform realises TRL 7, as a demonstrator model in an operational environment, particularly aimed at meeting the transfer objectives of PP, such as environmental and structural safety for territories and communities and for the HUB's objectives on Open Innovation, Start UP, Patent and Start up.

Impact Value

Ecophigital Open Platform (3), adds to the values the final levels of socio-technical and socio-environmental impact, new value for the expected socio-economic impact of technology transfer to territories and communities for quality of life and security in climate change scenarios.

2.2 Selection of Prototype Cases

This section explains the reasons behind the selection of the two prototype cases for the experiment, which involved the team working with university laboratories and on site with the Living Lab.[5]

The topic referred to in spoke 4 refers to the care and enhancement of "cultural and natural heritage", which the European Landscape Convention already defines as "protected assets", referring to "architectural and environmental assets".[6]

This category includes immovable property and parts of the built environment, complex buildings also considered in relation to their surroundings and their characteristic and traditional appearance; areas with unique characteristics and landscape, geological, natural and environmental value, such as parks and reserves, panoramic views and viewpoints, coastal areas, areas surrounding lakes, rivers and watercourses, mountain peaks, volcanoes, glaciers, woods and forests' (Treccani glossary).

The PP 4.7.1 team identifies the Grecanica Area of the Metropolitan City of Reggio Calabria as a territory and community with a strong cultural and natural identity, recognised by UNESCO as intangible heritage and recognised for the identity of the Greek minority of Calabria.[7]

The following case studies have therefore been identified:

[5] G. Mangano wrote about the Living Lab method and implementation in "The Living Lab project and community engagement in the pilot project 4.7.1"

[6] "property protect" Legislative Decree No. 42 of 22 January 2004, 'Code of Cultural Heritage and Landscape, pursuant to Article 10 of Law No. 137 of 6 July 2002' and by the European Landscape Convention, 2000.

[7] GRECANICA AREA – Cultural heritage - *Grecanica Area/ SNAI and SRAI Pilot Area _ National and Regional Strategy for Internal Areas*The administrative Grecanica area consists of 16 municipalities: Motta S. G., Montebello Jonico, San Lorenzo, Cardeto, Bagaladi, Roghudi, Roccaforte del Greco, Condofuri, Bova Marina, ***Bova, Palizzi,*** Staiti, Brancaleone, Samo, Africo

Prototype 1 – Municipality of Palizzi Marina – coastal territory/natural asset with testing area at the former slaughterhouse (municipal area) (Fig. 6)

Prototype 2 – Municipality of Bova – Promontory territory and historic centre/artistic and monumental asset with testing area at Palazzo Mesiani (municipal area) (Fig. 7)

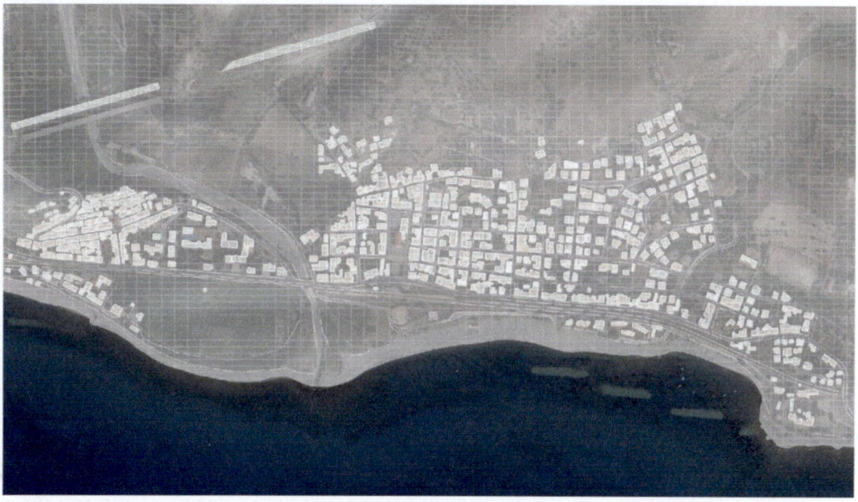

Fig. 6 . Settlement map of the municipality of Palizzi

Fig. 7 . Settlement map of the municipality of Bova and Mesiani Palace

The investigation of prototype cases.

and finally, Reggio Calabria. However, the Greek-speaking island, i.e. the area where the Greek language is still spoken, can be narrowed down to the municipalities of Bova, Bova Marina, Roghudi, Ghorio di Roghudi, Gallicianò, with traces also preserved in municipalities or parts of them such as Roccaforte del Greco and its Ghorio, Arangea and San Giorgio Extra di Reggio Calabria.

1. The "scale" knowledge levels for information and context areas may differ in the two prototype cases:

Prototype 1 – Municipality of Palizzi Marina – coastal area/natural asset

LT _ territorial level with reference to the Ionian coast of the Grecanico area

LSA_ environmental system level: coastal strip in the section up to 300 metres from the coastline, including the EX Mattatoio testing area (water/sea system, soil, vegetation, buildings, etc.)

Prototype 2 – Municipality of Bova – Promontory territory and historic centre/artistic and monumental heritage with Palazzo Mesiani testing area (municipal area)

LUCs_ urban level/historic centre (exposed fronts, related areas, roof area, soil, networks, etc.)

LAgM_ aggregate level of buildings/heritage and structures Palazzo Mesiani (masonry structures, roofs, etc.)

2. The knowledge activity on prototype cases will use
   a. A level of indirect information and study (bibliographies, studies, research, data, etc.)
   b. Direct information through survey activities of a practcal/instrumental nature – on cases, with equipment, mobile labs, etc. of a Technical-interpretative nature – with scenarios and simulation models, tests, etc.

Therefore these activities will implement the "open construction site" phase in the first year of research (M1) on which to work for the deliverables/M1, the LIVING LAB development area.

The data, information, studies, etc. will be subject to in-depth analysis and study for the preparation of material useful for transfer to the digital platform and for activities related to dissemination to users in the area and dissemination of scientific research products.

2.2.1 The Prototypes and the 'Data-Information-Resources' System

The focus of the application activity is to work on the identified cases to create prototypes from which to draw data, resources and information for the purpose of "managing information on structural and environmental safety" issues (together) of cultural and natural heritage, through open knowledge systems (physical transfer and digital technologies for multiple user profiles), in the context of climate change, in the three scenarios 2030, 2050 and 2085 (IPCC 2018–19), with the possibility of identifying an existing "asset" and a new one as a field of application. It should be noted that, as reported by T4Y, the aim is to outline a profile for improving the "quality of the environment and life".

This means that:

- The case studies are producers of methodologies, models, measures, etc. and therefore should be considered as such and are not the subject of experimentation, but rather the field of application for experimentation and integration into the demonstration case (the platform).
- The intention is to work on the concept of "risk" associated with related and interfering issues between structural safety (not only vulnerability and seismic risk, etc.) and

environmental safety (climate, biodiversity, etc. etc.), where climate change scenarios are the contextual and scientific systems from which to start and to which to refer, and in the specificity of the climate data set that will be produced on the case studies (see action 8).

- The work has been done on defining this multi-level risk, after conducting the initial exploratory phases of the action, and that we will do so with a co-design approach (as envisaged in the project).
- The activities have produced methodologies and results that have identified any differences, characteristics, etc. related to the three scenarios and across multiple models.
- In all experimental activities relating to deliverables, the aim was to produce "information useful for the integrated procedure with the platform", which therefore requires a specific programme for the prototypes.
- Not all actions produced activities on both prototype cases, specifically on the Bova prototype (actions 1, 2, 4, 5, 6, 7), on the Palizzi prototype (3, 8, 9), and on both prototypes (action 10).
- The activity had a digital information level that dealt with the transfer of information to the user profiling platform (digital platform) and a physical information level with the activities of the Living Lab (physical platform) for transfer to users (institutional, professional, community). That this scenario achieves what is referred to in the PP objective as "open knowledge".

3 Discussion– The Pilot Demonstrator *Ecophigital Platform*

3.1 Level of Innovation of the Demonstrator and User Profiles

The digital user profiling platform, with its highly transferable "open" and "interoperable" content, is the product of the pilot project, the Demonstrator, which has been awarded TRL 7 technological maturity level. The level of technological maturity achieved by the activities related to the actions/deliverables (the different TRLs) must be capable of "informing" the knowledge levels on information, content and interoperability of the platform, related to the themes of "structural and environmental safety" for heritage.

In summary and specifically, not only have advanced results been achieved from a scientific point of view, but the real challenge has been to make them "readable and interoperable" for multiple user profiles (Fig. 8).

The platform takes on different values for "open knowledge", responding as:

- A highly communicative and effective system for users/citizens, communicating levels of information on topics and simulations and models to help them understand the phenomenon of risks, but also possible "safety" scenarios with integrated visions of quality of life, the future, etc.
- A support system for users/technicians, to understand the scenarios, models, measures and methodologies that can be adopted in studies and activities on topics of research interest, illustrating simplified processes to identify feasible paths for interventions on heritage and its assets.

- A user/public body support system to understand, with regard to the studies and methodologies conducted and the work carried out with the Living Lab, scenarios of change in contexts, possible resources and opportunities for "safe" territorial investments.

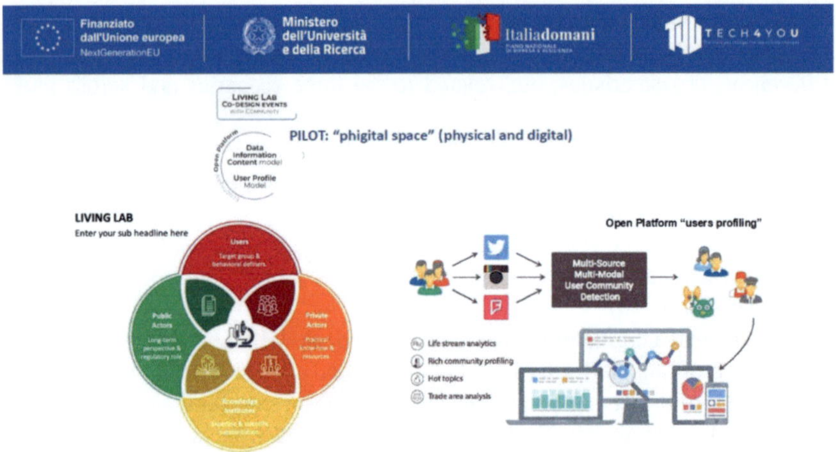

Fig. 8 . Structure and contents of the 'phigital space': platform and living lab

All activities and Pilot results with products to be included in the platform have a "filter" that selects according to access to the three types of users for the level of information most "dedicated" to them.

3.1.1 Integrated Procedure Between Actions/Activities and Platform

The platform has a communicative interface and content that is updated with the transfer of results and the outcome of actions. It is open because it has levels of interactivity and interoperability programmed by IT developers (Tab. 2).

Table 2. Content structure outline for the platform, useful for populating information.

AREA	Section 1 Contents	Section 2 Contents	Section 3 Contents
INFORMATION AREA	How to use the demonstration platform Guidelines and the 3 user profile filters	Presentation of TOPICS on structural and environmental safety and CC Texts, data, images, references	Presentation of prototype cases Dossiers, photos, videos, graphic tables, reference links

(*continued*)

Table 2. (*continued*)

AREA	Section 1 Contents	Section 2 Contents	Section 3 Contents
CONTENT AREA	Definitions and descriptions of RISK TYPES (references to the 2 prototype cases) data, texts, graphics Interoperability with access to other databases, platforms, etc.	MEASURES AND INTERVENTIONS Data, texts, graphics, models, simulations Interoperability with access to other databases, platforms, etc.	METHODOLOGIES USED WITH REFERENCE TO SECTION 1.2 Data, texts, graphics, processes, workflows, references, examples Interoperability with access to other databases, platforms, etc.
INTERACTIVITY AND INTEROPERA BILITY AREA	Interactivity Section dedicated to the filters of the 3 user profiles Platform testing activities planned in the PP	Interactivity Presentation of protocols to assist specific cases Platform testing activities planned in the PP	LIVING LAB Physical and digital activities With open consultation Platform testing activities planned in the PP

3.1.2 Platform Population Activities _WOP (Workflow of Platform) Document

The platform has a descriptive-demonstrative (non-processing) purpose, but interoperable only on models, data, hypertextually traceable information on the sections populated by the work of the PP actions for different profiles (user, scholar, institution) (Tab. 3)

1. With reference to what has been described for the activities in the university laboratory and in situ (LL Programme and Agenda document), correspond to these activities "products" transferable to the platform as indicated in the sections and areas.
2. Briefly describe only "specifying the product" in the type (data, texts etc…) and also in its "definition" (e.g. climate scenario models etc.).
3. Report products according to A and B and on several products use A1, A2.
4. More extended and referenced descriptions with explanations and sources will concern the document attached to the workflow.

Table 3. Workflow matrix for information provided by actions

PERIOD	UNIVERSITY LABORATORY ACTIVITIES	LABORATORY ACTIVITIES IN SITU (c/o municipalities)	OPEN PLATFORM – sect 1	OPENPLATFORM – Sect 2	OPENAPLATFORM – sect 3
	Report on Corresponding Deliverable/indicate site (Palazzo Bova, Costa Palizzi)	Report on Corresponding Deliverable/indicate site (Palazzo Bova, Costa Palizzi)	(content area) Definitions and descriptions of RISK TYPES (references to the 2 case-prototypes) data, text, graphics Interoperability with access to other databases, platforms, etc.	(information area) A- Presentation of the Themes on structural and environmental safety and CC Texts, data, images, references	(information area) A- Contribution to the Presentation of Prototype Cases (Dossiers, photos, videos, graphic tables, reference links
				(content area) B-MEASUREMENTS AND INTERVENTIONS Data, texts, graphics, models, simulations Interoperability with access to other databases, platforms, etc.	(content area) A- METHODOLOGIES USED WITH REFERENCE TO SECTION 1,2 Data, text, graphics, processes, workflows, references, examples Interoperability with access to other databases, platforms, etc.
I M period	*document*	*document*		***A-*** ***B-***	***A-*** ***B-***

Attach more extensive descriptive .doc document Workflow activities, as expressed in the matrix (Fig. 9), indicating:

1. Period of the activities and explanation if in LAB or SITU
2. Section and Area of reference
3. Content and more extensive description of the product/activity, with possible references to data type and sources used
4. Type of data to be allocated to the interface by User profile also multi-use_ indicating if (citizen/student/body)
5. Interoperability of the data/information- indicating if connected to other types of products/data expressed in other activities (Fig. 10)

PERIODO	ATTIVITA' IN LABORATORIO UNIVERSITARIO *Riferire sul Deliverable Corrispondente/ indicare sito (Palazzo Bova, Costa Palizzi)*	ATTIVITA' IN LABORATORIO IN SITU (c/o i comuni) *Riferire sul Deliverable Corrispondente/ indicare sito (Palazzo Bova, Costa Palizzi)*	PIATTAFORMA— **sezione 1**	PIATTAFORMA— **sezione 2**	PIATTAFORMA— **sezione 3**
			(area dei contenuti) Definizioni e descrizioni di TIPI DI RISCHIO (riferimenti ai 2 casi prototipi) *dati, testi, grafiche* *Interoperabilità con accesso a altri database, piattaforme, etc*	*(area delle informazioni)* A--Presentazione dei TEMI sulla sicurezza strutturale e ambientale e CC *Testi, dati, immagini, riferimenti* *(area dei contenuti)* B--MISURE E INTERVENTI *Dati, testi, grafiche, modelli, simulazioni* *Interoperabilità con accesso a altri database, piattaforme, etc*	*(area delle informazioni)* A- → Contributo alla Presentazione dei casi prototipi *(Dossier, foto, Video, Tavole grafiche, link refenze* *(area dei contenuti)* B- → METODOLOGIE UTILIZZATE CON RIFERIMENTO ALLA SEZIONE 1,2 *Dati, testi, grafiche, processi, workflow, riferimenti, esempi* *Interoperabilità con accesso a altri database, piattaforme, etc*
I milestone Trimestre OTTOBRE-NOVEMBRE-DICEMBRE 2023	**Action 8—Activity 5—D 7.1.2** (Palazzo Bova, Costa Palizzi): Elaboration of the Atlas structure for the organization of climate scenario data.		***Palazzo Bova, Costa Palizzi*** **Type of Content:** Comprehensive research report on climate change mechanisms impacting cultural and natural heritage. • **Risk Category:** Climate-induced deterioration and damage to heritage sites; Increased erosion due to extreme weather events; risks to structural integrity from temperature fluctuations. ***Palazzo Bova, Costa Palizzi*** **Type of Content:** Clear understanding of prototype requirements and specifications; early stage designs of 8 prototypes. • **Risk Category:** Inadequate design response to climate change; inefficiency in energy use leading to higher CO2 emissions. **Data/Product Transfer:** reports and climate scenario models; data visualization tools to represent climate change impacts	**INPUT** **A--Presentation of Themes on Environmental Safety and Climate Change** • **Content (A1):** Overview of climate change impacts on cultural and natural heritage sites, highlighting vulnerabilities. • **References (A2):** Studies and reports on climate change effects specific to heritage structures. **B--Measures and Interventions for Environmental Safety** • **Data and Texts (B1):** Strategies for mitigating climate change impacts on cultural and natural heritage sites. • **Models and Simulations (B2):** Climate models showing potential future impacts on these sites. • **Interoperability (B3):** Connection with climate databases and platforms for extended data access. **OUTPUT**	**INPUT** **A--Contribution to the Presentation of case studies Prototypes** • **Content (A1):** Dossiers, photos, and videos of climate change impacts on cultural and natural heritage. • **Graphic Tables and Links (A2):** Visual aids and external references to studies on climate change and heritage conservation. **B--Methodologies Used (Referencing Sections 1, 2)** • **Data and Texts (B1):** Documentation of research methods and findings on climate impacts. • **Graphics, Processes, Workflow (B2):** Visual representations of the research and analysis process. • **Interoperability (B3):** Integration with climate research databases and platforms. **OUTPUT** • **DM--Digital 3D Model for Case Studies:**

Fig. 9. Example of workflow matrix population (action 9)

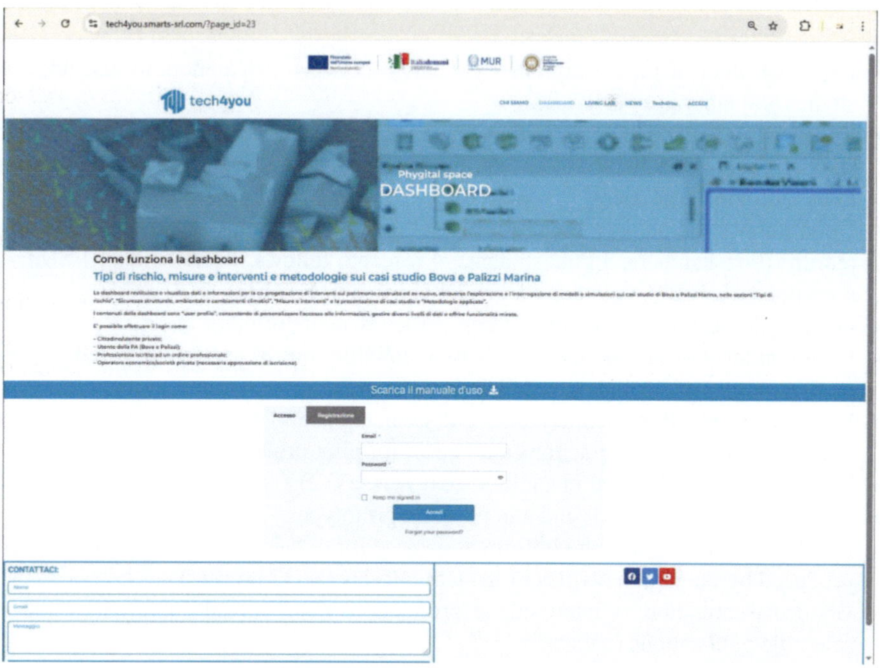

Fig. 10. Image of the Ecophigital platform (PP 4.7.1)

4 Conclusion

The PP4.7.1 pilot project carried out activities related to a three-year exploitation research programme, with great rigour in achieving the objectives and with reference to the activities of the 10 project actions. The experience gained by the team of senior and junior researchers highlighted a number of important conditions for the success of the competitive proposal and identified some critical issues during the course of the activities. Both conditions may serve as a reference for possible future research trends. Some points for discussion are presented below.

Barriers:

- At the start of the programme, a lack of communication between HUB T4Y and Spoke 4 impacted the progress of PP4.7.1.
- The availability of the budget for purchasing equipment for the laboratories was delayed in relation to the activity plan for the second milestone.
- Poor knowledge of the PNRR T4Y programme pilot projects for possible discussions on topics of common interest or similar research trajectories.
- Bureaucratic complexity of the procedural activities required.

Values:

- Excellent level of interdisciplinary exchange between researchers in carrying out studies and laboratory activities.
- Excellent and effective operational agenda for the team to exchange information and test the objectives achieved in an "agile process" (over 15 meetings in total, 4 Living Labs with municipalities).
- Good impact and balance between researchers from different departments and new recruits (9 researchers, 3 PhD students, 4 research fellows, 1 scholarship, 1 SME, 5 consultants, 2 BAC), with a good gender balance (9 F; 7 M).
- Good dialogue with the municipalities of the selected prototype cases.
- Excellent level of innovation for scientific activities aimed at open knowledge (new methodologies and collaboration between researchers).
- Excellent production of intermediate documents useful for semi-annual scientific reporting (draft paper for actions and report for pilot project).
- Excellent results obtained in audits with reviewers (2).
- Excellent production of digital and physical prototypes.
- Creation of a user profiling platform as a space for transferring data and information produced by the joint activities of the laboratories (WOP process).
- Good dissemination of intermediate products at international conferences and in scientific publications (papers).
- Achievement of measurable objectives with impact assessment as required by the proposal (paragraphs 1.2.3 and 1.2.4).

Acknowledgment. This study was carried out within the Innovation Ecosystem Tech4You – "Technologies for climate-change adaptation and quality-of-life improvement," Mission 4, Component 2, Investment 1.4 of Italy's National Recovery and Resilience Plan (PNRR), funded by the European Union—NextGeneration EU (project code ECS_00000009, CUP C33C22000290006). The author, is pilot project lead, Pilot Project 4.7.1 "Open platform phygital space for advanced, dynamic co-design of built-environment interventions" and scientific responsible for the activities carried out in the Design and Prototyping sections of ABITAlab, Department of Architecture and Design, Mediterranean University of Reggio Calabria. We would like to thank the entire PP4.7.1 team for their important contribution to the activities of PP 4.7.1, demonstrating scientific expertise and great agility in overcoming barriers and adding value to the initial proposal, achieving very promising scientific and cultural objectives for future experiences. Project team – Professors from the dAeD and Diceam departments, Mediterranean University of Reggio Calabria: P. Fuschi, A. Pisano, D. Porcino, A. Sofi, E. Chioccarelli, G. Failla, R. Pucinotti, F. Giglio; Young Researchers: M.C.Lasorella, G.Tomasello, A.Russillo, D.Lucanto, G.Mangano, G.Percolla, M.Paone, A.Hanida.

References

Gregor, S., Hevner, A.R.: The Knowledge Innovation Matrix (KIM): a clarifying lens for innovation. Inform Sci Inter J Emerg Transdiscip. **17**, 217–239 (2014) Retrieved from http://www.inform.nu/Articles/Vol17/ISJv17p217-239Gregor0800.pdf

Holste, D., Roche, I., Hörlesberger, M., Marianne, D., Besagni, T., Scherngell, F., Claire, P., Cuxac, P., Schiebel, E.: A concept for inferring "Frontier Research" in Research Project Proposals', ResearcheGate (2011)

IPCC.: Summary for policymakers. In: Climate Change 2022: Mitigation of Climate Change. Contribution of Working Group III to the Sixth Assessment Report of the Intergovernmental Panel on Climate Change [Shukla, P.R., Skea, J., Slade, R., Al Khourdajie, A., van Diemen, R., McCollum, D., Pathak, M., Some, S., Vyas, P., Fradera, R., Belkacemi, M., Hasija, A., Lisboa, G., Luz, S., Malley, J. (eds.)]. Cambridge University Press, Cambridge, UK and New York, NY, USA. (2022a). https://doi.org/10.1017/9781009157926.001

IPCC.: Summary for policymakers. In: Climate Change 2022. Impacts, Adaptation and Vulnerability. Contribution of Working Groups II to the Sixth Assessment Report of the Intergovernmental Panel on Climate Change [Pörtner, Hans-O., Roberts, D. C., Adams, H., […] and Ibrahim, Z. Z. (eds.)]. Cambridge University Press, Cambridge, United Kingdom and New York, NY, USA (2022b)

Lionello, P., Naumann, G.: Il rapporto IPCC spiegato dagli esperti italiani con i contenuti principali su Europa, Mediterraneo e Italia. CMCC for IPCC focal point for Italy (2022)

Lucanto, D., Nava, C., Mangano, G.: "Digital prototyping and regenerative design towards a carbon-neutrality and climate resilient built environment: a Multi-Scale Assessment of Environmental Multi-Risks. Buildings. **14**(12). Special Issue- Best Practices of Resilient Buildings (and Districts) and Post-disaster Reconstruction E-ISSN:2075-5309 -open access Q1-, 1–27 (2024). https://doi.org/10.3390/buildings14123934

Nava, C.: Tecnologie emergent per il progetto rigenerativo. 5 questioni teoriche su innovazione e sostenibilità dalla prassi della Ricerca di Frontiera, Aracne ed., Roma (2023)

Nava, C., Melis, A.: A radical bioecological paradigm for design technologies with a transdisciplinary approach. TECHNE J Tech Arch Envir. **21**, 103–111 (2021). https://doi.org/10.13128/techne-9938

Santagata, W.: Il Governo della Cultura. Promuovere sviluppo e qualità sociale, Il Mulino ed., Bologna (2014)

Stokes, D.E.: Pasteur's (TM's) Quadrant. Basic Science and Technological Innovation, Brookings Institutions (1997)

Structural Vulnerability Assessment in Historic Masonry Settlements Using Drone Survey

Paolo Fuschi(✉), Giulia Percolla, Mariaceleste Lasorella, and Aurora Angela Pisano

Department of Architecture and Design – dAeD, University Mediterranea of Reggio Calabria, Reggio Calabria, Italy
{paolo.fuschi,giulia.percolla,mceleste.lasorella, aurora.pisano}@unirc.it

Abstract. This work proposes a systematic technique to carry out a structural vulnerability assessment of historic masonry structures using Unmanned Aerial Vehicle (UAV) photogrammetry. The proposed approach aims to improve structural risk mitigation efforts in small historic towns by overcoming the limits of existing on-site approaches, which are often intrusive and resource expensive, by using drone-based image data to detect peculiar structural parameters. The proposed technique formalizes observable visual parameters in orthophotos and 3D models, transforming them into a decision support system. For this purpose, 23 parameters are considered, which are divided into four macro-categories. These parameters are borrowed from official Italian Civil Protection forms, adapting them for remote analysis. Weights to the parameters are assigned using the Analytic Hierarchy Process (AHP), which is guided by experts in pairwise comparisons. The approach is implemented in an Excel-Visual Basic Application (Excel-VBA) environment, allowing users to input visual data, calculate scores, and generate a vulnerability index ranging from 1.00 (low vulnerability) to 5.00 (high vulnerability) for each individual building. The index is then framed into three vulnerability classes. This method provides a scalable, non-invasive, and reproducible methodology for guiding early vulnerability mapping and prioritization, which is especially appropriate for ancient urban centers.

Keywords: UAV Photogrammetry · Structural Vulnerability Assessment · Analytic Hierarchy Process (AHP) · Structural Vulnerability Map

1 Introduction

In the context of heritage building safeguarding, the structural safety evaluation is a needed task, especially in small towns where the preservation of heritage and safety must coexist. These urban aggregates—particularly present in southern Italy—consist of dense, irregular clusters constructed with different materials and frequently lacking structural documentation. Such situations interfere with the implementation of traditional structural safety assessment methods, which generally depend on on-site evaluations, numerical modelling, and invasive diagnostic examinations (Julià & Ferreira, 2021).

C. Nava et al. (Eds.): *Climatic and Structural Safety in Multi-Hazard Regime of Cultural and Natural Heritage*, GREEN, pp. 29–40, 2026.
https://doi.org/10.1007/978-3-032-11926-1_2

To address the need for rapid and effective evaluation tools, recent studies have explored the use of remote sensing techniques—particularly UAV-based photogrammetry for documenting and analyzing existing buildings (Ulvi, 2021). These methods enable the production of high-resolution orthophotos and textured 3D models that can support preliminary diagnostic evaluations (Guo et al., 2022). However, despite their increased application in conservation and survey campaigns, few methodologies have formalized the obtainable outputs into structured frameworks for structural vulnerability assessment.

Some efforts have also combined photogrammetric data with GIS platforms or spatial decision-support systems to generate vulnerability maps (Leggieri et al., 2022). Others have introduced scoring-based approaches that rely on qualitative expert judgement. In this direction, Multi-Criteria Decision Making (MCDM) methods, and in particular the Analytic Hierarchy Process (AHP), have represented a systematic way to assign importance to multiple factors while ensuring logical consistency (Saaty, 1989; Saaty, 2002). Applications of AHP to vulnerability assessment have shown promising results in terms of transparency and adaptability (Shadmaan, M. S., & Popy, S. 2023; Sergiacomi, C., & Fagarazzi, C. 2020), yet often depend on field-derived data or semi-quantitative forms.

Nevertheless, in the authors' opinion, there remains a methodological gap; indeed, no current procedure provides a fully visual, photogrammetry-driven structure safety assessment system that translates image-based indicators into consistent vulnerability scores without the need for interior access or invasive surveys.

This work presents a systematic approach for the preliminary structural vulnerability evaluation of historic masonry structures, based just on using UAV photogrammetry. The methodology adapts official Italian Civil Protection forms: AeDES, GNDT and CARTIS (Baggio et al., 2007; GNDT, 1986; Zuccaro et al., 2023) for non-invasive applications and employs AHP to establish the relative significance of judgment criteria based on the visually detected parameters. The resulting operative tool, developed in a Microsoft Excel–VBA environment, is intended for use by local administrations and technical staff, facilitating the creation of vulnerability maps and digital decision-support tools for risk mitigation and building heritage preservation.

2 Methodology

The complete methodological workflow is sketched in Fig. 1. This flowchart summarizes the operational structure of the proposed method which starts with a UAV survey of the analyzed historic urban center followed by the construction of a 3D digital twin (DT) of the building aggregate. By visual inspection of the DT, the peculiar parameters affecting the structural vulnerability are then extracted for each single building. The extracted data are processed in an Excel-VBA environment where an AHP-based procedure also assigns a weight to each single parameter. Eventually, a vulnerability map of the aggregate is obtained, along with a vulnerability index associated with each single building. No comments are given on the survey procedure and the consequent construction of the DT model, both being standard procedures.

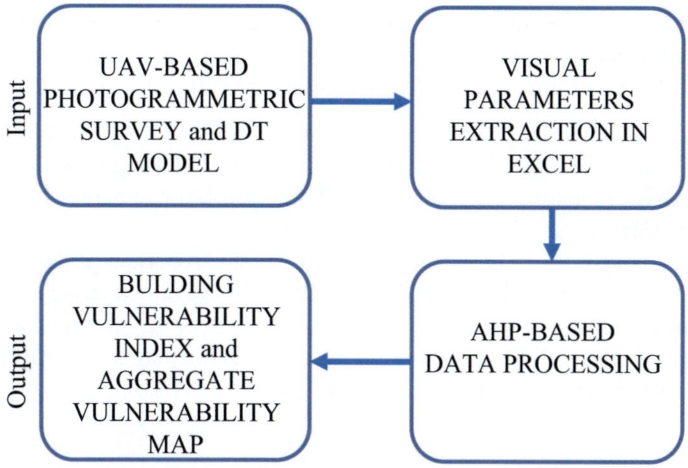

Fig. 1. Workflow of the proposed methodology, from UAV-based data acquisition to vulnerability index calculation via AHP and Excel–VBA automation.

2.1 Selection and Structuring of the Visual Detectable Parameters

The selection of the peculiar parameters is based on official documentation issued by the Italian Civil Protection Department for seismic damage and vulnerability assessment. The reference sources include the forms AeDES, GNDT and CARTIS (Baggio et al., 2007; GNDT, 1986; Zuccaro et al., 2023). Only parameters that are externally observable and reliably interpretable through UAV-based photogrammetry were considered. Parameters requiring invasive inspections, interior access, or unavailable documentation have been excluded.

The selected significant parameters have resulted in 23, and they have been grouped in 4 *evaluation criteria*, precisely:

- **General Characteristics:** typological position; site morphology; number of total floors; average building height; regularity in plan; regularity in elevation; type of aggregate; type of roof.
- **Structural Characteristics:** type of masonry; tie rods; structural bow windows; asymmetrical perforations.
- **External Envelope Feature:** surface damage; irregular vertical extension; presence of dangerous elements; different materials at different levels; presence of consolidation.
- **Structural Weaknesses:** visible crack; type of crack; crack orientation; out-of-plane walls deformation; damage at the connection between buildings; roof damage.

All the selected parameters were verified for compatibility with visual analysis conducted on orthophotos, 3D textured meshes, and oblique/nadir views generated via Structure-from-Motion (SfM) photogrammetry.

2.2 Hierarchical Weighting Model Based on AHP

The weighting process was structured according to a three-level hierarchy:

- Level 0: *Overall objective* that is the structural vulnerability assessment.
- Level 1: 4 criteria.
- Level 2: 23 Sub-criteria (distributed within the criteria).

Pairwise comparisons were conducted following Saaty's 1–9 scale (Saaty, 1989; Saaty, 2002). Five pairwise comparison matrices were constructed using expert judgment.

One matrix comparing the four criteria; Four matrices comparing the sub-criteria within each criterion.

For each matrix, the Consistency Index (*CI*) and Consistency Ratio (*CR*) were computed to verify internal coherence. All matrices yielded *CR* values < 0.10, confirming the reliability of the judgments. The *CI* and *CR* were calculated according to Saaty's method (Saaty, 1989) to assess the reliability of the judgments provided in each pairwise comparison matrix. The *CI* is defined as:

$$CI = \frac{\lambda_{max} - n}{n - 1} \tag{1}$$

Where n is the number of elements in the matrix and $\lambda_{\max}$ is the principal eigenvalue. The *CR* is computed as:

$$CR = \frac{CI}{RI} \tag{2}$$

Where *RI* is the Random Consistency Index, which depends on the order of the matrix. A *CR* value less than 0.10 is generally considered acceptable. If *CR* exceeds this threshold, the pairwise judgments are considered inconsistent and should be reviewed. Table 1 shows *RI* for $n = 1, 2, ..., 12$.

Table 1. Random Consistency Index

n	1	2	3	4	5	6
RI	0	0	0.58	0.90	1.12	1.24
n	7	8	9	10	11	12
RI	1.32	1.41	1.45	1.49	1.52	1.54

The hierarchical structure adopted in this study is illustrated in Fig. 2. It reflects the decomposition of the overall objective into four criteria, each further divided into sub-criteria.

This hierarchical structure enabled the calculation of a *global weight* (i. e., the weight of the single criterion within the four) and a *local weight* (i. e., the weight of each single sub-criterion within the criterion to which it belongs). A *final weight* of each sub-criterion (i. e., its weight within the 23) is simply obtained by multiplying the local by the global weight and it is used to compute the vulnerability index for each assessed building based on its visually detected features.

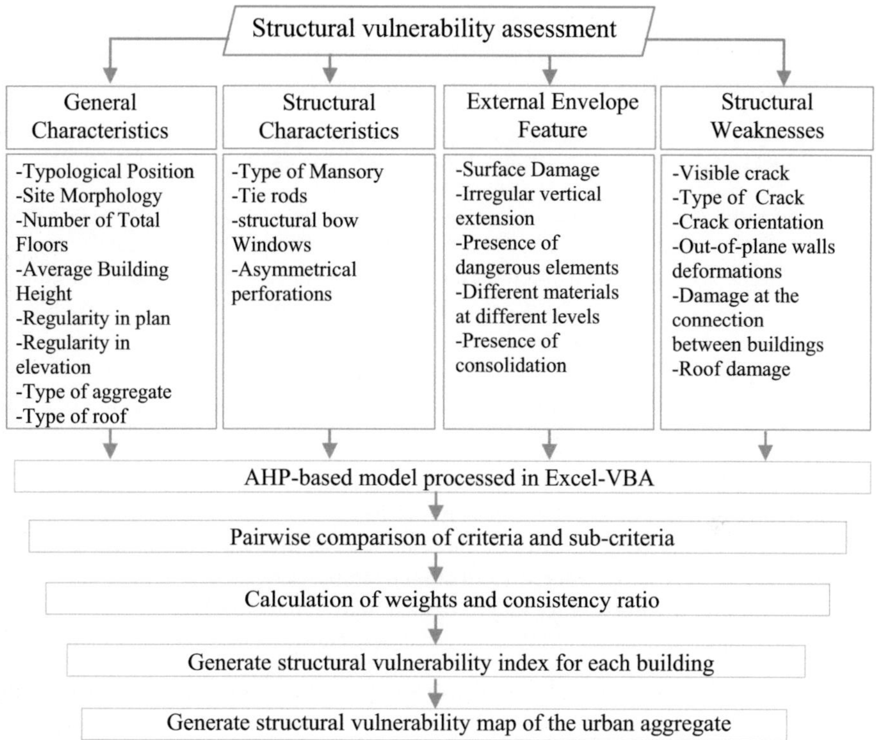

Fig. 2. AHP hierarchical structure used for criteria weighting and individual building vulnerability index computation.

2.3 Implementation in Excel-VBA

The methodology was implemented in Microsoft Excel, choosing it for its structured interface, accessibility and integrated automation support through Visual Basic for Applications (VBA). The spreadsheet is modular and includes the following core components:

- Pairwise comparison modules for entering the AHP judgement matrices at both criteria and sub-criteria levels.
- Automated processing blocks, executing matrix normalization, consistency checks, and hierarchical weight propagation.
- Building evaluation forms are used to input the observed peculiar characteristics (i. e., the discussed parameters or sub-criteria) of each examined building.

Each sub-criteria is associated with a closed list of possible answers, defined according to visual attributes identifiable in the 3D model or orthophotos. Each answer is assigned a score on a 1–5 ordinal scale as shown by Alizadeh et al. (2018), the validity of this scale stems from its capacity to convert subjective perceptions into consistent numerical values within a defined weighting system. The choice to implement a 1–5 scale was driven by the variability in the number of answer possibilities across sub-criteria: a few provide just two or three alternatives, while others present four. Employing a five-point scale facilitates a more equitable differentiation in the scoring system and mitigates excessive result variability that may develop with extended scales, where:

- 1 indicates a favorable condition (a low contribution to structure vulnerability).
- 5 indicates a critical condition (high contribution to structure vulnerability).

The scoring criteria follow two predefined classification frameworks based on national technical standards once again, reference is made to the forms: AeDES GNDT and CARTIS above quoted. The method facilitates the concurrent processing of numerous buildings within a single worksheet. For each building, the selected answers for each sub-criteria are converted into numerical scores, which are subsequently combined utilizing the weighted AHP model to get the building vulnerability index.

The VBA system automates the fundamental computational processes of the AHP-based evaluation methodology, including:

- Normalization of judgement matrices and computation of local priority vectors.
- Calculation of λ_{max}, CI and CR for each matrix.
- Application of the weights through the hierarchy and calculation of the weighted scores which are aggregated across all sub-criteria.
- Computation of a vulnerability index for each individual building.

The final output is a synthetic building vulnerability index, ranging from 1.00 (minimum vulnerability) to 5.00 (maximum vulnerability). The obtained values of the vulnerability index are finally grouped into three vulnerability classes:

- Low Vulnerability: 1.00–2.33
- Medium Vulnerability: 2.34–3.66
- High Vulnerability: 3.67–5.00

This classification, via the resulting vulnerability maps of the entire urban aggregate, supports decision-making by enabling intuitive comparison, spatial mapping, and prioritized technical inspections.

3 Results, Preliminary Validation and Discussion

The Analytic Hierarchy Process (AHP) indicated that the two criteria. “Structural Characteristics” and “Structural Weaknesses” have the greatest impact on the overall vulnerability assessment. Among the 23 sub-criteria, the most significant ones were out-of-plane walls deformation, type of masonry, damage at the connection between buildings and type of crack; such results correspond indeed with established literature on failure mechanisms in historic masonry structures.

The results of the AHP weighting process are summarized in the following tables. In particular:

- Table 2 reports the pairwise comparison matrix for the 4 criteria assumed. All values were assigned using Saaty’s fundamental scale. According to this scale, a score of 1 indicates equal importance between two criteria, while scores of 3, 5, 7, and 9 represent increasing levels of importance (moderate, strong, forceful, and extreme,

Table 2. Pairwise comparison matrix for the 4 criteria, with derived global weights and consistency indicators.

Criteria	General Characteristics	Structural Characteristics	External Envelope Feature	Structural Weaknesses	Global Weight
General Characteristics	1	1/3	1/2	1/4	10%
Structural Characteristics	3	1	2	1/2	28%
External Envelope Feature	2	1/2	1	1/3	16%
Structural Weaknesses	4	2	3	1	46%
λ_{max}			4,03		
CI			0,01		
CR			0,01		

Table 3. Pairwise comparison matrix for the sub-criteria within the "Structural Characteristics" Local and Final weights of the sub-criteria.

Structural Characteristics	Type of Masonry	Tie Rods	Structural Bow Windows	Asymmetrical Perforations	Local Weight
Type of Masonry	1	2	4	5	48,96%
Tie rods	1/2	1	3	4	30,54%
Structural Bow Windows	1/4	1/3	1	2	12,64%
Asymmetrical perforations	1/5	1/4	1/2	1	7,86%
Final Weight	13,70%	8,55%	3,54%	2,20%	
λmax			4,04		
CI			0,01		
CR			0,01		

respectively). Reciprocal values (e.g., 1/3, 1/5) are applied when the compared criterion is considered less important. The weights are expressed as percentages. The Consistency Ratio (*CR*) confirms the matrix is within acceptable consistency limits ($CR < 0.10$).

- Table 3 presents the sub-criteria comparison matrix for the "Structural Characteristics" criterion, just as a representative example. The same pairwise comparison procedure and scoring system based on Saaty's scale was applied to derive the relative weights.

Table 4. Final weights for each sub-criteria.

Criteria and Sub-Criteria	Local Weight	Final Weight
General Characteristics (10%)		
Typological position	9,77%	0,98%
Site morphology	8,38%	0,84%
Number of total floors	18,87%	1,89%
Average building height	20,76%	2,08%
Regularity in plan	13,09%	1,30%
Regularly in elevation	14,98%	1,50%
Type of aggregate	7,72%	0,77%
Type of roof	6,43%	0,64%
Structural Characteristics (28%)		
Type of masonry	48,96%	13,70%
Tie rods	30,54%	8,55%
Structural bow windows	12,64%	3,54%
Asymmetrical perforations	7,86%	2,20%
External Envelope Feature (16%)		
Surface damage	6,50%	1,04%
Irregular vertical extension	40,00%	6,40%
Presence of dangerous elements	25,10%	4,01%
Different materials at different levels	15,00%	2,40%
Presence of consolidation	13,40%	2,14%
Structural Weaknesses (46%)		
Visible crack	5,10%	2,35%
Type of crack	15,30%	7,04%
Crack orientation	10,40%	4,78%
Out-of-plane walls deformation	36,20%	16,65%
Damage at the connection between buildings	23,00%	10,58%
Roof damage	10,00%	4,60%

- Table 4 summarizes the local and final weights for all sub-criteria.
- Table 5 illustrates the calculation procedure for determining the vulnerability index combining the AHP-derived weights with the answer scores assigned to each sub-criterion. For every sub-criteria, the selected answer corresponds to a predefined score from 1 to 5. This value is multiplied by the sub-criteria's final weight to obtain the *sub-criteria weighted score.* For instance, the choice "neighboring" under the criterion Typological Position corresponds to a score of 3; multiplied by its final

weight of 0.98%, the result yields a weighted score of 0.03. The vulnerability index of the building is finally obtained as the sum of all sub-criteria weighted scores. The table reports only a selection of parameters for clarity, while the complete evaluation considers the entire set of criteria.

Table 5. Example of weighted score calculation for a sample building. The vulnerability index results from the sum of each sub-criteria's weight multiplied by the assigned answer score.

Criteria	Final Weight	Answer	Answer Score	Sub-criteria weighted score
Typological Position	0,98%	Isolated	1	0,03
		neighboring	3	
		connection	5	
Type of Masonry	13,70%	I don't know	0	0,00
		Regular masonry	1	
		Rough masonry	3	
		Irregular masonry	5	
Number of Total Floors	1,89%	1	1	0,09
		2	3	
		≥3	5	
…	…	…	…	…
		…	…	
Roof damage	4,60%	Yes	5	0,05
		No	1	
Sub-criteria weighted score = Final Weight X Answer Score				
Building vulnerability index = ∑ (Sub-criteria weighted score)				

Table 5 also includes cases where the selected response is "I don't know". In these situations, the corresponding sub-criteria is excluded from the evaluation, and its weight is proportionally redistributed among the remaining sub-criteria within the same criterion. For instance, when the type of masonry is not identifiable—frequent in drone-based surveys—the weight associated with that sub-criteria is reallocated across the other sub-criteria within the structural characteristics criterion.

To facilitate a cross-comparison among buildings of different typologies and state of preservation within the examined urban aggregate, the vulnerability index has been classified into *three vulnerability levels,* or *classes*. To this end the index range was divided into three equal intervals corresponding to *low*, *medium*, and *high vulnerability classes*, each associated with a specific color code and interpretation, as shown in Table 6.

Table 6. Classification of the vulnerability index. The index range is divided into three vulnerability classes, each associated with a color and a qualitative interpretation to support decision-making and prioritization of actions.

Vulnerability Class	Index Range	Color Code	Interpretation
Low	1.00 - 2.33	Green	The building has no visible structural weaknesses. No immediate intervention is required.
Medium	2.34 - 3.66	Yellow	Some vulnerabilities are present. Further inspection is recommended.
High	3.67 - 5.00	Red	The structure shows significant damage or instability. Priority intervention and detailed assessment are advised.

A colored structural vulnerability map of the entire urban aggregate can then be provided for a quick and easy consultation.

The preliminary application of the promoted tool has been implemented on a limited dataset that included masonry structures with diverse morphologies, materials and damage states. Precisely, as pilot test, the village of Bova, Reggio Calabria, Italy, was chosen. The village of about 500 inhabitants. The tool exhibited reliable performance, categorizing well-preserved and geometrically regular structures as low vulnerability while recognizing buildings with structural deterioration or geometric irregularities as medium or high vulnerability. The test affirmed the methodological effectiveness, structural vulnerability, and scalability for expedited assessment in historic urban environments.

4 Concluding Remarks

This study proposes a systematic and reproducible approach for evaluating the structural vulnerability of historic masonry structures, using only UAV-derived photogrammetric data. The methodology combines visual parameters obtained from official Italian national forms AeDES, GNDT and CARTIS with a multi-criteria weighting mechanism using the Analytic Hierarchy Process. The model is executed in a completely automated Microsoft Excel-VBA environment, guaranteeing methodological transparency and accessibility for non-specialist users.

The proposed method allows a non-invasive evaluation of buildings' structural vulnerability at the urban level, helping in the first identification of vulnerable buildings and guiding the prioritization of subsequent inspections or interventions. A pilot test validated the consistency and operational reliability of the tool.

Subsequent efforts will concentrate on expanding the application of the promoted tool to more extensive case studies, also incorporating the methodology into geospatial platforms or digital twin systems for enhanced planning and structural vulnerability management in historic built environments.

References

Alizadeh, M., Hashim, M., Alizadeh, E., Shahabi, H., Karami, M.R., Beiranvand Pour, A., Pradhan, B., Zabihi, H.: Multi-criteria decision making (MCDM) model for seismic vulnerability assessment (SVA) of urban residential buildings. ISPRS Int J Geo-Inf. *7*(11), 444 (2018)

Baggio, C., Bernardini, A., Colozza, R., Corazza, L., Della Bella, M., Di Pasquale, G., Dolce, M., Goretti, A., Martinelli, A., Orsini, G., Papa, F., Zuccaro, G.: Field manual for post-earthquake damage and safety assessment and short term countermeasures (AeDES). European Commission—Joint Research Centre—Institute for the Protection and Security of the Citizen, EUR, 22868 (2007)

GNDT, Emilia-Romagna Region, Tuscany Region.: Field Manual for the completion of the building vulnerability GNDT form. Litografia della Giunta Regionale (in Italian) (1986)

Guo, Q., Liu, H., Hassan, F.M., Bhatt, M.W., Buttar, A.M.: Application of UAV tilt photogrammetry in 3D modeling of ancient buildings. Int J Syst Assur Eng Manag. **13**(Suppl 1), 424–436 (2022)

Julià, P.B., Ferreira, T.M.: From single-to multi-hazard vulnerability and risk in historic urban areas: a literature review. Nat Hazards. ***108***(1), 93–128 (2021)

Leggieri, V., Mastrodonato, G., Uva, G.: GIS multisource data for the seismic vulnerability assessment of buildings at the urban scale. Buildings. **12**(5), 523 (2022). https://doi.org/10.3390/buildings12050523

Saaty, T.L.: Decision making, scaling, and number crunching. Decis Sci. **20**(2), 404–409 (1989)

Saaty, T.L.: Decision Making for Leaders: The Analytic Hierarchy Process for Decisions in a Complex World. RWS Publications, Pittsburgh (2002)

Sergiacomi, C., Fagarazzi, C.: Modelli Decisionali Multi Criterio per l'analisi della vulnerabilità sismica a scala territoriale: il caso studio della Garfagnana (Toscana) 1. Aestimum, 171–197 (2020)

Shadmaan, M.S., Popy, S.: An assessment of earthquake vulnerability by multi-criteria decision-making method. Geohazard Mech. **1**(1), 94–102 (2023). https://doi.org/10.1016/j.ghm.2022.11.002

Ulvi, A.: Documentation, three-dimensional (3D) modelling and visualization of cultural heritage by using unmanned aerial vehicle (UAV) photogrammetry and terrestrial laser scanners. Int J Remote Sens. **42**(6), 1994–2021 (2021). https://doi.org/10.1080/01431161.2020.1834164

Zuccaro, G., Dolce, M., Perelli, F.L., De Gregorio, D., Speranza, E.: CARTIS: a method for the typological-structural characterization of italian ordinary buildings in urban areas. Front Built Environ. **9**, 1129176 (2023)

A Visual Programming Tool for Assessing Out-of-Plane Collapse in Masonry Walls

Aurora Angela Pisano(✉), Mariaceleste Lasorella, Giulia Percolla, and Paolo Fuschi

Department of Architecture and Design – dAeD, University Mediterranea of Reggio Calabria, Reggio Calabria, Italy

{aurora.pisano,mceleste.lasorella,giulia.percolla, paolo.fuschi}@unirc.it

Abstract. The present paper focuses on the assessment and preservation of the structural safety of traditional masonry constructions by predicting their most probable collapse mechanism. The aim is to propose an efficient numerical approach for evaluating collapse in historical masonry structures, based on kinematic limit analysis applied to common local failure mechanisms and implemented through a visual scripting computational tool. The method integrates linear programming optimization within a CAD environment, utilizing an interactive Python component to compute the collapse load multiplier. The structural model is based on rigid blocks, following Heyman's assumptions. The effectiveness of the visual scripting framework is validated through the systematic evaluation of out-of-plane collapse mechanisms of masonry walls, demonstrating its potential for rapid and accurate structural safety assessments.

Keywords: visual scripting · optimization algorithm · masonry macro-blocks · kinematic limit analysis

1 Introduction

The preservation of the Italian cultural heritage is a crucial and widely discussed topic, addressed from various perspectives. Among these, particular attention is given to the structural safety of historic buildings, the majority of which are constructed with masonry. The structural integrity of these masonry constructions is influenced by a variety of factors, which must be carefully considered to ensure their long-term safety. Some of these factors are related to aspects inherent to the construction, such as the properties of materials used (e.g., compressive strength, porosity, presence of defects) and the construction techniques employed (e.g., mortar type, masonry quality, connections between elements). On the other hand, further factors depend on external conditions and involve forces acting upon the building. These include applied loads, e.g., permanent weights, variable loads, seismic forces, but also non mechanical actions like the ones exerted by sunlight, humidity, climatic actions and climate change in general. Such factors can lead to degradation and reduced durability of the masonry. Moreover, the interaction between internal and external factors determines the actual load-bearing capacity and vulnerability of the structure (Giuffrè, 1996) (Szabó et al., 2023).

C. Nava et al. (Eds.): *Climatic and Structural Safety in Multi-Hazard Regime of Cultural and Natural Heritage*, GREEN, pp. 41–58, 2026.
https://doi.org/10.1007/978-3-032-11926-1_3

The strong interest in the preservation of the architectural heritage and the complexity of the problem led to the development of various methods to assess the structural safety of masonry buildings. The choice of the appropriate method requires evaluating its specific advantages and limitations in relation to the structural problem at hand. A fundamental classification of masonry assessment approaches is based on the type of analysis method used to study the mechanical behavior of masonry and on how the masonry material is modeled within the method itself. A detailed overview is provided in the review by D'altri et al. (2020). In particular, analysis methods are generally categorized as: *nonlinear incremental approaches* and *direct or limit analysis ones.* While material modeling approaches include: *continuum mechanics-based models* and *rigid-block models* representing masonry as assemblies of discrete blocks.

The first category of approaches comprises nonlinear incremental methods primarily aimed at numerically simulating the full mechanical response of masonry structures, incorporating assumptions across different material scales and accounting for damage and interface effects. Recent studies employ finite element methods (FEM), often coupled with probabilistic approaches to address uncertainties in parameters used in FE analyses (Mazzeo et al., 2024). For instance, Galassi and Zampieri (2023) present an automated nonlinear incremental procedure to predict failure and collapse mechanisms in masonry arches, evaluating internal forces at block joints under equilibrium and continuity constraints. Similarly, Portioli and Lourenço (2024) propose a rigid-block model with elastoplastic softening interfaces for nonlinear static analysis, validated experimentally and numerically, focusing on material softening and failure mechanisms.

Nonlinear incremental methods provide insight, for example, into the seismic behavior of masonry, but face practical constraints: accurately modeling the heterogeneity and anisotropy of masonry requires detailed mechanical, material, and geometric parameters, making simulations computationally demanding; moreover, their strong sensitivity to input uncertainties can lead to wide variability in the predicted responses, limiting their use in real-world applications. All these factors necessitate careful consideration for selecting an appropriate technique for the assessment of masonry structures (Kumar et al., 2022).

On the other hand, limit analysis methods, despite their own limitations, provides a robust efficient framework alternative to nonlinear approaches. Indeed, they result particularly attractive when detailed material modeling and simulation are not feasible, offering key benefits in speed, accuracy for specific failure modes, and adaptability to various geometries. Limit analysis is, in fact, effective in predicting collapse loads, without modeling the full failure process. It is essentially based on equilibrium conditions which enhances its applicability in both seismic and non-seismic scenarios. Numerous local collapse phenomena have been studied and classified in this context (Milano et al., 2009), providing case studies to identify potential triggers of structural failure. The Italian regulatory framework (MIT, 2019) explicitly permits the application of limit analysis when assessing local collapse mechanisms. Consequently, numerous methods for masonry safety assessment are grounded on limit analysis theory, utilizing both static and kinematic approaches. These methods stem from the recognition that structural collapse often initiates from the localized instability of macro-elements.

Moreover, the numerical limit analysis methods, which frequently incorporate optimization algorithms, offer another significant advantage that is the capability of determining collapse loads and failure mechanisms even in complex geometries. Notably, in Baggio and Trovalusci (2000) the authors have explored nonlinear programming with friction for modeling masonry as rigid blocks, considering interface dilatancy. This allows for a more complete understanding of structural behavior under stress, especially for systems with irregular shapes or complex boundary conditions. Another advantage of limit analysis is its flexibility in modeling frictional interfaces, which significantly influence collapse mechanisms. Methods proposed by Ferris and Tin-Loi (2001) utilize constrained optimization to account for non-associative friction and tension-free contact, while in (Gilbert et al., 2006) a linear programming approach is introduced to model non-associative frictional joints. In (Gilbert, 2007) an efficient kinematic-based numerical method is also developed for bearing capacity assessment, further emphasizing the versatility of limit analysis in a wide range of scenarios. Nodargi et al. (2019) implemented a variational fixed-point algorithm for limit analysis of dry-masonry block structures considering frictional interfaces. Kinematic limit analysis approaches have instead been applied by Casapulla et al. (2023) for seismic vulnerability of masonry walls with or without grouted anchors. Although efficient, many applications of limit analysis are still restricted to idealized geometries and oversimplified collapse scenarios, which reduces their applicability to complex real-world masonry structures.

In the authors' opinion, in recent years, limit analysis has gained more attraction in professional practice through its integration within visual scripting environments, which enables efficient, interactive safety assessments of masonry structures. Pioneering work by (Block, 2009) introduced Thrust Network Analysis in CAD software, further improved in (Rippmann et al., 2012). In (Stockdale & Milani, 2019) a similar technique is applied in AutoCAD for arch safety evaluations. More recently, Funari et al. (2021) developed a Grasshopper-based tool in Rhinoceros that allows interactive and parametric exploration of collapse mechanisms via genetic algorithms. Mousavian et al. (2021) implemented a Grasshopper plugin for out-of-plane collapse in multi-storey walls, using a macro-block limit analysis that accounts for geometric and construction parameters. These recent developments demonstrate that embedding limit analysis in visual scripting environments delivers rapid, user-friendly, and computationally efficient workflows for both design and heritage conservation.

While previous studies have already demonstrated the potential of integrating limit analysis with visual programming, the present work aims to further develop this potential by refining the approach and emphasizing its effectiveness and broader applicability. Therefore, this study contributes to the current trend of coupling limit analysis with CAD tools by employing Grasshopper for safety evaluations of masonry walls. The promoted approach merges limit analysis, carried on masonry walls treated as a system of rigid macro-blocks, with a linear-programming optimization to identify collapse mechanisms and determine the corresponding load multiplier in real time. This approach, as already outlined, is particularly advantageous when detailed material behavior modeling is impractical due to the inherent complexity and variability of masonry materials. Through parametric inputs, users can instantly estimate how material, geometric, and loading variations influence structural behavior and test restoration or emergency intervention strategies. In line with recent advances in structural engineering, this study contributes to bridging the gap between theoretical research and practical heritage conservation tools. The paper is organized into five sections: after this introduction, Sect. 2 covers the theoretical background; Sect. 3 examines eight benchmark collapse mechanisms; Sect. 4 details the Grasshopper implementation; Sect. 5 offers concluding observations and future directions.

2 Theoretical Background

Limit analysis is here applied in its kinematic formulation also assuming that a masonry wall is modeled as an assembly of rigid, interlocked blocks under Heyman's hypotheses of zero tensile strength, infinite compressive strength and absence of sliding between blocks (Heyman, 1966). These assumptions apply to good-quality masonry with pre-existing interface cracks between visible, or recognizable, macro-elements.

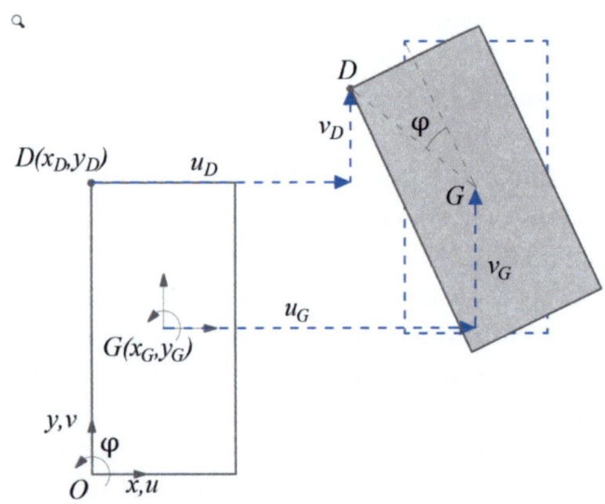

Fig. 1. Kinematic parameters related to the centroidal transverse section of the macro-element.

The procedure determines the collapse load multiplier, α, and the associated out-of-plane failure mechanism for each macro-element in the 2D plane of its centroidal transverse section (Fig. 1).

Referring to a global coordinate system (O, x, y), the displacement $\mathbf{u}_D = [\, u_D, v_D \,]^T$ of a point $D(x_D, y_D)$ on a macro-element can be written in terms of the horizontal and vertical displacements u_G and v_G of the macro-element centroid G, together with its rotation φ around G. The latter three parameters are assembled into the vector **d**:

$$\mathbf{d} = [u_G, v_G, \varphi]^T. \tag{1}$$

Displacements are taken positive when aligned with the global axes, as shown in Fig. 1. Moreover, the components u_D and v_D of the displacement vector can be defined by the kinematic matrix $\mathbf{S}$ defined as:

$$\mathbf{S} = \begin{bmatrix} 1 & 0 & -(y_D - y_G) \\ 0 & 1 & (x_D - x_G) \end{bmatrix}. \tag{2}$$

The displacement vector $\mathbf{u}_D$ can therefore be written as:

$$\mathbf{u}_D = \mathbf{Sd} \tag{3}$$

The kinematic description above enables characterization of the collapse mechanism via limit analysis. Specifically, the kinematic formulation leads to an optimization problem that minimizes the virtual work required to induce collapse. The application of the principle of virtual work yields the equilibrium condition equating the work of stabilizing dead loads, all collected in the vector $\boldsymbol{P}_d$, with that of destabilizing live loads, all collected in the vector $\boldsymbol{P}_l$; namely:

$$\boldsymbol{U}_d^T \boldsymbol{P}_d = \alpha \boldsymbol{U}_l^T \boldsymbol{P}_l \tag{4}$$

where $\boldsymbol{U}_d$ and $\boldsymbol{U}_l$ denote the displacement vectors of the application points of $\boldsymbol{P}_d$ and $\boldsymbol{P}_l$, respectively and α is the live loads multiplier. When the crack pattern is specified a priori, the collapse multiplier α can be obtained directly from the virtual-work balance (Eq. 4), as for the simple overturning mechanism of a masonry macro-element, see e.g. (Cifani, et al., 2006). Conversely, if the crack geometry is not known in advance, this explicit expression no longer holds and α must be determined by solving an optimization problem that searches for the most critical collapse configuration, as in the approaches proposed by (Ferris & Tin-Loi, 2001), (Gilbert, 2007) and (Nodargi et al., 2019).

Two collapse scenarios are considered for a masonry wall macro-element with fixed hinge A and roller B (Fig. 2(a,b)), reflecting crack patterns observed in situ. In the first scenario (Fig. 2(a)), the macro-element rotates about hinge A. In the second scenario (Fig. 2(b)), an internal hinge C appears at an undetermined height, dividing the macro-element into two sub-elements. This vertical bending mechanism often occurs in historic façade walls and the location of hinge C enters an optimization algorithm to identify the most critical collapse configuration. To formulate the optimization problem, a parameter $\mu > 1$ is introduced that defines the ratio between the unknown block heights h_1 and h_2, thereby fixing the internal hinge location. Specifically, for $\mu > 1$ the heights satisfy:

$$h_2 = \frac{h}{\mu}; \quad h_1 = \frac{\mu - 1}{\mu}; \quad \text{with } \mu > 1 \tag{5}$$

The optimization seeks the multiplier α from balance Eq. (4), subject to the constraint on μ, namely:

$$\min \alpha = \min_{\mu} \left[\frac{\boldsymbol{U}_d^T \boldsymbol{P}_d}{\boldsymbol{U}_l^T \boldsymbol{P}_l} \right] \quad \text{s.t.} \quad \mu > 1. \tag{6}$$

The optimization problem defined in Eq. (6) seeks the minimum collapse multiplier α under the constraint $\mu > 1$. This strict inequality guarantees that the internal hinge lies within the macro-element rather than at its extremity. The resulting linear-programming formulation is hereafter solved in Python using the *linprog* function.

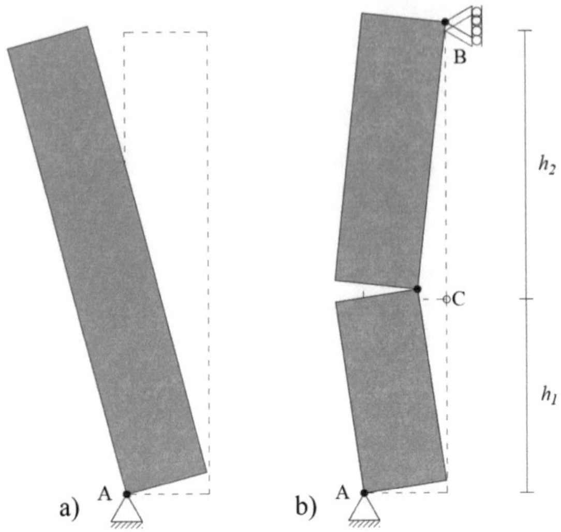

Fig. 2. Schematic representation of a masonry macro-element collapse mechanism: (a) simple overturning of a single macro-element around the external hinge A; (b) vertical bending of two interconnected macro-elements rotating around the internal hinge C, whose position is unknown.

3 Benchmark Collapse Mechanisms

Eight out-of-plane collapse mechanisms have been implemented, all falling under the two general categories introduced in the previous Sect. 2. Such mechanisms arise when external actions, most notably seismic forces, overcome the equilibrium of an entire wall or large wall segments. Their activation depends on factors governing out-of-plane response, including the quality of orthogonal wall connections, wall-to-wall interactions, roof configuration, opening layout, loads from upper stories, and even the building's position within the urban fabric. The considered mechanisms can be divided into *overturning mechanisms* and *bending mechanisms*. Overturning mechanisms may be *simple*, involving rotation of a single wall segment, or *composed*, if adjacent orthogonal wall portions are also dragged into motion; each mechanism can occur at one story or span multiple stories. Bending mechanisms are categorized by the orientation of the bending plane, *vertical* or *horizontal* and they too may affect a single level or multiple levels of the structure.

Tables 1, 2 and 3 schematically illustrate the eight out-of-plane collapse mechanisms addressed here, based on the classification of Milano et al. (2009), precisely: simple overturning of a single-story wall; simple overturning of a multi-story wall; composed overturning of a single-story wall; composed overturning of a multi-story wall; composed overturning of a corner wall; vertical bending of a single-story wall; vertical bending of a multi-story wall; and horizontal bending of a single-story wall. Each mechanism is represented in detail on the left column of Tables 1, 2 and 3, while in the right column the analytical expression of the corresponding collapse load multiplier is reported.

A simple description of each benchmark collapse mechanism is given next:

Simple Overturning of a Single-Story Wall

This mechanism affects a full single-story wall or a wide segment of it when out-of-plane forces cause it to move independently from adjacent sidewalls, as shown in Table 1(a). The wall rotates rigidly about a horizontal cylindrical hinge at its base, which spans the entire width of the macro-element. This simple overturning usually occurs at the topmost story or just below the roof. The analysis considers the centroidal transverse section of the wall, using the geometry, boundary conditions and loading depicted in Table 1(a). Specifically, W is the self-weight of the wall, F_v denotes the vertical thrust from arches or vaults, and P_s represents the floor or roof weight acting on the wall. Horizontal inertia forces arise by multiplying each of these vertical actions by the collapse multiplier α. The loads P_h and F_h account respectively for the static load from any head cover at the top of the macro-element and the horizontal thrust component of arches or vaults, while T represents the action of tie rods.

Simple Overturning of a Multi-Story Wall

This mechanism concerns a multi-story façade or large segments of it that, under out-of-plane forces applied at various levels, can detach from adjacent sidewalls (Table 1(b)). The wall undergoes rigid rotation about a known horizontal cylindrical hinge at its base, which spans the entire width of the macro-element. By analogy with the single-story case and using the same notation, the simple overturning of this multi-story macro-element is analyzed via its centroidal transverse section, with geometry, boundary conditions, and loads as shown in Table 1(b).

Composed Overturning of a Single-Story Wall

This mechanism concerns a single-story façade together with adjoining sidewall segments modeled as diagonal wedges (Table 1(c)). Under out-of-plane loading, the entire rigid macro-element rotates about a known horizontal cylindrical hinge at its base spanning its full width. The 2D analysis is carried out on the symmetry plane of this assembly, where the façade's centroidal section and the projected diagonal wedges, along with all applied forces, are treated as acting within that symmetry plane (Table 1(c)). This projection preserves the equilibrium of moments about the hinge.

Composed Overturning of a Multistory Wall

This mechanism addresses a multi-story façade and its adjoining sidewall segments, which extend across multiple levels and are idealized as paired diagonal wedges (Table 2(a)). Under out-of-plane loading, the entire rigid macro-element rotates about a fixed horizontal cylindrical hinge at its base spanning its full width. The analysis is conducted in 2D on the symmetry plane of this assembly, applying the same simplifying assumptions as before (Table 2(a)).

Composed Overturning of a Corner Wall

The corner-wedge mechanism is generated by two diagonal cracks along orthogonal walls, causing a rigid block to detach and rotate about a fixed hinge at the macro-element's base, as shown in Table 2(b). Typically observed in isolated buildings, this failure concentrates thrust at wall intersections. Assuming a 45° rotation plane (Table 2(b)), the analysis projects all loads, including the ridge-beam load P and the vertical and horizontal components P_{v1}, P_{v2}, and P_{h1} and P_{h2}, onto this plane (Table 2(b)).

Vertical Bending of a Single-Story Wall

The mechanism involves two rigid blocks in which the original macro-element divides. The mechanism is the one typical of a façade wall rotating out of plane about a horizontal cylindrical hinge that forms between successive levels, as shown in Table 3(a). Vertical cracks at the junctions with adjacent sidewalls define the detached wedge, which is typically restrained by tie rods or beams at the story's top and bottom. The 2D model on this symmetrical plane (Table 3(a)) includes a vertical load N applied at the macro-element's top. External hinges A and B are fixed a priori, while the intermediate hinge C height emerges from an optimization involving the blocks' weights W_1, W_2 and heights h_1, h_2 via the parameter μ.

Vertical Bending of a Multi-Story Wall

This mechanism mirrors the previous case but occurs in a multi-story façade wall—i.e., a wall spanning more than two levels where intermediate connections are ineffective—and involves out-of-plane rotation of the above mentioned two rigid blocks about a horizontal cylindrical hinge (Table 3(b)). The same considerations apply, and the 2D mechanical model is shown in Table 3(b).

Horizontal Bending of a Single-Story Wall

The mechanism ejects the top portion of a façade wall under out-of-plane loading, detaching wedge- or trapezoidal-shaped blocks via oblique vertical cylindrical hinges (Table 3(c)). This collapse mode typically affects the uppermost story when strong sidewall connections prevent overall overturning, but weak roof connections permit horizontal bending of the wall. The detached segment rotates as two rigid blocks, once again formed by the division of the original macro-element, about fixed hinges A and B at its base and an internal hinge C at an undetermined position (Table 3(c)), analyzed in the horizontal rotation plane of the block edges. The collapse multiplier is defined by balancing all applied loads—where H is the reaction from a bracing wall or tie rod, and P_{h_i} and P_{v_i} are the horizontal and vertical components of roof loads—and the hinge coordinates L_1, L_2 are functions of the optimization parameter μ, analogous to h_1 and h_2 in the vertical-bending case.

Table 1. Collapse mechanisms of overturning-type.

Overturning mechanism	
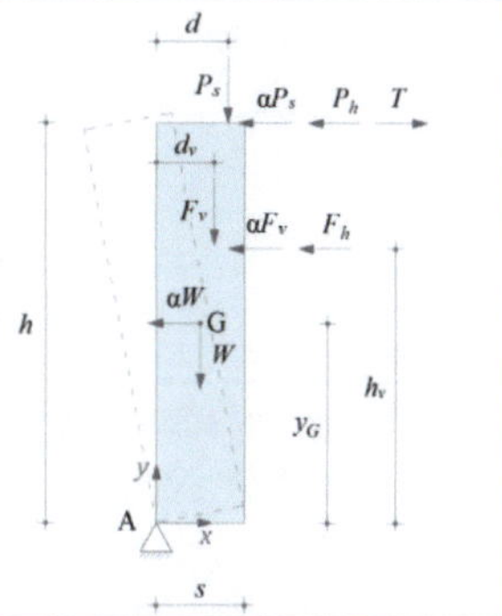	$\alpha = \frac{W\cdot\frac{s}{2}+F_v\cdot d_v+P_s\cdot d+T\cdot h-F_h\cdot h_v-P_h\cdot h}{W\cdot y_G+F_v\cdot h_v+P_s\cdot h}$
a) Simple overturning of a single-story wall	
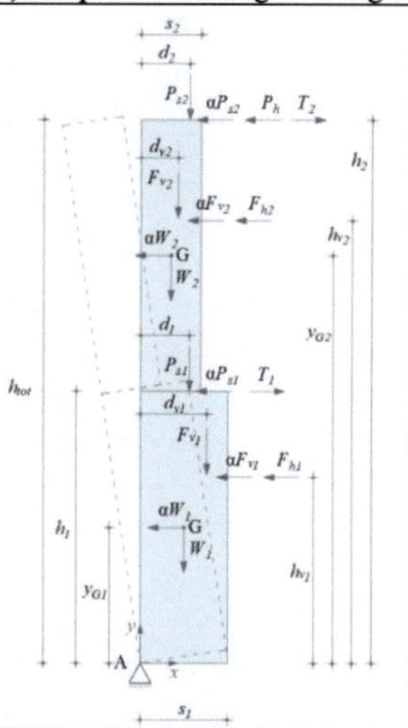	$\alpha = \frac{\left[\sum_i W_i\cdot\frac{s_i}{2}+\sum_i F_{vi}\cdot d_i+\sum_i P_{si}\cdot d_i+\sum_i T_i\cdot h_i-\sum_i F_{hi}\cdot h_{vi}-P_h\cdot h_{tot}\right]}{\left[\sum_i W_i\cdot y_{Gi}+\sum_i F_{vi}\cdot h_{vi}+\sum_i P_{si}\cdot h_i\right]}$
b) Simple overturning of a multi-story wall	
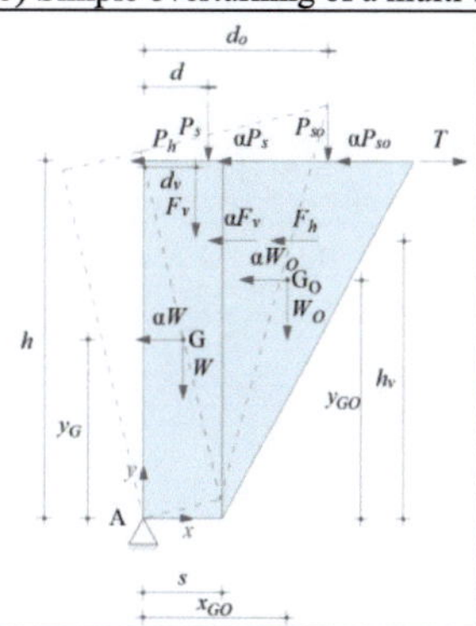	$\alpha = \frac{\left[W\cdot\frac{s}{2}+F_v\cdot d_v+W_O\cdot x_{GO}+P_s\cdot d+P_{so}\cdot d_o+T\cdot h-F_h\cdot h_v-P_h\cdot h\right]}{\left[W\cdot y_G+W_O\cdot y_{GO}+F_v\cdot h_v+P_s\cdot h+P_{so}\cdot h\right]}$
c) Composed overturning of a single-story wall	

Table 2. Collapse mechanisms of overturning-type.

Overturning mechanism	
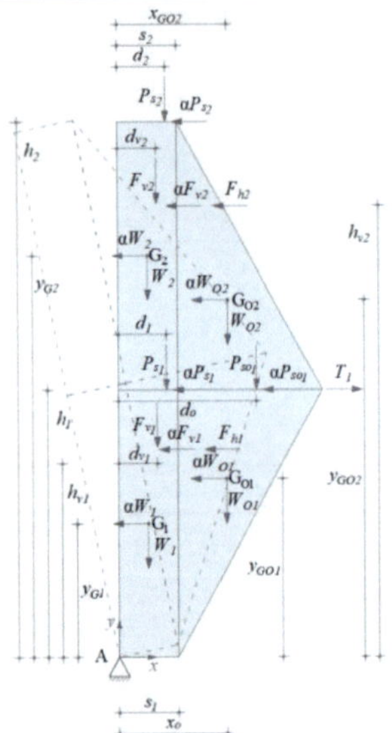	$\alpha = \dfrac{num}{\sum_i W_i y_{Gi} + \sum_i W_{Oi} y_{GOi} + \sum_i F_{vi} h_{vi} + \sum_i P_{si} h_i + \sum_i P_{soi} h_i}$ $num = \sum_i W_i \frac{s_i}{2} + \sum_i F_{vi}\, d_{vi} + \sum_i W_{Oi}\, x_{GOi} + \sum_i P_{si}\, d_i + \sum_i P_{soi}\, d_{oi} + \sum_i T_i\, h_i - \sum_i F_{hi}\, h_{vi}$
d) Composed overturning of a multistory wall	
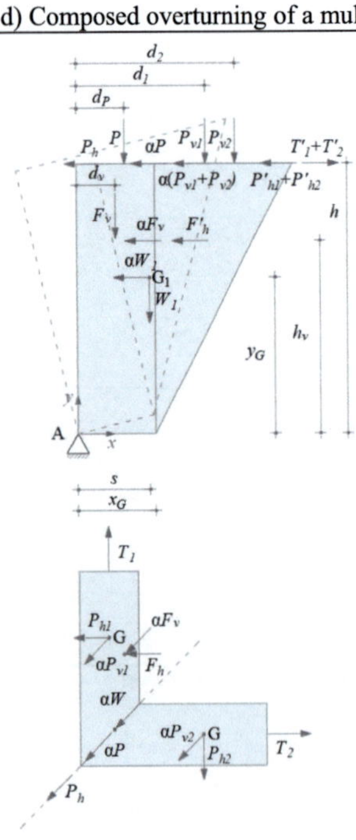	$\alpha = \dfrac{num}{W y_G + F_v h_v + (P+P_{v1}+P_{v2})\, h}$ $num = W x_G + F_v d_v + P\, d_p + P_{v1} d_1 + P_{v2} d_2 + (T'_1 + T'_2)\, h - F'_h h_v - (P_h + P'_{h1} + P'_{h2})\, h$
e) Composed overturning of a corner wall	

Table 3. Collapse mechanisms of bending-type.

Bending mechanism	
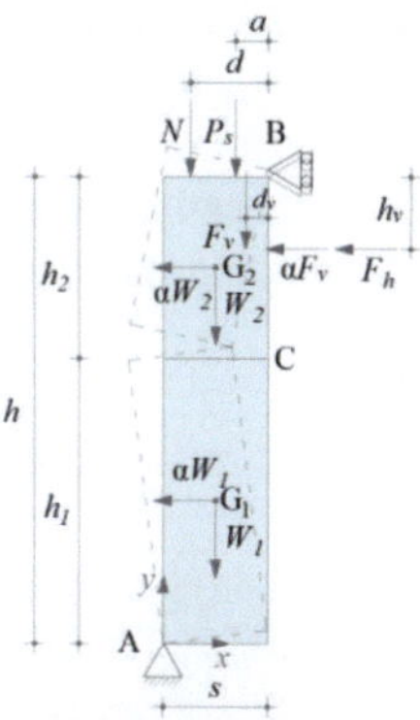	$\alpha = 2\dfrac{(\mu-1)(N\,d+P_s\,a+F_v\,d_v-F_h\,h_v)+s(W+N+P_s+F_v)}{(\mu-1)(W\,h/\mu+2F_v\,h_v)}$ $h_1=\dfrac{\mu-1}{\mu}\,h \quad h_2=\dfrac{h}{\mu} \quad W_2=\dfrac{W}{\mu} \quad W_1=\dfrac{\mu-1}{\mu}\,W$
f) Vertical bending of a single-story wall	
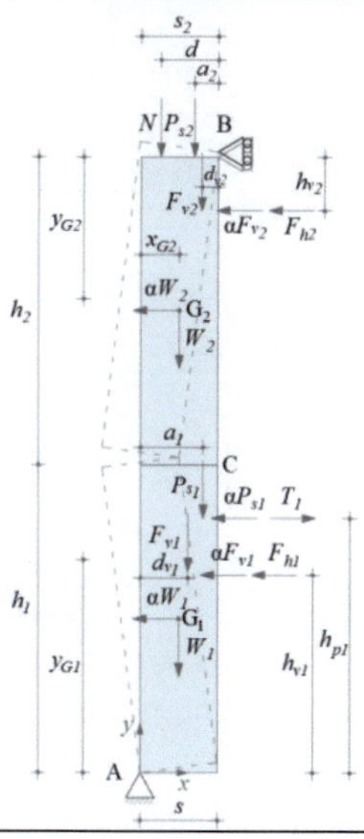	$\alpha = \dfrac{num}{W_1 y_{G1}+F_{v1}h_{v1}+P_{s1}h_{p1}+\left(W_2 y_{G2}+F_{v2}h_{v2}\right)(h_1/h_2)}$ $num = W_1 x_{G1}+W_2\left(s_2+x_{G2}\dfrac{h_1}{h_2}\right)+F_{v1}d_{v1}+F_{v2}s_2 +F_{v2}\dfrac{h_1}{h_2}d_{v2}+P_{s1}a_1+P_{s2}\left(s_2+a_2\dfrac{h_1}{h_2}\right) +N\left(s_2+d\dfrac{h_1}{h_2}\right)+T_1h_{p1}-F_{h1}h_{v1}-F_{h2}h_v\dfrac{h_1}{h_2}$ $h_1=\dfrac{\mu-1}{\mu}h,\ h_2=\dfrac{h}{\mu},\ W_2=\dfrac{W}{\mu},\ W_1=\dfrac{\mu-1}{\mu}W$
g) Vertical bending of a multi-story wall	
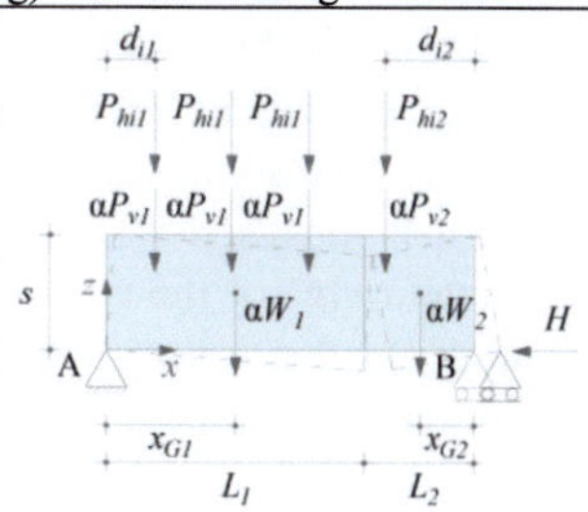	$\alpha = \dfrac{H\,s\left(1+\dfrac{L_1}{L_2}\right)-\sum_i P_{hi1}\,d_{i1}-\sum_i P_{hi}\dfrac{L_1}{L_2}d_{i2}}{W_1 x_{G1}+W_2\dfrac{L_1}{L_2}x_{G2}+\sum_i P_{vi1}\,d_{i1}+\sum_i P_{vi}\dfrac{L_1}{L_2}d_{i2}}$ $L_1=\dfrac{\mu-1}{\mu}L,\ L_2=\dfrac{L}{\mu},\ W_2=\dfrac{W}{\mu},\ W_1=\dfrac{\mu-1}{\mu}W$
h) Horizontal bending of a single-story wall	

4 Visual Programming Framework for Limit Analysis

The limit analysis optimization is implemented within Grasshopper, the visual scripting environment of Rhinoceros, enabling direct input of CAD geometries for masonry walls. This delivers an interactive workflow in which the kinematic limit analysis automatically identifies the most probable collapse mechanism and computes the corresponding load multiplier in real time. Figure 3 outlines the workflow as a series of Grasshopper components, organized into colored sections. The process is illustrated using one of the benchmark cases, addressed previously, namely the vertical bending mechanism in a two-story masonry wall.

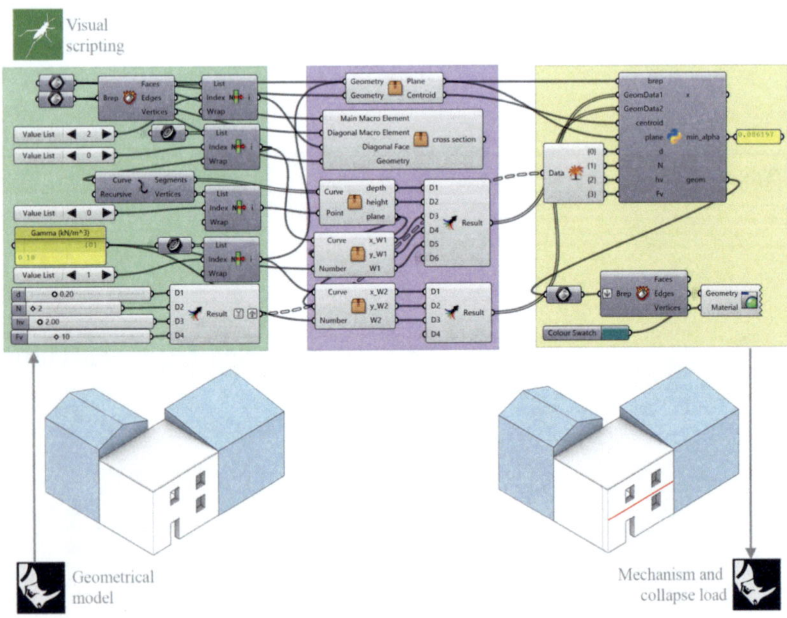

Fig. 3. Sketch of the workflow of the visual scripting approach for performing limit analysis on a double-story masonry façade with a vertical bending mechanism.

The process begins by defining the geometry of the two macro-elements and importing the CAD model into Grasshopper. The user then selects the relevant wall cross-section, i.e., the plane of potential mechanism rotations and specifies boundary conditions, applied loads, and material properties as shown in the green section (on the left) of Fig. 3. In the lilac section (central) the script automatically detects and extracts all necessary geometric information: the cross-section centroid locations, self-weight distributions, and dimensions. Finally, the yellow section (on the right) of the workflow visualizes the applied loads and contains the core Python script that joins all inputs and executes the limit-analysis optimization. The tool then outputs the collapse-load multiplier and furnishes the predicted collapse mechanism, including the location of the internal horizontal hinge. Users can adjust input parameters in real-time, with immediate feedback, allowing for rapid evaluation of how changes in geometry, material properties, or loading conditions impact the risk of collapse.

5 Results and Validation

The method previously presented is validated through the study of the eight benchmark collapse mechanisms, along with the related collapse multipliers, referred to a generalized wall, where the presence of openings is neglected. The thickness, height, and specific weight of the masonry wall will be kept constant and equal to respectively 0.4 m, 3 m for each level and 18 kg/m^3. The graphs showing the results obtained from the visual scripting tool are given next in terms of collapse multiplier α versus the peculiar actions of each benchmark collapse mechanism.

The first example analyzed is the simple overturning mechanism involving a single level of the structure. In this case, the thickness was reduced from 0.4 m to 0.2 m for the analysis. As shown in Fig. 4(a), values of the collapse multiplier α were obtained by varying T, representing the presence of a tie rod, and P_s, the weight of the floor/roof acting on the wall, within the range from 0 to 10 kN and from 0 to 20 kN, respectively. It can be observed that T has a stabilizing effect as P_s increases. The presence of the tie rod, when combined with an increase in the floor weight, allows the α value to increase. In the absence of P_s, and with only T, this value would have more consistently increased. Moving on to the second benchmark case in Fig. 4(b) of multi-story simple overturning in the façade, the input parameters T and P_s are also considered variable. The trend is similar to the previous case, with the key difference being that the maximum value of the collapse multiplier is reduced by half. This is due to the increased slenderness of the macro-element, as it concerns a two-story façade. In the third case, shown in Fig. 4(c), involving a composed overturning mechanism, the varying parameters are still T and P_s. The trend closely follows that of the previous case, with higher values of the collapse multiplier due to the presence of the sidewall. The sidewall, being involved in the overturning mechanism, counteracts the collapse with its own weight. The fourth case in Fig. 4(d) involves the entire height of the wall, with the sidewalls participating in the overturning mechanism. In this scenario, in addition to T, the action exerted by a vault on the top level, F_v, varies. As shown by comparing the trend of the graph with that in Fig. 4(b) although the sidewall is involved and counteracts the overturning with its own weight, the collapse multiplier is much lower. This highlights how the presence of the vault is a determining and critical factor in triggering the mechanism, thus affecting the vulnerability of the structure. In this case as well, T helps increase the collapse multiplier. Figure 4(e) shows the trend of the collapse multiplier for the corner overturning case. In this case, once again the parameters T and P_s were varied, resulting in a graph trend consistent with the other overturning cases.

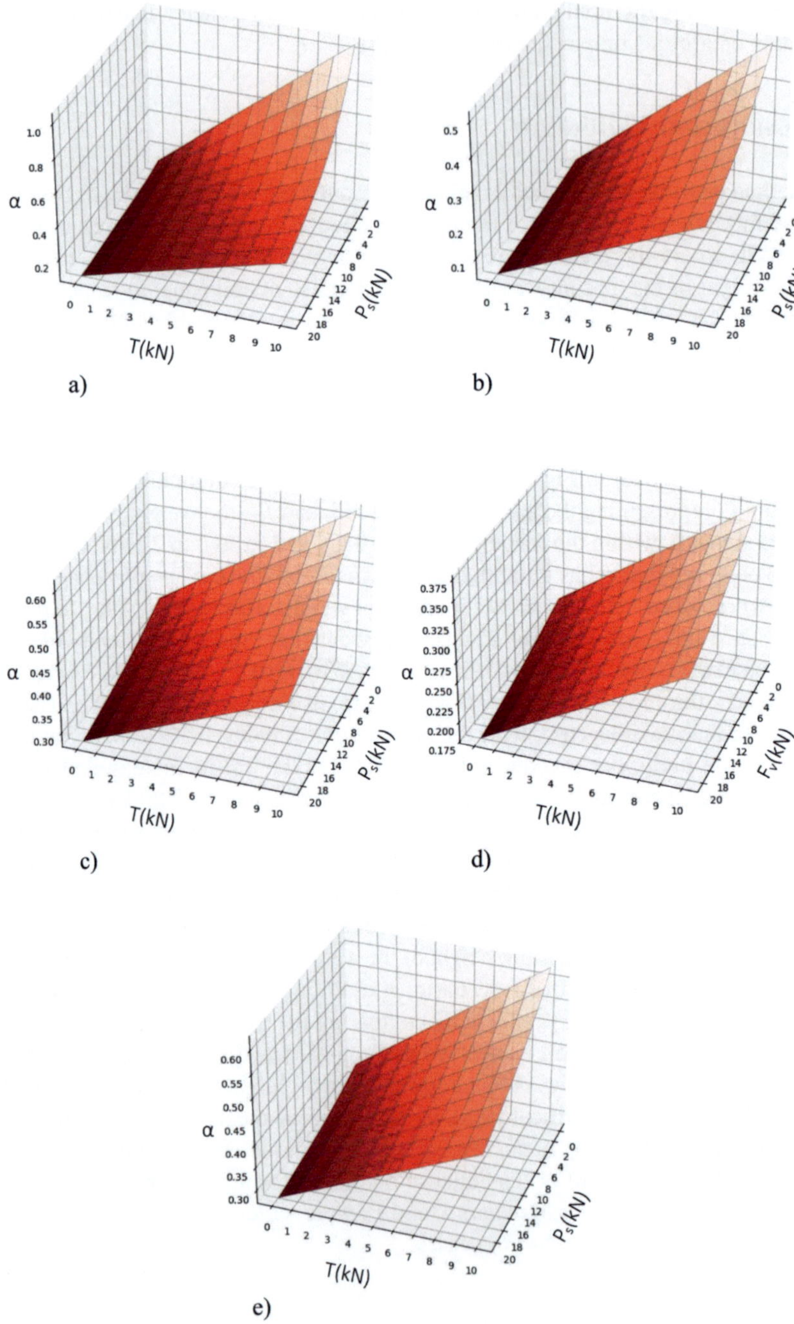

Fig. 4. Collapse multiplier versus the peculiar actions of each benchmark mechanism: the overturning mechanisms: a) simple overturning of a single-story wall, (b) simple overturning of a multi-story wall, (c) composed overturning of a single-story wall, (d) composed overturning of a multi-story wall, (e) composed overturning of a corner wall.

Figure 5 analyzes cases involving vertical or horizontal bending. In the first case, vertical bending (Fig. 5(a)) is considered for a single story, with F_v and N as varying parameters. The trend differs significantly from the overturning cases, with the mechanism being highly sensitive to changes in the values F_v and N. In this case, N plays a stabilizing role against F_v, ensuring a significant increase in the collapse multiplier.

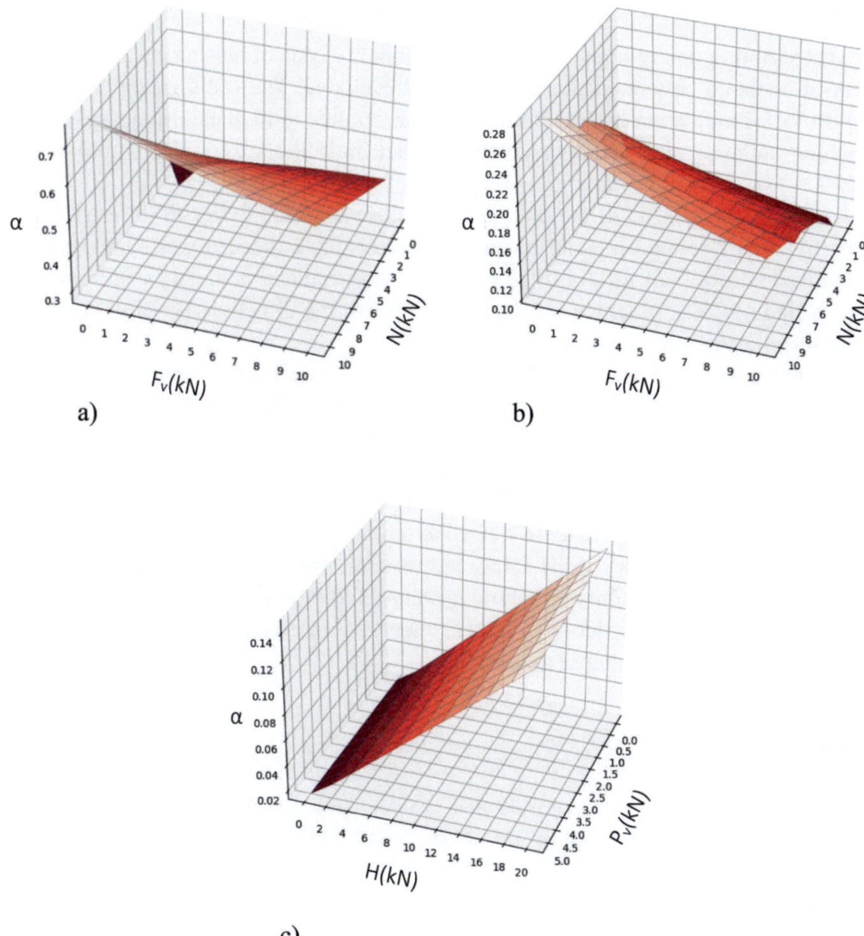

Fig. 5. Collapse multiplier versus the peculiar actions of each benchmark mechanism: the bending mechanisms: (a) vertical bending of a single-story wall, (b) vertical bending of a multi-story wall, and (c) horizontal bending of a single-story wall.

Fig. 5(b) represents the same mechanism for multiple stories. Here, the collapse multiplier is reduced, highlighting the structure's increased vulnerability compared to the single-story case, as the bending mechanism is more likely to develop along the entire height of the wall rather than being confined to a single level.

The last Fig. 5(c) shows the collapse mechanism due to horizontal bending, with H and P_v varying as the reaction force and vertical load, respectively. As observed, the presence of H increases the multiplier significantly, confirming its stabilizing effect in preventing the activation of the collapse mechanism.

6 Concluding Remarks

A visual scripting environment for kinematic limit analysis of masonry walls subjected to out-of-plane collapse mechanisms has been developed. The walls are modeled using a rigid-block-based approach, with limit analysis applied in its standard kinematic form to predict collapse mechanisms and related load multipliers through an optimization process.

A sensitivity analysis of the computed collapse load multiplier to different values of the live loads acting on eight benchmark out-of-plane mechanisms examined on a generalized (simplified) wall of fixed geometry, material and boundary conditions, has been carried out. The obtained results seem to prove the effectiveness of the proposed visual-scripting-based limit analysis tool.

The promoted user-friendly interactive approach allows quick identification of key factors, such as geometry, material properties, and applied load intensity, that influence the activation or prevention of collapse mechanisms in cracked walls. The tool also enables real-time evaluation of how adjustments, corresponding to structural restoration interventions, can improve safety, especially in post-seismic scenarios.

The results presented have to be viewed as a first step toward developing an efficient predictive tool for masonry walls safety assessment, with an ongoing research focusing on more complex masonry structural systems, sliding rigid-block mechanisms, and refining the rigid-block modeling to better reflect real masonry behavior including also the analysis of in-plane collapse mechanisms.

References

Baggio, C., Trovalusci, P.: Collapse behaviour of three-dimensional brick-block systems using non-linear programming. Struct Eng Mech. **10**(2), 181 (2000)

Block, P.P.: Thrust Network Analysis: Exploring Three-Dimensional Equilibrium. Massachusetts Institute of Technology: Doctoral Dissertation (2009)

Casapulla, C., Maione, A., Ceroni, F., Prota, A., Di Ludovico, M.: Limit analysis and design-oriented approach for out-of-plane loaded masonry walls strengthened by grouted anchors. Eng Struct. **285**, 115991 (2023)

Cifani, G., Petrucci, G., Lazzaro, D., D'Alessandro, S., Cialone, G., Martinelli, A., et al.: Repertorio dei meccanismi di danno, delle tecniche di intervento e dei relativi costi negli edifici in muratura. ITC-CNR (2006)

D'altri, A.M., Sarhosis, V., Milani, G., Rots, J., Cattari, S., Lagomarsino, S., …, De Miranda, S.: Modeling strategies for the computational analysis of unreinforced masonry structures: review and classification. Archives of computational methods in engineering (2020)

Ferris, M.C., Tin-Loi, F.: Limit analysis of frictional block assemblies as a mathematical program with complementarity constraints. Int J Mech Sci. **43**(1), 209–224 (2001)

Funari, M.F., Mehrotra, A., Lourenço, P.B.: A tool for the rapid seismic assessment of historic masonry structures based on limit analysis optimisation and rocking dynamics. Appl Sci. **11**(3), 942 (2021)

Galassi, S., Zampieri, P.: A new automatic procedure for nonlinear analysis of masonry arches subjected to large support movements. Eng Struct. **276**, 115359 (2023)

Gilbert, M.: Limit analysis applied to masonry arch bridges: state-of-the-art and recent developments. ARCH'07 – 5th International arch bridges conference (2007)

Gilbert, M., Casapulla, C., Ahmed, H.: Limit analysis of masonry block structures with non-associative frictional joints using linear programming. Comput Struct. **84**(13–14), 873–887 (2006)

Giuffrè, A.: A mechanical model for statics and dynamics of historical masonry buildings. In: Protection of the Architectural Heritage Against Earthquakes, pp. 71–152. Springer (1996)

Heyman, J.: The stone skeleton. Int J Solids Struct, 249–279 (1966)

Kumar, N., Barbato, M., Rengifo-López, E.L., Matta, F.: Capabilities and limitations of existing finite element simplified micro-modeling techniques for unreinforced masonry. Res. Eng. Struct. Mater. **8**(3), 463–490 (2022)

Mazzeo, M., Laudani, R., Santoro, R.: Uncertainty effect on seismic capacity assessment in the out-of-plane failure mechanisms of masonry structures by probabilistic and non-probabilistic approaches. Dev Built Environ. **17**, 100366 (2024)

Milano, L., Mannella, A., Morisi, C., Martinelli, A.: Illustrative sheets of the main local collapse mechanisms in buildings existing masonry and related kinematic models of analysis. Annex to the Repair and Strengthening Guidelines of structural elements, infills and partitions (2009)

MIT. Ministero delle Infrastrutture e dei Trasporti. CNTC19 – Circolare applicativa delle Norme Tecniche delle Costruzioni di cui al D.M. 17-01-2018 (NTC 2018). Gazzetta Ufficiale N. 42 del 20-02-2018. 2019 (in Italian) (2019)

Mousavian, E., Argiento, L. U., Casapulla, C.: Visual programming for macro-block analysis of multi-storey masonry buildings. 8th International Conference on Computational Methods in Structural Dynamics and Earthquake Engineering, COMPDYN 2021 (2021)

Nodargi, N.A., Intrigila, C., Bisegna, P.: A variational-based fixed-point algorithm for the limit analysis of dry-masonry block structures with non-associative Coulomb friction. Int J Mech Sci. **161**, 105078 (2019)

Portioli, F.P., Lourenço, P.B.: Nonlinear static analysis of masonry structures with mortar joints and cracking units by optimization-based rigid block models. Earthq Eng Struct Dyn. **53**(13), 3963–3982 (2024)

Rippmann, M., Lachauer, L., Block, P.: Interactive vault design. Int J Space Struct. **27**(4), 219–230 (2012)

Stockdale, G., Milani, G.: Diagram based assessment strategy for first-order analysis of masonry arches. J Build Eng. ***22***, 122–129 (2019)

Szabó, S., Funari, M.F., Lourenço, P.B.: Masonry patterns' influence on the damage assessment of URM walls: current and future trends. Dev Built Environ. **13**, 100119 (2023)

Exploring Earthquake-Induced Liquefaction Hazard Linked to Climate Change Effects in Sandy Soils

Daniela Dominica Porcino(✉) and Giuseppe Tomasello

University Mediterranea, Reggio Calabria, Italy
{daniela.porcino,giuseppe.tomasello}@unirc.it

Abstract. Climate change is increasingly affecting the stability of geo-structures by altering groundwater regimes, intensifying drought-rainfall cycles, and causing sea-level rise (*SLR*). Seismic liquefaction due to *SLR*, among others, poses a critical issue for low-lying coastal areas, as evidenced by the analysis of several case histories which report the expected impact of liquefaction in terms of in situ potential damages. These evidences can be explained considering the increased saturation of soils due *SLR*, as it leads to higher development of excess pore water pressure (*PWP*) during seismic events. In this context, it is relevant to predict in a reliable way the increase of excess *PWP* during seismic events taking into account future climate scenarios. In the present study, a series of 16 high-quality undrained cyclic simple shear tests, performed on Ticino clean sand prepared at relative densities ranging from 43 to 86% under a vertical effective stress of 100 kPa, were interpreted to calibrate the energy-based *PWP* model proposed by Berrill and Davis (*BD*) providing empirical correlations for the model's parameters. The performance of the *BD* model, implemented in DEEPSOIL 1D nonlinear response analysis software, was ascertained through numerical simulations of a centrifuge test, revealing the good agreement between experimental *PWP* measurements and numerical predictions.

Keywords: Climate change · Soil liquefaction · Simple shear · Energy-based pore water pressure model · Centrifuge test · Numerical analysis

1 Introduction

Climate change is affecting weather patterns, resulting in more frequent and intense precipitation events, as well as an increase in sea-level rise [1, 2]. These changes pose significant risks to the stability and serviceability of natural and artificial geo-structures, such as slopes, dams, and building foundations. These risks arise under both static conditions (e.g. slope instability, settlements) and dynamic conditions, particularly due to the seismic liquefaction geohazard.

Liquefaction occurs when excess pore water pressure (*PWP*) builds up in saturated sandy soils during seismic events, leading to a total or partial loss of soil strength and

C. Nava et al. (Eds.): *Climatic and Structural Safety in Multi-Hazard Regime of Cultural and Natural Heritage*, GREEN, pp. 59–76, 2026.
https://doi.org/10.1007/978-3-032-11926-1_4

development of critical settlements [3–5]. This geohazard is exacerbated in low-lying coastal areas due to sea-level rise (*SLR*) caused by climate changes [2, 6–9]. Therefore, an accurate prediction of the generation and dissipation of *PWP* during and after earthquakes is crucial for the evaluation of the seismic safety and resilience of structures and infrastructures. For this purpose, one-dimensional nonlinear numerical codes are generally devoted to predict excess *PWP* buildup during seismic events by various empirical *PWP* models, such as those based on stress, strain and energy approaches [10–12]. Among these, energy-based models offer the advantage of not requiring conversion of the seismic input motion into an equivalent number of loading cycles. The most widely adopted energy-based models are those proposed by Berrill and Davis [13] and Green et al. [11], both implemented into the DEEPSOIL software [14]. These models require the calibration of the constitutive parameters based on laboratory undrained cyclic tests. Most of the studies carried out in the literature referred to cyclic triaxial tests, while a limited number were based on cyclic torsional and simple shear tests [15–17], which better replicate in situ stress conditions before and during seismic loading.

The validation of excess *PWP* models by real site evidences is made complicated by the fact that it would require the installation of adequate in-situ instrumentation to measure *PWP* during real earthquake and only few case histories in the literature are available [18, 19]. In contrast, seismic geotechnical centrifuge tests provide a realistic response (at prototype scale) of in-situ soil conditions, allowing for the assessment of *PWP* development and susceptibility to liquefaction under controlled seismic loading conditions [20].

The present study, after examining the impact of climate change effects on geohazards, focuses on the increased risk of seismic liquefaction associated with *SLR*. The predictive capability of the Berrill and Davis [13] *PWP* energy-based model was validated through numerical simulations of a seismic centrifuge test performed to study the liquefaction response of Ticino sand. A comprehensive laboratory program, consisting in undrained cyclic simple shear tests, were carried out on Ticino clean sand to derive specific correlations for the empirical model parameters. The results gathered from nonlinear seismic response analysis of centrifuge test in terms of *PWP* variations with time and depth were compared with the corresponding experimental measurements.

This research is part of an ongoing Tech4You project ("Technologies for Climate Change Adaptation and Quality of Life Improvement"), which is funded by the National Recovery and Resilience Plan (*PNRR*) and supported by the European Union's NextGenerationEU initiative.

2 Geohazards and Climate Change Effects: An Overview

According to literature evidences [2, 6, 21–27], the climate signals which appear to have the most significant impact on building foundations and other structures interacting with soils (geo-structures) that are the aims of the Tech4You project are:

A. decreased precipitation/increased drought periods;
B. increased number of intense rain-drought cycles;
C. increased frequency of high rainfall events;
D. sea level rise.

These climate change signals can have multiple effects on geotechnical and geological processes and properties, affecting the ground, underlying soil or rock, groundwater, surface water and vegetation. They can therefore increase the existing known hazards as well as pose additional risks to geo-structures. The damages induced by a given climate event on a geo-structure are generally the result of various physical, chemical and mechanical processes whose study is a complex task that needs interdisciplinary knowledge and efforts. Table 1 reports a list of the geohazards and damages on existing buildings, along with the implications for the design and construction of the foundations of new buildings.

Among the listed geohazards, heave/subsidence phenomena in swelling/shrinkable soils are one of the most widely documented and detrimental risks, and they were observed in many countries located in tropical and temperate zones worldwide [28–30]. Soils that are prone to swelling or shrinkage do so in response to changes in moisture content, resulting in ground movement that can influence the safety and serviceability of structures built on them (Fig. 1).

Research into climate change effects predicts that a greater amount of moisture recharge can occur in a shorter period, thereby potentially exacerbating the shrink/swell effect [31]. Depending on the type of structure, local soil type, local climate, and expected future changes, the impact can vary in different scenarios. A first approximation estimate of the potential swelling risk of a soil is based on correlations with its liquid limit (*LL*) and plasticity index (*PI*) [32–34]. Nevertheless, there are many other parameters that can have an influence on the magnitude and rate of foundation heave [29], such as soil properties (composition of solids, concentration of pore fluid salts, composition of pore fluid, dry density, structure) and environmental conditions (i.e. climate, groundwater, drainage, vegetative cover, confinement, field permeability). A more rigorous evaluation of the swelling behaviour should be made through specific swell tests and suction tests, namely one-dimensional swell tests performed in traditional oedometers or consolidometers on undisturbed soil specimens [28, 35]; this is a good practice if a potential heave higher than 1.2 cm is estimated through the data provided by the correlations.

When groundwater regime naturally fluctuates in response to seasonal variations, it results in water table lowering during summer months and rising groundwater levels across the winter periods. On this regard, an increasing lowering of the groundwater level due to increasing deficit of rainfalls (A, Table 1) will increase the need of groundwater pumping from the deep aquifers by the communities. A change in groundwater level produces changes in the normal effective stresses of soils according to Terzaghi's principle of effective stresses and consequently a variation in soil mechanical properties. For example, groundwater level rise due to higher rainfall events (B, Table 1) can cause reduction in terms of strength of the soils and consequently, in regions characterised by slope stability problems, significant damages due to landslides can occur. On the other hand, in level ground conditions, settlements in compressible soils are expected to occur during groundwater lowering. Several case histories concerning subsidence phenomena induced by important long-term water exploitation projects occurred all over the world (including Italy) [25]. Figure 2 illustrates the link between intensive water extraction (water wells) and land subsidence (subsidence contours) and associated damages observed for a case history in Oltrepo Pavese (Italy) [36].

Table 1. Climate change signal and associated geohazards directly relevant to the design and construction of building foundations (modified from Insana et al. [22])

Climate change signals	Geohazard	Impact on existing buildings and foundation design for new buildings
A. Decreased precipitation/increased drought periods	Degradation of material strength due to shrinkage/desiccation and increased physical weathering Decreased surface and ground water level and flow	Potential damage to existing buildings constructed in accordance with current standards New buildings may require design and construction of deeper foundations to counter the effects or use of alternatives
B. Increased frequency of storms with high rainfall events	Flooding – direct damage, erosion and site inundation	Potential damage to existing buildings.
	Washout induced damage to foundations and structures	Design of foundations for buildings in flood-prone areas will have to consider the risk of washout.
	Slope stability – landslips	Potential damage to existing buildings.
	Reduction of strength associated with increase in pore water pressure on engineered structures and natural slopes	Potential increases in pore water pressure and potential loss of strength will have to be considered in stability assessments during design of new developments. Potential requirement for slope stabilisation.
	Dissolution and mining of chalk limestone and evaporites	Rainwater is slightly acidic and as a result, dissolves evaporities and calcareous rocks such as chalk and limestone. This process creates underground voids and collapse and potential catastrophic surface settlements and subsidence. All these features have the potential to be exacerbated by the effects of climate change.

(*continued*)

Table 1. (*continued*)

Climate change signals	Geohazard	Impact on existing buildings and foundation design for new buildings
	Instability of earthworks and fills (principally cohesive and weak rock fills) due to inundation	Potential damage to existing buildings. Potential requirement on new developments for pre-construction consolidation of fills.
	Mining areas – sinkholes and other erosion leading to subsidence and building damage	Potential damage to existing buildings. For new developments, careful assessment of old mine entries and associated backfill required in light of potential groundwater rise.
C. Increased number of intense rain/drought cycles	Degradation of material strength due to increased saturation/desiccation and increased weathering Increased shrink–swell behaviour of clay soils	Cracking and instability of engineered structures
D. Continued sea-level rise	Coastal flooding and storm surges – coastal erosion and site inundation leading to rockfalls, subsidence and building damage	Direct and indirect effects on existing buildings. Siting of new buildings will require careful assessment but then will not directly affect foundation design.
	Higher sea levels in coastal areas/corrosion of foundations from salt water	Potential damage to existing buildings. New buildings will require foundation design to consider corrosion potential.
	Increased risk of seismic liquefaction during seismic events in low lying coastal areas	Building damage due to ground surface subsidence, lateral spreading, etc.

Finally, sea-level rise is another impactful consequence of climate change which affects global communities living along coasts. In the following section, a detailed analysis concerning the linkage between sea-level rise and soil liquefaction geohazard, including an examination of the expected seismic-liquefaction induced damages through case histories, will be conducted.

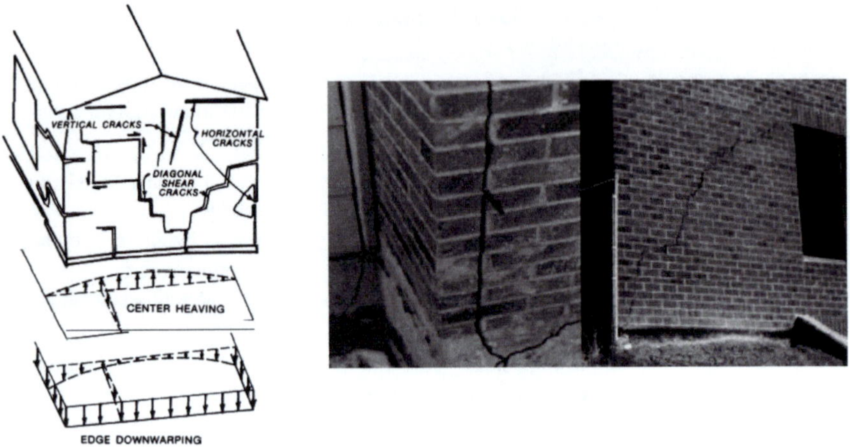

(a) (b)

Fig. 1. (a) Schematisation of wall fractures from doming heave of swelling and shrinking foundation soils [29], (b) photo of vertical and diagonal cracks on a structure rested on expansive foundations soils [30].

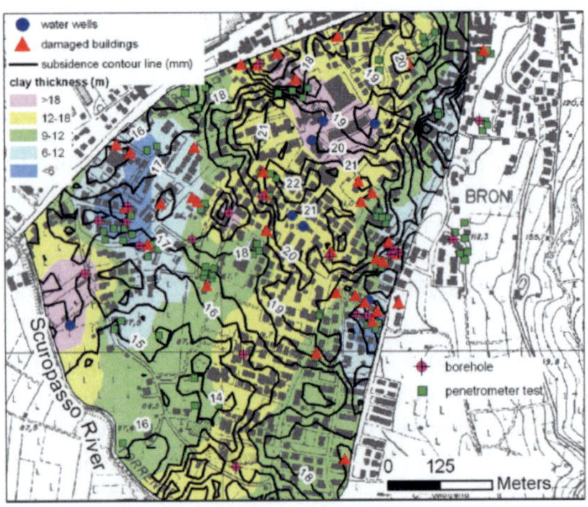

Fig. 2. Map of the ground displacements and damaged buildings in Oltrepo Pavese (Italy) during the drought period May 1998–November 2000 due to groundwater withdrawals [36].

3 Sea-Level Rise and Soil Liquefaction: Effects and Case Histories

Sea-level rise (*SLR*) is attributed to planetary-scale processes that contribute to the current sea-level trend. The influencing factors are global warming, which is driving the melting of ice and the thermal expansion of the oceans and finally, the vertical geological movements of the land along the coastal zones, which may accelerate (or decrease) the submersion of low-elevated coasts [37].

Between 1901 and 2010, various researchers have used different models to measure the rate of global *SLR* over different time scales, and the average rate of *SLR* was about 1.7 mm/year with a fluctuation range of 0.5 mm/year to 3 mm/year [38]. Since the end of the last century, more precise devices, such as the altimeter satellites, have been used to measure the rate of global *SLR*, showing average rates up to 3.3 mm/year from 1993 to 2010 [38]. With the passage of time, the value of global average *SLR* observed from the tide gauge and altimetry has become approximately 3.6 mm/year from 2006 to 2015 [39]. In recent years, satellite altimetry has shown that the increasing rate of the global average *SLR* occurs along with the disappearance of glaciers in many places [40, 41]. Glaciers melting is one of the major control factors of the *SLR* trend worldwide. It is linked to the global warming of the earth caused by the massive emission of CO_2 into the atmosphere due to human activities. *SLR* can influence underground and surface water levels in the near shore land in many ways. Firstly, when sea level rises, the runoff capacity of streams, rivers and underground fresh water in the near onshore land decreases. This will cause an uplift of rivers and groundwater table levels in the near shore areas. Such uplift can extend up to a significant distance from the coastline. Secondly, sea-level rise also increases the intensity of tidal action in estuary areas, enhancing the risk of coastal erosion. Finally, when the tidal level surpasses the underground water level, much seawater flows back into the underground fresh water. The interface between them is pushed towards the inland increasing the saline wedge intrusion. Work by Chang et al. [42] has shown that given enough time, aquifers and, as a result, the groundwater table position will respond to match the changed sea level position, assuming they are connected, and the aquifer is not artesian. In this regard, Fig. 3 shows the response of fresh and saline water to sea-level rise in an unconfined aquifer [43].

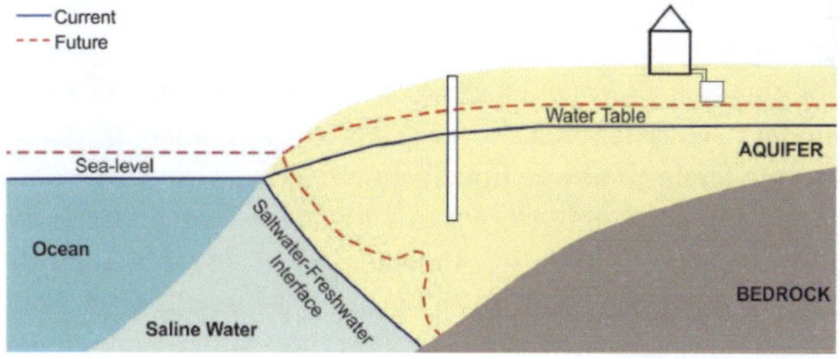

Fig. 3. Response of fresh and saline groundwater to sea-level rise in an unconfined aquifer (image from Jamaluddin et al. [43]).

Groundwater emerges from the earth where the water table intersects the land surface and may form new seeps and springs or provide additional flow to existing bodies of water. Sea level rise increases the risk of three potential geohazards: coastal erosion, corrosion of reinforced concrete foundations due to saline intrusion, and seismic liquefaction during seismic events in low-lying coastal areas. Seismic liquefaction is a common phenomenon associated with earthquakes, which causes severe building and infrastructure damage and loss of life and property (Fig. 4).

Liquefaction occurs under two conditions: (i) loose, granular, saturated soils such as young, sandy sediments, commonly found in reclaimed land or recent natural sandy-silty deposits; (ii) sufficiently strong and intense long-duration earthquake shaking. When sea level increases, groundwater level accordingly rises, especially in coastal areas, leading to a full saturation of previously unsaturated soils. During seismic events, this condition can cause high excess pore water pressures to develop, leading to a reduction of effective vertical stresses and, consequently, of shear strength and stiffness. As a result, large permanent shear strain up to a full collapse can occur with associated severe damages to structures and infrastructures.

Fig. 4. Effects of liquefaction phenomena after the 1989 Loma Prieta earthquake in coastal area (image from Poitras et al. [44]).

Several literature studies have accounted for the increased risk of liquefaction due to future *SLR* scenarios in coastal cities [2, 6–9]. By comparing liquefaction hazard maps before and after the sea-level rise in the range from 0.5 to 1 m, Quilter et al. [6] described a procedure for mapping liquefaction hazard based on liquefaction potential index [45]. The study highlights that *SLR* can exacerbate the liquefaction risk for earthquakes of varying magnitudes in Christchurch, New Zealand. In particular, the percentage of areas expected to have moderate-to-severe liquefaction manifestations when *SLR* was 0.5 and 1 m, resulted to be approximately from 2 to 4 times higher than those evaluated in the current sea level condition. Similarly, Abueladas et al. [8] evaluated the liquefaction probability in the Aqaba-Elat region when the sea level would risen from 0.5 m to 2 m using the approach proposed by Youd et al. [46]. The authors concluded that the risk of seismic liquefaction in this area would increase by 26% to 49%. United States Geological Survey (USGS) studied sea-level rise effects on liquefaction severity around San Francisco Bay [44]. Four hundred thirty-seven geotechnical borings were used throughout the Bay Area, primarily in San Francisco, Alameda, and Santa Clara Counties. It was found that although most sites were not sensitive to sea-level changes of less than 1 meter, some sites were susceptible to small changes in water levels. Therefore, for coastal cities, the risk of seismic liquefaction in this area will increase significantly under the effect of *SLR*.

4 Cyclic Pore Water Pressure Generation in Sandy Soils: From Experimental Study to Modelling

Given the increasing liquefaction risk due to the rise of sea level, it is relevant to predict in an accurate way the generation of excess pore water pressure (*PWP*) during earthquakes. For this reason, in the present research, an energy-based *PWP* model for sands was calibrated using high-quality undrained cyclic simple shear test results and subsequently validated through the numerical simulation of a seismic centrifuge test.

The analysis of a dataset of available laboratory undrained stress-controlled cyclic simple shear (*CSS*) tests previously performed at the University Mediterranea of Reggio Calabria [47] on Ticino sand (Table 2) was analysed for the purpose of the paper.

Table 2. Physical properties of Ticino sand

G_S	e_{max}	e_{min}	D_{50}	C_U
2.68	0.93	0.58	0.57	1.5

Note: G_S = specific gravity, e_{max}, e_{min} = maximum and minimum dry density, D_{50} = mean grain size, C_U = uniformity coefficient.

The testing program comprised 16 undrained *CSS* tests performed on specimens reconstituted with moist tamping method at different relative densities (D_R) ranging from 43% to 86%, and consolidated at the same value of vertical effective stress σ'_{v0} equal to 100 kPa, before the application of different cyclic stress ratios (*CSR*) varying from 0.12 to 0.25. To establish the onset of liquefaction, a criterion of 3.75% shear strain in single amplitude (γ_{SA}) was adopted.

Figure 5 shows a picture of the apparatus used to perform laboratory tests, i.e. a modified *NGI* simple shear device [48], and the typical results of a test ($D_R = 43\%$, $CSR = 0.14$) in terms of shear stress (τ), shear strain (γ) and excess pore water pressure (Δu) with the number of loading cycles (N), as well as the τ vs. γ relationship. It is worth noting that the simple shear apparatus replicates in a more realistic way the in situ ground conditions before and during earthquakes, compared to the conventional cyclic triaxial apparatus. Typical *CSS* results represented in Fig. 5b highlight that, under undrained conditions (typical of seismic events), sands tend to develop significant Δu, leading to reduced stiffness and increasing shear strains, which consequently cause severe damages to structures and infrastructures.

A “promising” approach, i.e. the energy-based approach, was considered for the interpretation of simple shear test results [11, 13, 15, 17, 49, 50]. Some energy-based models, namely Green et al. [11] (*GMP*) [Eq. (1)] and Berrill and Davis [13] (*BD*) [Eq. (2)], are implemented in the widely used DEEPSOIL 1D numerical code [14] to evaluate the earthquake site response of a site:

$$R_u = \sqrt{\frac{\left(W/\sigma'_{v0}\right)}{PEC}} \tag{1}$$

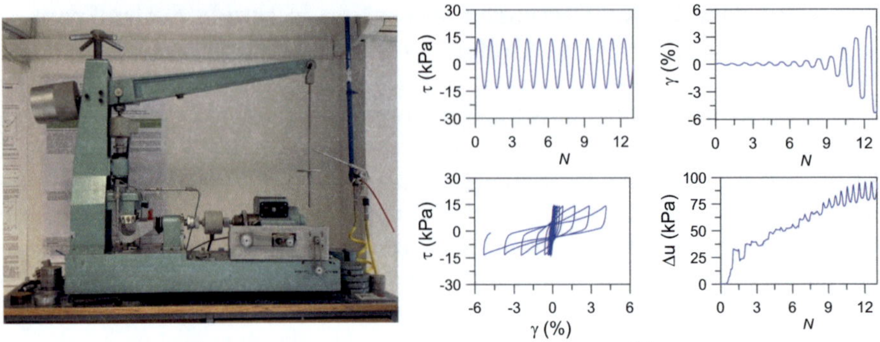

(a) (b)

Fig. 5. (a) Picture of the *NGI* simple shear apparatus used in the experimentation and (b) typical results of a test executed on Ticino sand.

$$R_u = \alpha \bullet \left(\frac{W}{\sigma'_{v0}}\right)^{\beta} \tag{2}$$

where $R_u = \Delta u/\sigma'_{v0}$ is the excess pore water pressure ratio, (W/σ'_{v0}) is the normalised dissipated energy per unit volume, *PEC*, α and β are positive calibration parameters. Setting the calibration parameters α and β in Eq. (2) equal to $1/PEC^{0.5}$ and 0.5, respectively, one can observe that the *GMP* model is a special case of the general model initially proposed by Berrill and Davis [13].

In previous studies [15, 17, 47] where the experimental data of undrained *CSS* tests were interpreted by using the *GMP* and *BD* energy-based *PWP* models, positive conclusions were drawn in terms of measured vs. predicted $R_u - W/\sigma'_{v0}$ trends.

In the present study, the dissipated energy per unit volume *W* was calculated from the hysteresis loops in τ-γ plane (Fig. 5b), as follows [Eq. (3)]:

$$W = \frac{1}{2} \bullet \sum_{i=1}^{n-1} (\tau_{i+1} + \tau_i) \bullet (\gamma_{i+1} - \gamma_i) \tag{3}$$

where n = number of load increments, τ_{i+1} and τ_i are the applied shear stress at load increment i + 1 and i, respectively, whereas γ_{i+1} and γ_i are the shear strains at load increment i + 1 and i, respectively. In the following analyses, the normalised dissipated energy was expressed as a percentage (i.e., γ in percent) in order to define empirical parameters to be used directly in numerical simulations, in accordance with the requirements of the DEEPSOIL code.

Figure 6a reports residual excess pore water pressure ratio $R_{u,res}$ vs. W/σ'_{v0} for Ticino clean sand with different D_R and applied *CSR*. It should be emphasised that, in Fig. 6a, $R_{u,res}$ represents R_u measured when the applied cyclic stress ratio is equal to zero.

It is evident that relative density of sand has a remarkable impact on the $R_{u,res} - W/\sigma'_{v0}$ relationship, whereas the applied cyclic stress ratio induced by earthquake loading resulted in a minor influence (Fig. 6a). It was also observed that the rate of excess pore water pressure generation increases as relative density decreases, and this trend was well captured by *DB* model. For practical utility, correlations relating model

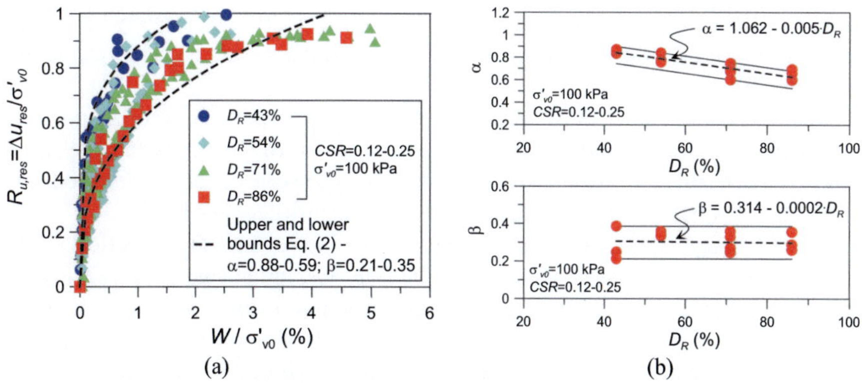

Fig. 6. (a) Residual excess pore water pressure as a function of normalised dissipated energy using the model proposed by Berrill and Davis [13] and (b) correlation for α and β of the *BD* model as a function of D_R for Ticino sand.

calibration parameters α and β of *BD* model with D_R (expressed in percentage) of Ticino sand were also determined, obtaining:

$$\alpha = 1.062 - 0.005 \bullet D_R \tag{4a}$$

$$\beta = 0.314 - 0.0002 \bullet D_R \tag{4b}$$

Such correlations [Eq. (4a and b)] are reported in Fig. 6b. The *BD* model does not appear particularly sensitive to the value of β, as suggested also by Berrill and Davis [13]. An "optimal" value of β for the tested sand is 0.314 (Fig. 6b), which results lower than the value recommended by Berrill and Davis [13], i.e. 0.5. As may be observed from Fig. 6b, α decreases with increasing D_R.

5 Validation of Energy-Based *PWP* Generation Models through Seismic Centrifuge Tests

In a previous study [47], the authors validated the *GMP* energy-based model for predicting both the generation and dissipation of excess pore water pressure in earthquake events through numerical simulations of a seismic centrifuge test, emphasising the importance of an accurate calibration of soil parameters. With the aim to provide an additional advanced tool for practitioners involved in liquefaction geohazard studies, in the present paper the authors applied the same methodological approach to assess the performance of the *BD* energy-based model.

It is worth mentioning that seismic geotechnical centrifuge tests utilise a high acceleration field, preserving the stress-strain response of the prototype soil, thus providing a more realistic prediction of field behaviour in terms of excess pore pressure buildup and onset of liquefaction at the prototype scale [20, 51, 52]. The selected geotechnical seismic centrifuge test, labeled as M1_GM17, was carried out at ISMGEO of Seriate

(Bergamo, Italy), which hosts the unique geotechnical centrifuge currently operative in Italy (Fig. 7a). All details of the analysed M1_GM17 test are reported in Fioravante et al. [20] (Fig. 7b).

Numerical simulations of M1_GM17 test at the prototype scale were conducted by the non linear 1D seismic code DEEPSOIL [14], where the *BD* energy-based *PWP* model proposed by Berrill and Davis [13] is implemented. The performance of this model was verified by comparing numerical vs. experimental (centrifuge) test results.

Figure 8 shows the main soil parameters required to perform the numerical simulations; in particular there are shown: soil profile (Fig. 8a), initial void ratio (e_0) (Fig. 8b), shear modulus at small strain (G_0) (Fig. 8c), coefficient of volumetric compression (*mv*) (Fig. 8d), permeability coefficient (k) (Fig. 8e) profiles and normalised shear modulus (G/G_0) and damping ratio (D) vs. shear strain curves (Fig. 8f). All these parameters were determined through high-quality laboratory tests specifically performed on Ticino sand, i.e. the same material used in the centrifuge test. The empirical parameters α and β of the *BD* model were derived from undrained cyclic simple shear tests as described in Section 4. Preliminary sensitivity analyses evidenced the need to adopt a depth-dependent profile for α and β parameters, in addition to that of D_R. Specifically, the upper bound of the α range (Fig. 6b) was assigned at the top of the centrifuge model and the lower bound at the base, while the opposite trend was adopted for β (Fig. 6b). This variation with depth is consistent with previous studies [47, 52], which reported an underestimation of excess pore water pressure at shallow depths when constant values of empirical parameters were used in energy-based models.

(a)

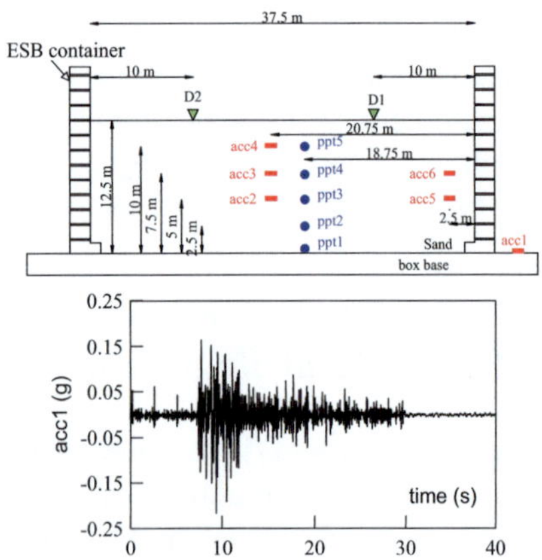

(b)

Fig. 7. (a) ISMGEO centrifuge test and (b) schematic view of centrifuge model test at prototype scale with the applied seismic input at the base of the model [53].

The coefficient of permeability (k) of Ticino sand during shaking was set equal to 2.5 times the static value (namely 0.005 m/s) (Fig. 8e) to better simulate the excess pore water pressure dissipation phase, which generally takes place during the shaking phase both before and after liquefaction occurs.

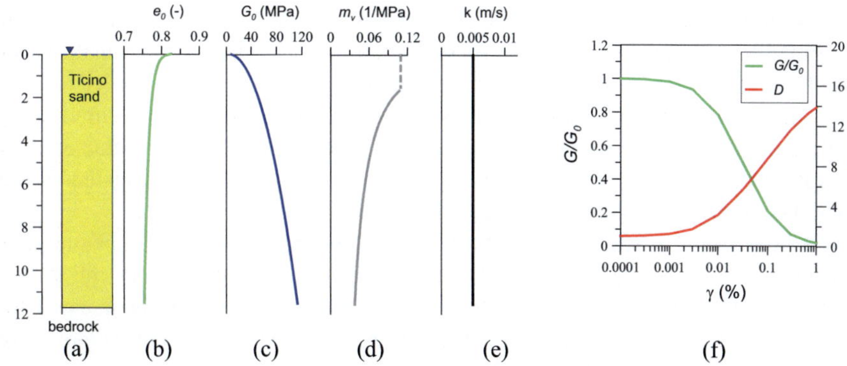

Fig. 8. Profiles of the soil input parameters used in the numerical analysis of centrifuge tests: (a) soil profile, (b) initial void ratio, (c) shear modulus at small shear strain, (d) coefficient of volumetric compression, (e) permeability coefficient; and (f) variation of the normalised shear modulus) (G/G_0) and damping ratio (D) with shear strain for Ticino sand.

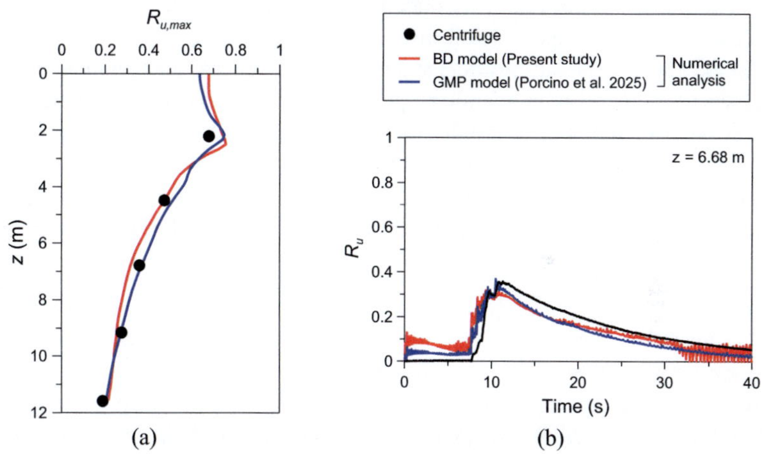

Fig. 9. Comparison between measured and predicted: (a) $R_{u,max}$ profiles, (b) R_u vs. time at $z =$ 6.68 m using *BD* and *GMP* models implemented in DEEPSOIL code.

The results of the numerical analyses carried out on the M1_GM17 centrifuge test by using the *BD* energy-based *PWP* model are reported in Fig. 9 in terms of maximum excess pore water pressure ratio ($R_{u,max}$) vs. depth and R_u vs. time for a specific depth ($z = 6.68$ m). For comparison purposes, the results obtained using the *GMP* model in the previously mentioned work conducted by the authors [47] were also reported. As it can be seen, the *BD* model reasonably predicts the peak excess pore water pressure ratio $R_{u,max}$ (Fig. 9a), with an accuracy that is fully comparable to that of the *GMP* model [47]. The differences between the two energy-based models were minimal, and both proved to be effective in simulating the experimental excess pore water pressures. Moreover, the build-up and the dissipation of excess pore pressures observed in the dynamic centrifuge tests were both accurately captured by the numerical simulations (Fig. 9b). A slight overestimation of the excess pore water pressure ratio, R_u, was observed during the initial phase of the seismic loading (i.e., $t < 8$ s), particularly in the *BD* model. This discrepancy can be attributed to the values of the empirical parameters adopted in the simulations. Notably, the *BD* model uses smaller values of β parameter than the fixed value equal to 0.5 employed in the *GMP* model. As a result, even small values of dissipated energy (W) can lead to appreciable values of R_u using the *BD* model formulation.

As a final result, it was verified that a careful and realistic selection of the input parameters in the numerical simulations, such as soil density state, compressibility, permeability, and shear modulus at small strain profile, is crucial for capturing in a realistic way the experimental measurements in terms of excess pore pressure, consistently with what reported by other authors [51, 52].

6 Conclusions

Seismic liquefaction induced by sea-level rise associated to climate change represents a relevant topic in coastal areas, as documented by several studies in the literature. In the present study, an experimental investigation, consisting of laboratory undrained cyclic simple shear tests and physical modelling in a seismic geotechnical centrifuge, was performed to analyse the seismic response of Ticino clean sand in terms of excess pore water pressure generation and dissipation. The aim of the research was to validate through numerical analyses of centrifuge tests at the prototype scale the energy-based model proposed by Berrill and Davis (*BD*) for predicting excess pore water pressures generated during earthquakes causing liquefaction phenomena. The main findings of the study are the following:

- Sea-level rise alters coastal groundwater regimes, leading to higher saturation levels that markedly increase the risk of seismic liquefaction in low-lying areas. This geohazard poses a threat to the stability of built environments, amplifying the potential for ground failure, infrastructure damage, and loss of life during seismic events. Accordingly, liquefaction risk must be evaluated in the light of ongoing climatic changes.
- The results of undrained cyclic simple shear (*CSS*) tests reported in R_u vs. normalised energy (W/σ'_{v0}) plane indicate that a key factor influencing the development of seismic excess pore water pressure (*PWP*) is the relative density (D_R) of sandy soils, whereas cyclic stress ratio (*CSR*) has a minor impact. Useful correlations for deriving α and β parameters of *BD* model were proposed as a function of D_R. Such correlations can be effectively used in numerical simulations of seismic response in potentially liquefiable sands.
- The comparison between the *PWP* response detected in the centrifuge test and the corresponding one predicted by the numerical analyses, carried out using the *BD* model implemented in the DEEPSOIL code, demonstrates good agreements at all depths.
- The numerical analyses of the seismic centrifuge test highlighted that an accurate and reliable selection of the input soil parameters, such as density state, compressibility, permeability, and shear modulus at small strain at all depths, is a relevant issue to

predict the experimental measurements in terms of excess pore water pressures in a realistic way.

Acknowledgements. This research was funded by the National Recovery and Resilience Plan (PNRR), Mission 4, Component 2, Investment 1.5, funded by the European Union—NextGenerationEU for the realization of Innovation Ecosystem Tech4You "Technologies for climate change adaptation and quality of life improvement", Pilot Project 4.7.1—Open platform "phigital space" (physical and digital) of the type "user profiling" for the advanced and dynamic co-design of interventions on the built and ex novo (Project code: ECS_00000009, CUP: C33C22000290006).

References

1. Galassi, G., Spada, G.: Sea-level rise in the Mediterranean Sea by 2050: roles of terrestrial ice melt, steric effects and glacial isostatic adjustment. Glob. Planet. Chang. **123**, 55–66 (2014)
2. Dong, L., Cao, J., Liu, X.: Recent developments in sea-level rise and its related geological disasters mitigation: a review. J Mar Sci Eng. **10**, 1–20 (2022)
3. Kramer, S.L.: Geotechnical earthquake engineering. Prentice Hall, Upper Saddle River (1996)
4. Bray, J.D., Sancio, R.B., Durgunoglu, T., Ohap, A., Youd, T.L., Stewart, J., Seed, R.B., Cetin, O.K., Bol, E., Baturay, M.B., et al.: Subsurface characterization at ground failure sites in Adapazari, Turkey. J. Geotech. Geoenvironmental Eng. **130**, 673–685 (2004)
5. Porcino, D., Diano, V.: Laboratory study on pore pressure generation and liquefaction of low-plasticity silty Sandy soils during the 2012 earthquake in Italy. J. Geotech. Geoenviron. Eng. **142**, 04016048 (2016)
6. Quilter, P.W, van Ballegooy, S., Russ, M.: The Effect of Sea Level Rise on Liquefaction Vulnerability. In: Proceedings of the 6th International Conference on Earthquake Geotechnical Engineering, Christchurch (New Zealand), 1–4 November (2015)
7. Monk, C.B., van Ballegooy, S., Hughes, M., Villenueve, M.: Liquefaction vulnerability increase at North New Brighton due to subsidence, sea level rise and reduction in thickness on the non-liquefying layer. Bull N Z Soc Earthq. Eng. **49**(4), 334–340 (2016)
8. Abueladas, A.-R.A., Niemi, T.M., Al-Zoubi, A., Tibor, G., Kanari, M., Ben-Avraham, Z.: Liquefaction susceptibility maps for the Aqaba-Elat region with projections of future hazard with sea-level rise. Q. J. Eng. Geol. Hydrogeol. **54**, 1–12 (2021)
9. Tomasello, G., Porcino, D.D., Lucanto, D., Mangano, G.: The impact of sea-level rise on seismic liquefaction susceptibility of coastal areas and risk mitigation. In: Sànchez, A., Tunio, M.N., Longato, D., Mohammad, P. (eds) Mediterranean landscapes Infrastructure: A Data-Driven Analysis of Housing and Design. CROSSMED 2024. Springer Series in Sustainable Civil Infrastructure. Springer Nature Switzerland AG (in press)
10. Vucetic, M., Dobry, R.: Pore Pressure Build-Up and Liquefaction at Level Sandy Sites during Earthquakes [Research report CE-86-3]. Department of Civil Engineering. Rensselaer Polytechnic Institute, Troy (1986)
11. Green, R.A., Mitchell, J.K., Polito, C.P.: An energy-based excess pore pressure generation model for cohesionless soils. In: Proceedings developments in theoretical geomechanics–the john booker memorial symposium, Sydney (Australia) (2000)
12. Park, T., Park, D., Ahn, J.K.: Pore pressure model based on accumulated stress. Bull. Earthq. Eng. **13**, 1913–1926 (2015)
13. Berrill, J.B., Davis, R.O.: Energy dissipation and seismic liquefaction of sands: revised model. Soils Found. **25**(2), 106–118 (1985)

14. Hashash, Y.M.A., Musgrove, M.I., Harmon, J.A., Ilhan, O., Xing, G., Numanoglu, O., Groholski, D.R., Phillips, C.A., Park, D., Deepsoil. A nonlinear and equivalent linear seismic site response of 1-D soil columns. 2024, V7.1. User manual. Urbana, IL: Board of Trustees of University of Illinois at Urbana-Champaign [Software], Version 7.1 (2024)
15. Porcino, D.D., Tomasello, G., Farzalizadeh, R.: Pore-pressure generation of sands subjected to cyclic simple shear loading: an energy approach. In: Wang, L., Zhang, J.M., Wang, R. (eds) Proceedings of the 4th international conference on performance based design in earthquake geotechnical engineering (Beijing 2022). PBD-IV 2022. Geotechnical, geological and earthquake engineering, vol 52. Springer, Cham (2022)
16. Khashila, M., Karray, M.K., Hussien, M.N., Ramirez, J., Chekired, M.: Validation of strain/energy-based pore pressure model in one-dimensional response analyses using centrifuge tests. Soil Dynam Earthq Eng. **173**, 108096
17. Tomasello, G., Porcino, D.D.: Energy-based pore pressure generation models in silty sands under earthquake loading. Geosciences. **114**, 166 (2024)
18. Ishihara, K., Shimizu, K., Yamada, Y.: Pore water pressures measured in sand deposits during an earthquake. Soils Found. **21**(4), 85–100 (1981)
19. Zimmaro, P., Kwak, D.Y., Tsai, Y.T., Stewart, J.P., Brandenberg, S.J., Mikami, A., Kataoka, S.: Database on seismic response of instrumented flood control levees. Earthquake Spectra. **36**(2), 924–938 (2020)
20. Fioravante, V., Giretti, D., Airoldi, S., Moglie, J.: Effects of seismic input, fine crist and existing structure on liquefaction from centrifuge model tests. Bull. Earthq. Eng. **19**, 3807–3833 (2021)
21. Vardon, P.J.: Climatic influence on geotechnical infrastructure: a review. Environ Geotechnics. **2**, 166–174 (2015)
22. Insana, A., et al.: Climate change adaptation of geo-structures in Europe: emerging issues and future steps. Geosciences. **11**, 1–24 (2021)
23. Gariano, S.L., Guzzetti, F.: Landslides in a changing climate. Earth Sci. Rev. **162**, 227–252 (2016)
24. Dixit, M.S.: Damage mechanism in problematic soils. Int J Civil Eng Tech. **7**, 232–241 (2016)
25. Tzampoglou, P., Ilia, I., Karolis, K., Tsangaratos, P., Zhao, X., Chen, W.: Selected worldwide cases of land subsidence due to groundwater withdrawal. Water. **15**, 1–29 (2023)
26. Toll, D.G., Abedin, Z., Buma, J., Cui, Y., Osman, A.S., Phoon, K.K.: The Impact of Changes in the Water Table and Soil Moisture on Structural Stability of Buildings and Foundation Systems: Systematic Review CEE10–005 (SR90). Technical Report 2012. Collaboration for Environmental Evidence (2012)
27. Vahedifard, F., Williams, J.M., Aghakouchak, A.: Geotechnical engineering in the face of climate change: role of multi-physics processes in partially saturated soils. IFCEE GSP. **295**, 353–364 (2018)
28. Chen, F.H.: Foundations on Expansive Soils. Elsevier Scientific Publishing Company, Amsterdam (1975)
29. TM 5–818-7 (Technical Manual). Foundations in Expansive Soils. Department of the Army, Headquarters, Washington (1983)
30. Arya, A., Choudhary, A., Khan, A., Ram, H.: Foundations of expansive soils. Int J Eng Res Tech (IJERT). **2**(3), 340–343 (2014)
31. Karim, M.R., Rahman, M.M., Nguyen, K., Cameron, D., Iqbal, A., Ahenkorah, I.: Changes in Thornthwaite moisture index and reactive soil movements under current and future climate scenarios—a case study. Energies. **14**, 6760 (2021)
32. Bowles, J.E.: Foundation Analysis and Design. McGraw-Hills Inc., New York (1988)
33. Vijayvergiya, V.N., Ghazzaly, O.I.: Prediction of swelling potential for natural clays. In: Proceedings 3rd International Conference on Expansive Soils, Haifa, Israel (1973)

34. McKeen, G.R., Lytton, R.L.: Expansive Soil Pavement Design Using Base Studies. In: International Conference on Case Histories in Geotechnical Engineering, Rolla, Mo., May (1984)
35. Gaspar, T.A.V., Jacobsz, S.W., Heymann, G., Toll, D.G., Gens, A., Osman, A.S.: The mechanical properties of a high plasticity expansive clay. Eng. Geol. **303**, 106647 (2022)
36. Meisina, C., Zucca, F., Fossati, D., Ceriani, M., Allievi, J.: Ground deformation monitoring using the permanent scatterers technique: the example of the Oltrepo Pavese (Lombardia, Italy). Eng. Geol. **88**, 240–259 (2006)
37. Antonioli, F., De Falco, G., Lo Presti, V., Moretti, L., Scardino, G., Anzidei, M., Bonaldo, D., Carniel, S., Leoni, G., Furloni, S., Marsico, A., Petitta, M., Randazzo, G., Scicchitano, G., Mastronuzzi, G.: Relative Sea-level rise and potential submersion risk for 2100 on 16 Coastal Plains of the Mediterranean Sea. Water. **12**, 2173 (2020)
38. Micheal, O,, Bruce, G., Hinker, J., Roderik, V., Frederikse, T.: Sea level rise and implications for low Lying Islands, coast and communities. In: IPCC special report on the ocean and cryosphere in a changing climate, Geneva (Switzerland) (2019)
39. Fox-Kemper, B., Hewitt, B.H.T, Xiao, C., Aðalgeirsdóttir, G., Drijfhout, S.S., Edwards, T.L., Golledge, N.R., Hemer, M., Kopp, R.E., Krinner, G., et al.: Climare change 2021: the physical science bases. In: a contribution of working Group of sea Level to the sixth assessment report of the Intergovernmental Panel on Climate Change, Intergovernmental Panel on Climate Change (IPCC), Geneva (Switzerland) (2021)
40. Chen, X., Zhang, X., Church, J.A., Watson, C.S., King, M.A., Monselesan, D., Legresy, B., Harig, C.: The increasing rate of global mean sea-level rise during 1993–2014. Nat Clim Change. **7**, 492–495 (2017)
41. Zemp, M., Huss, M., Thibert, E., Eckert, N., McNabb, R., Huber, J., Barandun, M., Machguth, H., Nussbaumer, S.U., Gartner-Roer, I., et al.: Global glacier mass changes and their contributions to sea-level rise from 1961 to 2016. Nature. **568**, 382–386 (2019)
42. Chang, S.W., Clement, T.P., Simpson, M.J., Lee, K.-K.: Does sea-level rise have an impact on saltwater intrusion? Adv. Water Resour. **34**, 1283–1291 (2011)
43. Jamaluddin, U.A., Yaakub, J., Suratman, S., Pereira, J.J.: Threats faced by groundwater: a preliminary study in Kuala Selangor. Bull Geol Soc. Malaysia. **62**, 65–72 (2016)
44. Poitras, T., Grant, A., Wein, A., Knudnes, K., Befus, K., Erdman, M., Petersen, K.: Liquefaction and sea-level rise. Sensitivity to liquefaction hazards from sea-level rise in the San Francisco Bay area, California. Available online: https://geonarrative.usgs.gov/liquefactionandsealevelrise/. Accessed 07.02.2022
45. Iwasaki, T., Tatsuoka, F., Tokida, K., Yasuda, S.: A practical method for assessing soil liquefaction potential based on case studies at various sites in Japan. In: Proceedings of the Second International Conference on Microzonation for Safer Construction Research and Application, San Francisco (California), 26 November – 1 December (1978)
46. Youd, T.L., et al.: Liquefaction resistance of soils: summary report from the 1996 NCEER and 1998 NCEER/NSF workshops on evaluation of liquefaction resistance of soils. J. Geotech. Geoenviron. **127**(10), 817–833 (2001)
47. Porcino, D.D., Tomasello, G., Giretti, D., Fioravante, V.: Numerical simulations of seismic centrifuge tests: validation of pore water pressure models in sand. Soil Dyn. Earthq. Eng. **197**, 109459 (2025)
48. Porcino, D., Caridi, G., Malara, M., Morabito, E.: An automated control system for undrained monotonic and cyclic simple shear tests. In: GeoCongress 2006: geotechnical engineering in the information technology age, Atlanta (Georgia) (2006)
49. Jafarian, Y., Towhata, I., Baziar, M.H., Noorzad, A., Bahmanpour, A.: Strain energy based evaluation of liquefaction and residual pore water pressure in sands using cyclic torsional shear experiments. Soil Dynam Earthq Eng. **35**, 13–28 (2012)

50. Zhou, G.Y., Pan, K., Yang, Z.X.: Energy-based assessment of cyclic liquefaction behavior of clean and silty sand under sustained initial stress conditions. Soil Dynam Earthq Eng. **164**, 107609 (2023)
51. Dobry, R., El-Sekelly, W., Abdoun, T.: Calibration of non-linear effective stress code for seismic analysis of excess pore pressure and liquefaction in the free field. Soil Dyn. Earthq. Eng. **107**, 374–389 (2018)
52. Pervaiz, U., Park, D., Hashash, Y., Xing, G.: Testing performance of pore pressure models implemented in one-dimensional site response analysis program against centrifuge test data measured in mildly sloping ground. Soil Dyn. Earthq. Eng. **149**, 106867 (2021)
53. Fioravante, V., Giretti, D.: Physical Modeling of Liquefaction in Various Granular Materials. In: Earthquake Ground Motion, IntechOpen (2024)

Interval Modelling and Propagation of Uncertainties in Masonry Structures

Alba Sofi[(✉)]

University Mediterranea of Reggio Calabria, Reggio Calabria, Italy
alba.sofi@unirc.it

Abstract. The ultimate capacity of masonry structures is affected by several factors, such as geometry, external actions, material properties, and environmental conditions. The accurate definition of the mechanical properties is quite challenging due to various sources of uncertainties arising from workmanship techniques, block shape and texture, properties of single components, environmental conditions, existing damage, deterioration, and other factors. A major drawback is the lack of sufficient experimental data for the accurate probabilistic characterization of the uncertain properties. In this Chapter, uncertainty analysis of masonry structures is addressed within a non-probabilistic framework using the interval model, which is suitable to describe uncertainties when only range information is available. To limit overestimation affecting interval computations, the uncertain parameters are described by means of the *Improved Interval Analysis* via *Extra Unitary Interval*. As a practical application, the bounds of the in-plane resistance of unreinforced masonry walls involving interval uncertainties are estimated by interval extension of the analytical formulations provided by the Italian Building Code.

Keywords: Masonry · Uncertainty · Interval Analysis

1 Introduction

The structural analysis of masonry buildings is a very complex task which is performed nowadays using sophisticated numerical approaches able to deal with the heterogeneous nature and highly nonlinear behaviour of masonry (see e.g., [1]). The ultimate capacity of masonry structures is affected by several factors, such as geometry, external actions, material properties, and environmental conditions (e.g., humidity and temperature). The proper definition of the mechanical properties of masonry is quite challenging, even in the case of regular assemblages. Different sources of uncertainty arise from workmanship techniques, block shape and texture, properties of single components, environmental conditions, past damage, deterioration, etc. Furthermore, existing buildings are characterized by different types of masonry. For historic masonry, the mechanical characterization of materials is particularly difficult due to history-related phenomena and the prolonged exposure of buildings to mechanical and environmental loads. As a result, the mechanical properties of masonry can vary in a wide range. Italian Building Codes

The original version of the chapter has been revised. A correction to this chapter can be found at https://doi.org/10.1007/978-3-032-11926-1_21

C. Nava et al. (Eds.): *Climatic and Structural Safety in Multi-Hazard Regime of Cultural and Natural Heritage*, GREEN, pp. 77–89, 2026.
https://doi.org/10.1007/978-3-032-11926-1_5

and Guidelines for existing masonry buildings (see e.g., [2–4]) provide typical ranges of values for the mechanical properties of recurrent masonry typologies. Furthermore, estimates of the possible range of variability of material properties can be obtained by applying visual inspection approaches for the mechanical characterization of historic masonry, such as the Masonry Quality Index [5].

To obtain reliable predictions of the behaviour of masonry structures, uncertainties need to be taken into account using suitable non-deterministic approaches. The traditional probabilistic model has been extensively adopted in the literature to describe uncertainties in masonry structures (see e.g., [6–9]). A major drawback is the lack of sufficient experimental data for the accurate probabilistic characterization of the uncertain mechanical properties of masonry, mainly due to the need to preserve the cultural and artistic value of historical buildings. Therefore, classical probabilistic approaches often rely on subjective assumptions about the probability density function of the uncertain properties, which may lead to inaccurate predictions of structural performance. Indeed, failure probabilities are highly sensitive to the assumed probabilistic distribution of input parameters, especially in the tails [10].

Criticism regarding the use of probabilistic approaches when available data are insufficient, vague or imprecise has led to the development of alternative mathematical descriptions of uncertainty based on non-probabilistic concepts, such as the interval model [11], convex models [12] or fuzzy-sets [13].

Within the framework of the Pilot Project 4.7.1 of the PNRR Project Tech4You, the present Chapter addresses uncertainty analysis of historic masonry structures, proposing the use of the interval model as a suitable approach to deal with limited data availability. This model, originally developed from *Interval Analysis* [11], describes the uncertain parameters as interval variables with assigned lower bound (LB) and upper bound (UB) with no need for a full probabilistic characterization, which may be unfeasible in practical cases. To limit overestimation typically arising in interval computations, the uncertain parameters are described using the *Improved Interval Analysis* (*IIA*) via *Extra Unitary Interval* (*EUI*) [14]. The propagation of interval uncertainties in structural analysis involves the evaluation of the LB and UB of the response quantity of interest. The main purpose, indeed, is to identify the most and least favourable values of the response as the uncertain parameters vary within their respective intervals.

To illustrate the main features of interval uncertainty analysis of masonry structures, the in-plane resistance of unreinforced masonry walls involving interval parameters is addressed. The bounds of the interval flexural and shear resistance are estimated relying on the interval extension of the analytical formulations provided by the Italian Building Code [3, 4]. Different uncertainty scenarios are considered to investigate the relative importance of material, geometrical and loading parameters.

The rest of the Chapter is organized as follows: Section 2 introduces some basic notions of *Interval Analysis* and the main features of the *Improved Interval Analysis* via *Extra Unitary Interval*; Section 3 presents the interval model of masonry properties and their propagation in structural analysis; Section 4 addresses the assessment of the in-plane resistance of unreinforced masonry walls involving interval uncertainties; in Section 5, some concluding remarks are provided.

2 Interval Model of Uncertainty

2.1 Mathematical Representations of Uncertainty Beyond Classical Probability

The classical probabilistic model is the most widely used approach for uncertainty quantification in engineering problems. It provides the basis for the development of specialized software and design codes, and is employed to perform reliability analysis, reliability-based optimization and robust design. Despite well-established methods have been developed to assess the influence of random uncertainties on the performance of structural systems, available data are often insufficient to define reliable probabilistic distributions of the uncertain parameters. The analysts then rely on subjective assumptions, which may lead to inaccurate results in the context of reliability analysis due to the high sensitivity of the probability of failure to the assumed probabilistic model of the uncertain parameters, especially in the tails [10].

Since the second half of the 20th century, alternative mathematical representations of uncertainty beyond the pure probabilistic confines, such as fuzzy sets theory, evidence theory, possibility theory, interval model, and imprecise probabilities, have been proposed (see e.g., [15]). Nowadays, uncertainty models are commonly cast within a generalized framework [16] including probabilistic, non-probabilistic and imprecise probability approaches. The selection of the most suitable mathematical representation of uncertainty should rely on the source of non-determinism and available information [17].

The interval model is a popular non-probabilistic approach, originally derived from Moore's *Interval Analysis* [11]. The key idea is to describe the generic uncertain parameter as an interval variable with assigned lower bound (LB) and upper bound (UB). No information on the probability of occurrence of values between the bounds is needed. Therefore, the interval model is suitable to describe uncertainties in engineering problems when only the range of variability of the uncertain parameters (or tolerance) is known, while available information is insufficient to provide a reliable definition of the probability distribution within such a range.

2.2 Basic Notions of Interval Analysis

In this section, some basic notions of *Interval Analysis* [11] are briefly summarized.

A real interval variable $x^I = \left[\underline{x}, \overline{x}\right] \in \mathbb{IR}$, denoted by the apex I, is defined as:

$$x^I \equiv \left[\underline{x}, \overline{x}\right] = \left\{x | \underline{x} \leq x \leq \overline{x}, x \in \mathbb{R}\right\} \tag{1}$$

where $\{x | P(x)\}$ means " the set of x such that the proposition $P(x)$ holds"; $\underline{x}$ and $\overline{x}$ denote the LB and UB of the interval; $\mathbb{IR}$ is the set of all real interval numbers.

The interval variable x^I is fully defined by its LB and UB. Alternatively, it can be characterized by the midpoint value, x_0, and the deviation amplitude (or radius), Δx, given by:

$$x_0 = \text{mid}\left\{x^I\right\} = \frac{\underline{x} + \overline{x}}{2}; \quad \Delta x = \frac{\overline{x} - \underline{x}}{2} \tag{2a,b}$$

where mid{•} is an operator which yields the midpoint of the interval quantity between curly brackets.

In the framework of *Interval Algebra*, the basic operations of ordinary algebra are substituted by the corresponding operations on some combinations of the endpoints of the operands. In particular, for the two intervals $x^I = [\underline{x}, \overline{x}]$ and $y^I = [\underline{y}, \overline{y}]$, the basic algebraic operations read as follows:

$$\begin{aligned} x^I + y^I &= \left[\underline{x} + \underline{y}, \overline{x} + \overline{y}\right]; \\ x^I - y^I &= \left[\underline{x} - \overline{y}, \overline{x} - \underline{y}\right]; \\ x^I \times y^I &= \left[\min\left(\underline{x}\underline{y}, \underline{x}\overline{y}, \overline{x}\underline{y}, \overline{x}\overline{y}\right), \max\left(\underline{x}\underline{y}, \underline{x}\overline{y}, \overline{x}\underline{y}, \overline{x}\overline{y}\right)\right]; \\ x^I / y^I &= [\underline{x}, \overline{x}] \times \left[1/\overline{y}, 1/\underline{y}\right] \text{ if } 0 \notin y^I. \end{aligned} \tag{3a-d}$$

Let f be a real-valued function of a single variable x and assume that x varies over a given interval x^I. The image of the set x^I under the mapping f is an interval function:

$$f\left(x^I\right) = \left\{f(x) | x \in x^I\right\} \tag{4}$$

given by the interval extension of the real function $f(x)$.

The algebraic properties that hold for real numbers are not always valid for intervals. Often, they hold in a weaker form. The *Classical Interval Analysis* (*CIA*) [11] is unable to recognize the multiple occurrence of the same interval variable in an expression i.e., all the operand interval numbers are assumed independent. As a result, when an expression contains multiple instances of one or more interval variables, overestimation of the interval solution width occurs. This drawback, known as *dependency phenomenon* [11], may lead to overly conservative results, which hinder the application of the *CIA* to real engineering problems.

2.3 Improved Interval Analysis via Extra Unitary Interval

In the literature, several approaches have been proposed to limit the overestimation affecting interval computations in engineering problems. In this section, the *Improved Interval Analysis* (*IIA*) via *Extra Unitary Interval* (*EUI*) [14], developed in the context of interval structural analysis, is briefly described.

Based on the *IIA*, the $i-$ th interval variable x_i^I is expressed in the following *affine form*:

$$x_i^I = x_{0,i} + \Delta x_i \hat{e}_i^I \tag{5}$$

where $x_{0,\,i}$ and Δx_i are the midpoint and deviation amplitude, respectively; $\hat{e}_i^I = [-1, 1]$ is a particular unitary interval, called *EUI,* which differs from the *Classical Unitary Interval* (*CUI*) $e^I = [-1, 1]$. Specifically, the *EUI* satisfies the following properties [14]:

$$\begin{aligned} &\hat{e}_i^I - \hat{e}_i^I = 0; \quad \hat{e}_i^I \times \hat{e}_i^I = \left(\hat{e}_i^I\right)^2 = [1, 1]; \\ &\hat{e}_i^I / \hat{e}_i^I = [1, 1]; \quad \hat{e}_i^I \times \hat{e}_j^I = [-1, 1], \quad i \neq j; \\ &x_i \hat{e}_i^I \pm y_i \hat{e}_i^I = (x_i \pm y_i) \hat{e}_i^I; \\ &x_i \hat{e}_i^I \times y_i \hat{e}_i^I = x_i y_i \left(\hat{e}_i^I\right)^2 = x_i y_i [1, 1] \end{aligned} \tag{6a-f}$$

where $[1, 1] = 1$ is a degenerate interval. The subscript i in $\hat{e}_i^I$ means that, unlike the *CUI*, the *EUI* is associated with the $i-$ th interval variable x_i^I. This feature enables the *IIA* to keep track of the dependencies between interval variables throughout calculations. In this way, the overestimation due to the *dependency phenomenon* can be drastically reduced.

If x_i^I is a symmetric interval variable, Eq. (5) reduces to $\Delta x_i \hat{e}_i^I$ since $\underline{x}_i = -\overline{x}_i$ and, therefore, $x_{0,\,i} = 0$.

3 Masonry Structures with Interval Parameters

3.1 Masonry Mechanical Properties Modelled as Interval Variables

The mechanical properties of masonry exhibit high variability due to several sources of uncertainty. Only a limited amount of experimental data is often available for historic masonry buildings, mainly due to the need to preserve their cultural and artistic value. To deal with limited data availability, in this Chapter, the interval model is adopted as a suitable approach to describe uncertainties affecting masonry mechanical properties. Though this section focuses on the mechanical properties, other relevant parameters involved in masonry structural analysis, such as geometrical properties or loads, can also be modelled as interval variables [18].

Following the *IIA*, the $i-$ th uncertain parameter is described as an interval variable expressed in the following *affine form* [14]:

$$d_i^I = \left[\underline{d}_i, \overline{d}_i\right] = d_{0,i}\left(1 + \alpha_i^I\right) = d_{0,i}\left(1 + \Delta\alpha_i \hat{e}_i^I\right) \tag{7}$$

where $\hat{e}_i^I$ is the associated *EUI*; $\alpha_i^I = \left[\underline{\alpha}_i, \overline{\alpha}_i\right] \in \mathbb{IR}$ is the fluctuation around the nominal value $d_{0,\,i}$, which is modelled as a dimensionless symmetric interval variable, i.e., $\overline{\alpha}_i = -\underline{\alpha}_i$, with deviation amplitude $\Delta\alpha_i$:

$$\Delta\alpha_i = \frac{\Delta d_i}{d_{0,i}} = \frac{\overline{d}_i - \underline{d}_i}{2d_{0,i}} \tag{8}$$

which represents the normalized deviation amplitude of d_i^I and must satisfy the condition $\Delta\alpha_i < 1$ to guarantee positive values of the uncertain properties.

A crucial issue is selecting the bounds of the uncertain mechanical properties based on available information. Ideally, such bounds should be defined relying on experimental data. When no information is available from experimental tests, the range of the interval mechanical properties can be defined based on Building Codes and Guidelines (see e.g., [2–4]) and visual inspection approaches, such as the Masonry Quality Index method [5]. Once the masonry typology has been identified, a reference value for each uncertain mechanical parameter is chosen as the midpoint $d_{0,\,i}$ of the pertinent interval. Then, the deviation amplitude $\Delta\alpha_i$ of the interval fluctuation α_i^I around the midpoint value needs to be defined based on the degree of uncertainty affecting the mechanical properties. The latter is related to many factors, such as workmanship techniques, block geometry and texture, properties of mortar and bricks, environmental conditions, and existing damage.

It is worth mentioning that the mechanical properties of masonry structures often exhibit significant spatial variability due to inherent heterogeneity, varying degrees of degradation, restoration interventions, different construction techniques, environmental conditions, and other factors. Spatially dependent interval mechanical properties of masonry can be described using the *interval field* model (see e.g., [19]).

3.2 Bounds of the Interval Response

Let us consider a masonry structure involving N interval parameters ($i = 1, 2, \ldots, N$), see Eq. (7), whose fluctuations around the midpoint value are collected into the interval vector $\boldsymbol{\alpha}^I = [\underline{\boldsymbol{\alpha}}, \overline{\boldsymbol{\alpha}}] \in \mathbb{IR}^N$.

Focusing on the generic response quantity, for each realization of the interval parameters $\boldsymbol{\alpha} \in \boldsymbol{\alpha}^I$, structural analysis provides the associated solution through a deterministic mapping $f : \mathbb{R}^N \to \mathbb{R}$, i.e., $y = f(\boldsymbol{\alpha}) \in \mathbb{R}$. Thus, a solution set containing all possible responses obtained as the interval parameters vary within their range is derived:

$$\Sigma_y = \left\{ y | y = f(\boldsymbol{\alpha}), \boldsymbol{\alpha} \in \boldsymbol{\alpha}^I = [\underline{\boldsymbol{\alpha}}, \overline{\boldsymbol{\alpha}}] \in \mathbb{IR}^N \right\}. \tag{9}$$

The exact evaluation of the solution set is quite challenging. Within the framework of interval structural analysis, the main purpose is to search for the LB and UB of the response quantity of interest by solving a global optimization and anti-optimization problem [20] under the constraint that the uncertain parameters vary over the pertinent intervals:

$$\underline{y} = \min_{\boldsymbol{\alpha} \in \boldsymbol{\alpha}^I} \{f(\boldsymbol{\alpha})\}; \quad \overline{y} = \max_{\boldsymbol{\alpha} \in \boldsymbol{\alpha}^I} \{f(\boldsymbol{\alpha})\}. \tag{10a,b}$$

In other words, the main goal is to identify the best and worst scenarios. To this aim, procedures based on different approaches, such as perturbation analysis, *Affine Arithmetic*, *Improved Interval Analysis*, optimization, and surrogate modelling have been proposed (see e.g., [21]).

The application of the interval model of uncertainty to engineering problems has been enhanced by the development of the Interval Finite Element Method (IFEM) by interval extension of the classical FEM (see e.g., [21–[22]). Over the last decades, several versions of the IFEM have been proposed with the main goal of limiting the effects of the *dependency phenomenon* to find sharp bounds of the solution and guarantee computational efficiency (see e.g., [23]).

4 In-Plane Resistance of Unreinforced Masonry Walls with Interval Properties Based on Code Formulations

4.1 Italian Building Code Formulations

The in-plane behaviour of masonry walls has been investigated through extensive experimental and numerical studies, which have highlighted the influence of key parameters, such as [24–26]: mechanical properties of the material; constraint conditions; vertical compression levels; slenderness of the wall; and masonry texture. The in-plane failure modes of unreinforced masonry walls are generally classified as follows: flexural failure (rocking or toe crashing), diagonal shear failure, and sliding shear failure.

In the literature, several simplified analytical formulations are available to estimate the in-plane capacity of masonry walls with reference to both flexural and shear failure modes. In this Chapter, the formulations provided by the Italian Building Code [3] are considered. The Commentary to the Italian Code [4] distinguishes between walls made of regular and irregular masonry. To illustrate the main features of the interval model of uncertainty in the assessment of in-plane resistance, attention is restricted to irregular masonry walls commonly found in historic buildings, i.e., made of irregular arrangements of units and mortars not organized in regular layers. The geometry of the masonry walls is defined by the base width L, height H, and thickness t. Reference is made to walls subjected to a horizontal force V and average normal stress σ_0 (see Fig. 1).

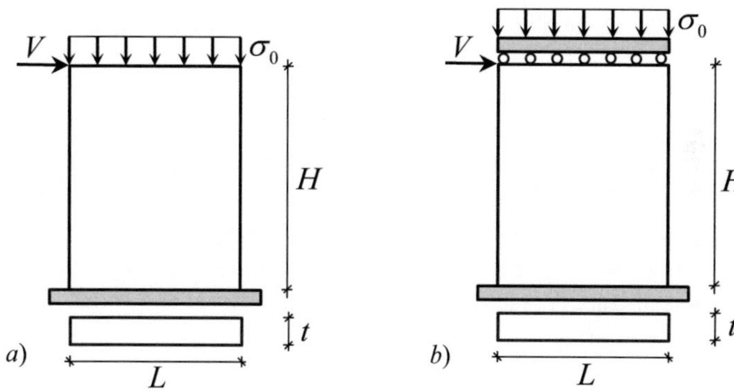

Fig. 1. Sketch of the masonry wall: *a*) cantilever wall; *b*) fixed-fixed wall.

The Italian Building Code [3] provides the following expression of the flexural capacity (F):

$$V_F = Lt\frac{\sigma_0}{2\psi\lambda}\left(1 - \frac{\sigma_0}{0.85 f_{c,d}}\right) \tag{11}$$

where $f_{c,\,d}$ is the design value of the compressive strength of masonry; $\psi\lambda = H_{eff}/L$ is the aspect ratio, with $H_{eff} = H$ for cantilever (Fig. 1a) and $H_{eff} = 0.5H$ for doubled-fixed constraint (Fig. 1b).

As far as the shear capacity is concerned, irregular masonry walls can only exhibit a diagonal shear failure (DS) characterized by diagonal cracks. The Commentary to the Italian Code [4] predicts the DS capacity of irregular masonry walls by the following equation:

$$V_{DS} = Lt\frac{f_{t,d}}{b}\sqrt{1 + \frac{\sigma_0}{f_{t,d}}} \tag{12}$$

where $f_{t,\,d}$ is the design value of the tensile strength of masonry; $b = H/L$ is the shape factor, which ranges between 1.0 and 1.5.

The in-plane resistance of the masonry wall is the minimum between the values provided by Eqs. (11) and (12).

4.2 In-Plane Resistance of Unreinforced Masonry Walls with Interval Parameters

In this section, uncertainty analysis of the flexural and shear resistance of irregular masonry walls is carried out by applying the interval model of uncertainty in conjunction with the above code-based formulations. The base width L and thickness t are assumed to be deterministic, while the remaining parameters are modelled as interval variables. The reference values of the mechanical and geometrical properties are listed in Table 1 [24]. Furthermore, a double-fixed constraint condition is assumed (Fig. 1b).

Table 1. Reference values of the mechanical and geometrical properties of the masonry wall.

L [m]	H [m]	t [m]	σ_0 [MPa]	$f_{c,d}$ [MPa]	$f_{c,t}$ [MPa]
1.00	1.00	0.25	0.60	5.00	0.50

Following the *IIA* via *EUI*, the generic uncertain parameter is expressed in *affine form* (see Eq. (7)), with the nominal value set equal to the reference value in Table 1. The same deviation amplitude of the interval fluctuation around the nominal value, $0 \leq \Delta\alpha \leq 0.5$, is assumed for all uncertain properties. Under these assumptions, by interval extension of Eqs. (11) and (12), the interval flexural and shear resistances take the following expression:

$$V_F^I \equiv V_F\left(f_{c,d}^I; H^I; \sigma_0^I\right) \equiv V_F\left(\boldsymbol{\alpha}_F^I\right) = \frac{L^2 t}{H^I}\sigma_0^I\left(1 - \frac{\sigma_0^I}{0.85 f_{c,d}^I}\right) \tag{13}$$

and

$$V_{DS}^I \equiv V_{DS}\left(f_{t,d}^I; b^I; \sigma_0^I\right) \equiv V_{DS}\left(\boldsymbol{\alpha}_{DS}^I\right) = Lt\frac{f_{t,d}^I}{b^I}\sqrt{1 + \frac{\sigma_0^I}{f_{t,d}^I}} \tag{14}$$

where $b^I = H^I/L$ satisfies the restriction $1.0 \leq b \leq 1.5$ as H varies; $\boldsymbol{\alpha}_F^I$ and $\boldsymbol{\alpha}_{DS}^I$ denote the interval vectors collecting the fluctuations of the interval parameters appearing in Eqs. (13) and (14).

The bounds of the interval flexural and shear resistance can be evaluated by solving the following optimization and anti-optimization problem:

$$\underline{V}_Q = \min_{\boldsymbol{\alpha}_Q \in \boldsymbol{\alpha}_Q^I}\left\{V_Q\left(\boldsymbol{\alpha}_Q\right)\right\}; \quad \overline{V}_Q = \max_{\boldsymbol{\alpha}_Q \in \boldsymbol{\alpha}_Q^I}\left\{V_Q\left(\boldsymbol{\alpha}_Q\right)\right\}, \quad Q = F, DS. \tag{15a,b}$$

For each failure mode, the LB of the associated interval resistance defines the worst-case scenario.

A measure of the dispersion of the interval resistance around the midpoint value is provided by the so-called *coefficient of interval uncertainty* (*c.i.u.*), defined as:

$$c.i.u.\left[V_Q^I\right] = \frac{\Delta V_Q}{\mathrm{mid}\left\{V_Q^I\right\}}, \quad Q = F, DS \tag{16}$$

where $\mathrm{mid}\left\{V_Q^I\right\}$ and ΔV_Q denote the midpoint and deviation amplitude of the interval capacity.

The bounds of the interval flexural resistance V_F^I are evaluated by solving Eqs. (15 a,b) considering three different uncertainty scenarios: all the parameters uncertain (Full Uncertainty, FU) i.e., $\boldsymbol{\alpha}_F^I = \left[f_{c,d}^I \; H^I \; \sigma_0^I\right]^{\mathrm{T}}$; geometrical and material properties uncertain (Structural Uncertainty, SU) i.e., $\boldsymbol{\alpha}_F^I = \left[f_{c,d}^I \; H^I\right]^{\mathrm{T}}$; uncertainty affecting only the normal stress σ_0^I (Loading Uncertainty, LU). The built-in functions "NMinimize" and "NMaximize" of Wolfram Mathematica [27] are used to solve Eq. (15a,b). It is worth mentioning that the LB and UB of V_F^I are achieved when the height of the wall H^I is set to its UB and LB, while the compressive strength of masonry $f_{c,d}^I$ and the normal stress σ_0^I attain their LB and UB, respectively.

Figure 2 shows the LB and UB of the interval flexural resistance V_F^I versus the normalized deviation amplitude of the uncertain parameters for the three different uncertainty scenarios. The nominal solution obtained considering the reference values of the parameters listed in Table 1 is also plotted. Inspection of Fig. 2 suggests the following main observations: the nominal solution overestimates the flexural resistance compared to the worst-case scenario (LB); the region enclosed by the resistance bounds becomes wider as the normalized deviation amplitude of the interval parameters increases; when only the vertical compressive stress σ_0 is assumed uncertain (LU), the amplitude of the capacity range is significantly reduced, which highlights the predominant role of SU.

Figure 3 displays the *c.i.u.* of the interval flexural resistance versus the normalized deviation amplitude of the uncertain parameters for the three different uncertainty scenarios. As expected, the dispersion of the flexural resistance around the nominal value increases when larger degrees of uncertainty are considered. In agreement with the results plotted in Fig. 2, it is observed that uncertainties cause significant dispersion around the midpoint value, especially when FU is considered. Furthermore, the predominant role of SU over LU is confirmed.

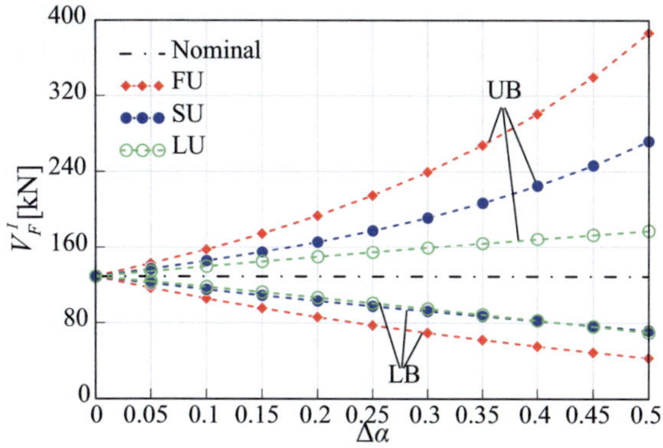

Fig. 2. LB, UB, and nominal value of the interval flexural resistance of the unreinforced masonry wall versus the normalized deviation amplitude of the uncertain parameters, considering three different uncertainty scenarios.

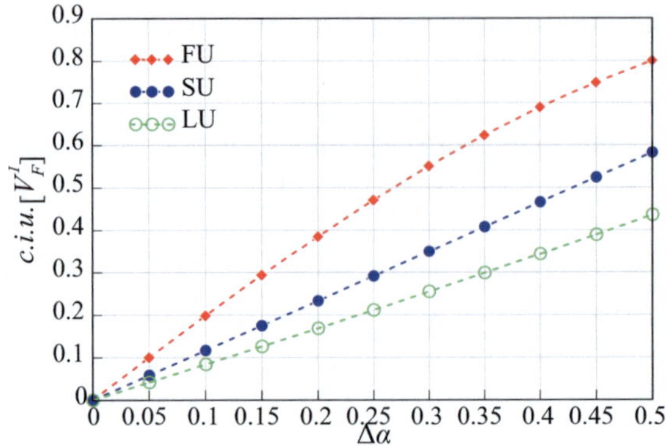

Fig. 3. *Coefficient of interval uncertainty* of the interval flexural resistance of the unreinforced masonry wall versus the normalized deviation amplitude of the uncertain parameters, considering three different uncertainty scenarios.

Similarly, the bounds of the interval diagonal shear resistance V_{DS}^I are evaluated by solving Eq. (15a,b) considering three different uncertainty scenarios: all the parameters uncertain (Full Uncertainty, FU) i.e., $\boldsymbol{\alpha}_{DS}^I = \left[f_{t,d}^I \; b^I \; \sigma_0^I\right]^{\mathrm{T}}$; geometrical and material properties uncertain (Structural Uncertainty, SU) i.e., $\boldsymbol{\alpha}_{DS}^I = \left[f_{t,d}^I \; b^I\right]^{\mathrm{T}}$; uncertainty affecting only the normal stress σ_0^I (Loading Uncertainty, LU). The LB and UB of V_{DS}^I are achieved when the shape factor b^I is set to its UB and LB, while the tensile strength of masonry $f_{t,d}^I$ and the normal stress σ_0^I attain their LB and UB, respectively.

Figure 4 shows the LB, UB and nominal value of the interval diagonal shear resistance V_{DS}^I versus the normalized deviation amplitude of the uncertain parameters for the three different uncertainty scenarios. The associated *c.i.u.* is plotted in Fig. 5. Figures 4 and 5 highlight features similar to those shown by Figs. 2 and 3 for the flexural resistance. Furthermore, it is observed that the shear resistance is less affected by LU than the flexural capacity. SU yields nearly the same *c.i.u.* of the interval flexural and shear resistance. Overall, the flexural resistance exhibits a larger dispersion around the midpoint value due to the more significant influence of LU.

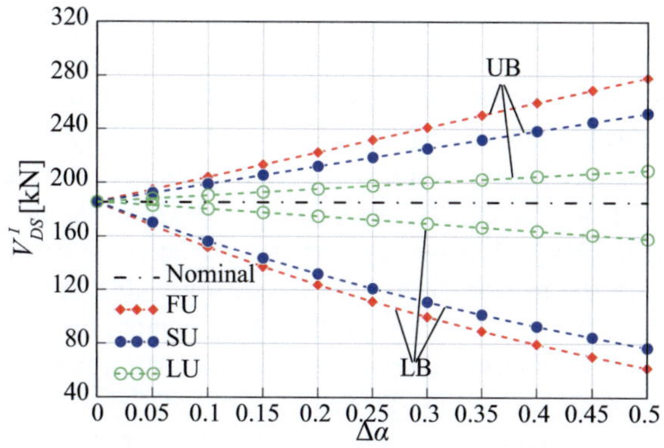

Fig. 4. LB, UB, and nominal value of the interval diagonal shear resistance of the unreinforced masonry wall versus the normalized deviation amplitude of the uncertain parameters, considering three different uncertainty scenarios.

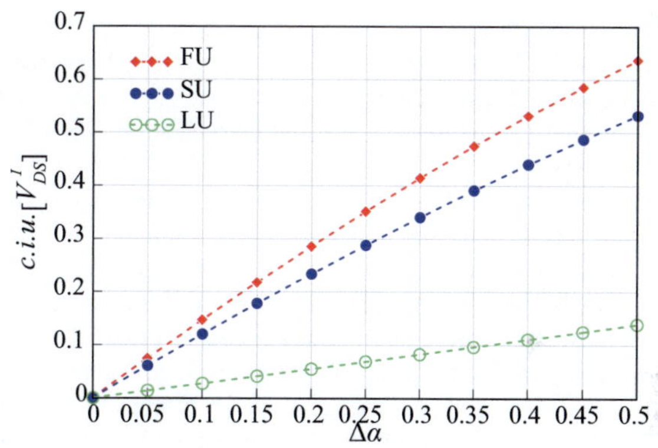

Fig. 5. *Coefficient of interval uncertainty* of the interval diagonal shear resistance of the unreinforced masonry wall versus the normalized deviation amplitude of the uncertain parameters, considering three different uncertainty scenarios.

5 Conclusions

Uncertainty analysis of masonry structures has traditionally been addressed using the classical probabilistic approach. However, available experimental data on masonry are often insufficient to obtain an accurate probabilistic characterization of the uncertain properties.

The interval model provides a suitable mathematical description of uncertainties able to deal with the common situation in which only limited, fragmentary or vague information on historic masonry structures is available. Indeed, this model describes the uncertain parameters as bounded intervals with assigned lower and upper bounds. Overestimation affecting interval computations can be reduced by applying the *Improved Interval Analysis* via *Extra Unitary Interval.*

In this Chapter, the in-plane resistance of unreinforced masonry walls with interval geometrical and mechanical properties has been investigated. The bounds of the interval flexural and shear resistance have been estimated by interval extension of the formulations provided by the Italian Building Code for irregular masonry walls. Three different uncertainty scenarios have been examined to analyze the influence of uncertainties associated with structural and loading parameters. Very wide ranges of the flexural and shear resistance have been obtained, demonstrating the crucial role played by uncertainties in the assessment of the in-plane capacity of masonry walls. In particular, assuming the nominal value of the structural and loading parameters may lead to a significant overestimation of the resistance compared to the worst-case scenario.

In line with the objectives of Pilot Project 4.7.1 of the PNRR Project Tech4You, the presented interval-based approach provides an effective means to increase confidence in the safety assessment of masonry structures, accounting for inherent uncertainties based on limited information typically available for historic buildings.

References

1. D'Altri, A.M., Sarhosis, V., Milani, G., Rots, J., Cattari, S., Lagomarsino, S., Sacco, E., Tralli, A., Castellazzi, G., de Miranda, S.: Modeling strategies for the computational analysis of unreinforced masonry structures: review and classification. Arch Comp Methods Eng. **27**(4), 1153–1185 (2020)
2. Linee guida per la valutazione e la riduzione del rischio sismico del patrimonio culturale con riferimento alle Norme Tecniche per le Costruzioni. Presidenza del Consiglio dei Ministri; Rome, Italy. Istituto Poligrafico e Zecca dello Stato (In Italian) (2011)
3. Aggiornamento delle Norme Tecniche per le Costruzioni. Ministero delle Infrastrutture e dei Trasporti; Rome, Italy, 2018. Decreto Ministeriale January 17, 2018, Gazzetta Ufficiale Serie Generale n.42 February 20, 2018 – Suppl. Ordinario n. 8 (In Italian)
4. Istruzioni per l'applicazione dell'aggiornamento delle 'Norme Tecniche per le Costruzioni' di cui al DM 17/01/2018. Ministero delle Infrastrutture e dei Trasporti; Rome, Italy. Circolare January 21, 2019, Gazzetta Ufficiale n. 35 February 11, 2019 (In Italian) (2019)
5. Borri, A., Corradi, M., Castori, G., De Maria, A.: A method for the analysis and classification of historic masonry. Bull Earthquake Eng. **13**(9), 2647–2665 (2015)
6. Benasciutti, D., Milani, G.: Limit analysis of masonry structures accounting for uncertainties in constituent materials mechanical properties. Open Civil Eng J. **2**, 51–57 (2008)
7. Sherafati, M.A., Sohrabi, M.R.: Probabilistic model for bed-joint shear-sliding strength of clay-brick walls based on field test data. ASCE-ASME J Risk Uncert Eng Syst Part A Civil Eng. **2**(4), 04016007 (2016)
8. Iannacone, L., Andreini, M., Gardoni, P., Sassu, M.: Probabilistic models and fragility estimates for unreinforced masonry walls subject to in-plane horizontal forces. J Struct Eng. **147**(6), 04021074 (2021)
9. Zeng, B., Cruz-Noguez, C., Li, Y.: Enhanced prediction and uncertainty analysis for in-plane and out-of-plane resistance of unreinforced masonry walls: a multifidelity approach. ASCE-ASME J Risk Uncert Eng Syst Part A Civil Eng. **10**(4), 04024053 (2024)
10. Ben-Haim, Y.: A non-probabilistic concept of reliability. Struct Safety. **14**(4), 227–245 (1994)
11. Moore, R.E., Kearfott, R.B., Cloud, M.J.: Introduction to Interval Analysis. Society for Industrial and Applied Mathematics, Philadelphia (2009)
12. Ben-Haim, Y., Elishakoff, I.: Convex Models of Uncertainty in Applied Mechanics. Elsevier, Amsterdam (1990)

13. Zadeh, L.A.: Fuzzy sets. Inform Control. **8**(3), 338–353 (1965)
14. Muscolino, G., Sofi, A.: Stochastic analysis of structures with uncertain-but-bounded parameters via improved interval analysis. Probabilistic Eng Mech. **28**, 152–163 (2012)
15. Corotis, R.B.: An overview of uncertainty concepts related to mechanical and civil engineering. ASCE-ASME J Risk Uncert Eng Syst Part B Mech Eng. **1**(4), 040801 (2015)
16. Ayyub, B.M., Klir, G.J.: Uncertainty Modeling and Analysis in Engineering and the Sciences. Chapman & Hall/CRC, Taylor & Francis Group, Boca Raton (2006)
17. Möller, B., Beer, M.: Engineering computation under uncertainty – Capabilities of non-traditional models. Comp Struct. **86**, 1024–1041 (2008)
18. Mazzeo, M., Laudani, R., Santoro, R.: Uncertainty effect on seismic capacity assessment in the out-of-plane failure mechanisms of masonry structures by probabilistic and non-probabilistic approaches. Dev Built Environ. **17**, 100366 (2024)
19. Sofi, A., Romeo, E., Barrera, O., Cocks, A.: An interval finite element method for the analysis of structures with spatially varying uncertainties. Adv Eng Software. **128**, 1–19 (2019)
20. Elishakoff, I., Ohsaki, M.: Optimization and Anti-optimization of Structures Under Uncertainty. Imperial College Press, London (2010)
21. Faes, M., Moens, D.: Recent trends in the modeling and quantification of non-probabilistic uncertainty. Arch Comp Methods Eng. **27**(3), 633–671 (2020)
22. Moens, D., Hanss, M.: Non-probabilistic finite element analysis for parametric uncertainty treatment in applied mechanics: Recent advances. Finite Elem Anal Design. **47**(1), 4–16 (2011)
23. Sofi, A., Romeo, E.: A novel Interval Finite Element Method based on the improved interval analysis. Comp Methods Appl Mech Eng. **311**, 671–697 (2016)
24. Celano, T., Argiento, L.U., Ceroni, F., Casapulla, C.: Literature review of the in-plane behavior of masonry walls: theoretical vs. experimental results. Materials. **14**(11), 3063 (2021)
25. Celano, T., Argiento, L.U., Ceroni, F., Casapulla, C.: In-plane behaviour of masonry walls: numerical analysis and design formulations. Materials. **14**(19), 5780 (2021)
26. Tariq, H., Najafgholipour, M.A., Sarhosis, V., Milani, G.: In-plane strength of masonry wall panels: a comparison between design codes and high-fidelity models. Structures. **47**, 1869–1899 (2023)
27. Wolfram Research, Wolfram Mathematica 14.2 (2025)

Exploring the Impact of Climate Change on the Progressive Collapse Resistance of an Existing Reinforced Concrete Building

Nicola Scattarreggia[1] and Eugenio Chioccarelli[2](✉)

[1] Department of Structural and Geotechnical Engineering, Sapienza University of Rome, Rome, Italy
nicola.scattarreggia@uniroma1.it

[2] Department of Civil, Energy, Environmental and Materials Engineering, University "Mediterranea" of Reggio Calabria, Reggio Calabria, Italy
eugenio.chioccarelli@unirc.it

Abstract. The growing impact of climate change on several fields and human activities has prompted the need for specific research about its potential consequences on structures. Climate change causes variations in temperature, atmospheric humidity, and carbon dioxide concentration levels, three factors that play a role in material degradation of reinforced concrete (RC) structures. This study explores the impact of climate change on the progressive collapse resistance of a real RC structure. Specifically, the corrosion of steel bars is modelled, with and without climate change, and its consequences on the structural capacity under vertical loads is discussed. It is assumed that the process of steel corrosion is induced by chloride ion penetration due to atmospheric environment; such a corrosion leads to a reduction in the reinforcement cross-sectional area, ductility, and strength. These parameters are assumed to vary over time based on a commonly adopted corrosion propagation model that explicitly accounts not only for chloride concentration, but also for temperature and humidity, which are both forecasted in alternative future scenarios via a climate change prediction model. To assess the gravity load capacity of the case-study building over time, a series of pushdown analyses are carried out. Over a 55-year period (lasting from today to the year 2080), the analysis reveals an approximate 35 % reduction in the vertical collapse load compared to the as-is condition (that is, today in 2025). The impact of climate change appears negligible on this result which is comparable to the one obtained for constant climate conditions. Larger consequences of potential climate change are observed when the considered period is extended to 150 years. The results also indicate that structural aging not only decreases the overall capacity but can also alter the collapse mechanism over time, underscoring the long-term structural risks associated with corrosion progression.

Keywords: Climate change · Corrosion · Reinforced concrete · Collapse analysis · Applied element method

---

The original version of the chapter has been revised. A correction to this chapter can be found at https://doi.org/10.1007/978-3-032-11926-1_21

C. Nava et al. (Eds.): *Climatic and Structural Safety in Multi-Hazard Regime of Cultural and Natural Heritage*, GREEN, pp. 90–107, 2026.
https://doi.org/10.1007/978-3-032-11926-1_6

1 Introduction

Climate change is increasingly reshaping the environmental conditions under which structures and infrastructure operate. Rising temperatures, elevated carbon dioxide (CO_2) levels, and fluctuating humidity patterns are not only affecting natural ecosystems but are also posing significant challenges to build structures. These environmental shifts accelerate the degradation processes in structure materials, especially reinforced concrete, which is widely used. Given that climate change is projected to worsen in the coming decades (e.g., Masson-Delmotte et al. 2021), understanding and quantifying its long-term effects on structures aging is essential for an accurate assessment of risk over time.

Climate change can cause the modification of structural loads (e.g., Mishra and Sadhu 2023) as well as accelerate the degradation of concrete structures (e.g., Bastidas-Arteaga et al. 2013 and Dimova et al. 2024), specifically by intensifying reinforcement corrosion, which affects both structural safety and serviceability (Stewart et al. 2012). Variation of environmental conditions associated with the mentioned climate change not only weaken the structural capacity of reinforced concrete increasing the risk but also contribute to substantial repair costs globally, which are projected to reach hundreds billions of dollars annually (Wang et al. 2014). For example, Wang et al. (2010) estimated that the damage risk from carbonation in Australia may rise by up to 16 % by 2100, meaning that additional infrastructure (building, bridges, powerplants, wharves, etc.) will require costly repairs. With over $1.8 trillion in global annual costs attributed to corrosion, managing these accelerated risks is imperative. As result, understanding and adapting to the climate-induced acceleration of reinforced concrete structures aging is essential for long-term infrastructure sustainability.

In recent decades, the structural engineering community has devoted increasing attention to the progressive collapse of buildings, following tragic events such as the Ronan Point collapse (1968), the Alfred P. Murrah Federal Building bombing (1995), and the World Trade Center attacks (2001). These incidents have highlighted the catastrophic consequences that can arise from the local failure of a primary load-bearing element, triggering disproportionate structural damage. Understanding and enhancing the collapse resistance of existing structures has therefore become a critical task, particularly in the context of aging infrastructure. This issue is further exacerbated by the above-described effects of climate change, which may accelerate deterioration. Numerous studies have investigated the progressive collapse resistance of reinforced concrete (RC) structures through numerical analysis (e.g., Brunesi et al. 2015; Li et al. 2016; Xu et al. 2022) also with the inclusion of reinforcement corrosion (e.g., Bao et al. 2023, 2025), with most of them relying on continuum-based techniques as the Finite Element modelling (FEM). However, FEM may encounter limitations in simulating large-scale collapse scenarios involving severe cracking, separation and collisions of components (Grunwald et al. 2018). In such cases, the Applied Element Method (AEM) (Meguro and Tagel-Din 2000, 2001, 2002), a discrete element-based approach, represents a valuable alternative, as demonstrated in several recent applications to RC structures (e.g., Grunwald et al. 2018 and Scattarreggia et al. 2022a).

Building on these considerations, this study, developed in the context of the Tech4You PNRR research project, investigates how variations in temperature and humidity influence the progression of corrosion and, consequently, the evolving vulnerability over time of a case-study RC frame building. The structure, built in the 1980s, is located in a marine atmospheric environment in Southern Italy, where the progression of corrosion due to its exposure to chloride-laden air may be further exacerbated by the changing local climatic conditions. The study focuses on the analysis of the ground floor, estimating its vertical load capacity up to collapse and the corresponding failure mechanism using AEM-based models.

2 Corrosion Modelling

As discussed in prior works, e.g. by Hanjari et al. (2011), the impact of corrosion on RC elements can be simulated by considering several factors. The effects on concrete include the removal of spalled concrete, and a reduction in mechanical properties of the concrete surrounding corroded reinforcement (e.g., Scattarreggia et al., 2021). In the case of steel reinforcement, reductions in cross-sectional area as well as in ductility, yield, and ultimate strengths should be taken into account (e.g., Biondini and Vergani, 2015). Finally, the interaction between these modified geometrical and mechanical properties of both concrete and reinforcement influences the bond between them, leading to its degradation (e.g., Wang and Liu, 2004).

The corrosion rate of steel reinforcement is directly related to the corrosion current density (i_{corr}), with higher values indicating more rapid deterioration. Accurately estimating i_{corr} is therefore critical for evaluating the durability and service life of reinforced concrete structures. However, characterizing i_{corr} is challenging due to the complex nature of the corrosion process, which is influenced by numerous interacting factors. These include the pH of the concrete, the availability of oxygen and moisture at the steel surface, and the surrounding environmental conditions (e.g., Vu and Stewart, 2000). In particular, concrete quality and cover thickness significantly affect the transport properties that control the ingress of oxygen and water to the corrosion cell (e.g., Guo et al., 2015a).

Several models have been proposed in the literature to describe i_{corr} during the propagation phase, each incorporating different assumptions and influencing parameters (see, the review by Raupach, 2006). For the purposes of this study, the model introduced by Vu and Stewart (2000) to calculate the corrosion current density as a function of time, t, that is, $i_{corr}(t)$ (in μA/cm^2), is applied according to Eq. 1 in which w/c is the water-to-cement ratio, x is the concrete cover thickness, t is the time (in years) and t_i is the corrosion initiation time:

$$i_{corr}(t) = \left[\frac{37.8 \cdot \left(1 - \frac{w}{c}\right)^{-1.64}}{x} \cdot 0.85 \cdot (t - t_i)^{-0.29} \right] \cdot f_T \cdot f_{RH} \cdot f_{Cl} \tag{1}$$

The baseline model in Eq. 1, proposed by Vu and Stewart (2000), was developed for an ambient relative humidity of 75 % and temperature of 20 °C. Therefore, to account for the local site-effects as proposed by Guo et al. (2015b), the equation incorporates the modification factors for temperature (f_T) (DuraCrete, 2000), relative humidity (f_{RH}) (Balafas and Burgoyne, 2010, and Guo et al., 2015b), and chloride concentration (f_{Cl}) (Liu and Weyers, 1998, Guo et al., 2015b). This model has been selected given that it has been widely used in other previous studies (e.g., Mortagi and Ghosh, 2020, and Nasr et al., 2022).

The temperature factor f_T (Eq. 2) adjusts for temperature deviations from the 20 °C reference temperature (where K_T is a factor estimated according to DuraCrete, 2000 guidelines, and T is the site temperature), while the humidity factor f_{RH} (Eq. 3) accounts for changes from the reference relative humidity (RH) of 75 %. Finally, the chloride concentration factor f_{Cl} (Eq. 4) accounts for the effect of increasing chloride surface concentration (C_s) during the propagation phase, relative to the critical threshold, C_{cr}.

$$f_T = \{1 + K_T \cdot [T - 20]\}, K_T = \begin{cases} 0.025 & if \quad T \leq 20°C \\ 0.073 & if \quad T > 20°C \end{cases} \tag{2}$$

$$f_{RH} = e^{-6000 \cdot \left(\frac{RH-75}{100}\right)^6} \tag{3}$$

$$f_{Cl} = \frac{C_s + C_{cr}}{2 \cdot C_{cr}} \tag{4}$$

The corrosion level of the reinforcement is typically quantified through steel rebar mass loss, expressed as a percentage of the pristine steel rebar mass (e.g., Shetty et al., 2015). This is a primary parameter used in combination with structural details to determine additional parameters affected by corrosion, such as reduced yield and ultimate strength and ductility of steel reinforcement, and degradation of concrete properties (Coronelli and Gambarova, 2004). As commonly observed in chloride-induced corrosion, steel reinforcement undergoes uniform cross-sectional area loss along its length, often accompanied by the development of localized deep pits. While pitting corrosion can significantly affect structural reliability under aggressive environmental conditions, such as direct marine splash or exposure to de-icing salts, its impact is generally limited in milder environments, such as marine atmospheric exposure (Mortagi and Ghosh, 2020; Shekhar et al., 2018). In line with these observations, the present study adopts a uniform corrosion model, considering the marine atmospheric exposure condition as representative of the degradation environment. This assumption is deemed possible by the location of the building, which lies at approximately 4 km from the coastline.

As above described, the corrosion-related phenomena influence both concrete and steel as well as their interaction. However, this study accounts only for the reduction of cross-section of rebars and steel ductility (the other related phenomena will be investigated in the future). Furthermore, although corrosion may slightly reduce the yield strength, its effect can be considered negligible compared to the most relevant effects in reduction in cross-sectional area and ductility (Apostolopoulos and Papadakis, 2008).

The time-dependent cross-sectional area, A_r, of one steel reinforcement bar due to uniform corrosion is evaluated as follows:

$$A_r(t) = \frac{\pi}{4} \cdot d_r(t)^2. \tag{5}$$

In the equation, d_r is the residual diameter which depends on $i_{corr}(t)$ as per Eq. 6, in which $d_{r,0}$ corresponds to the pristine rebar diameter (i.e., at $t = 2025$ yrs) (see, e.g., Vu and Stewart, 2000):

$$d_r(t) = d_{r,0} - 0.0232 \cdot \int_{t_i}^{t} i_{corr}(t) \cdot dt. \tag{6}$$

According to the model proposed by Biondini and Vergani (2015), the corresponding reduction of steel ultimate strain (ε_u) of a rebar subjected to uniform corrosion is evaluated as function of the steel mass loss (indicated with the parameter ψ):

$$\varepsilon_u(\psi) = \begin{cases} \varepsilon_u(0) & 0 \le \psi < 0.016 \\ 0.1521 \cdot (\psi)^{-0.4583} \cdot \varepsilon_u(0) & 0.016 \le \psi \le 1 \end{cases}, \tag{7}$$

where $\psi = \left(1 - \frac{A_r}{A_{r,0}}\right)$, in which $A_{r,0}$ and A_r represent the pristine and corroded area of the steel rebar, respectively.

As a result of the reduction of ε_u, the steel ductility (defined as $\mu = \varepsilon_u/\varepsilon_y$, where ε_y corresponds to the yielding steel strain deformation) is also reduced.

The models presented above are applied in a deterministic manner in the current case study. However, a more refined approach should involve a probabilistic framework to account for the inherent uncertainties in material properties, geometry, deterioration processes, and environmental and mechanical loading conditions.

3 Case-Study Data Collection

3.1 Structural Characteristics

The case-study structure focuses on a reinforced concrete frame building located in Bova (RC), Calabria, originally constructed in the 1980s as a Youth Hostel (Fig. 1). In 2005, the building's use was changed to a Carabinieri Station. In 2024, a seismic retrofitting work commenced, guided by a project designed in compliance with the Italian 2018 Technical Construction Standards (hereafter identified as NTC18, C.S.LL.PP., 2018) to enhance the building's structural performance/safety.

This study focuses on the pre-intervention building aiming at evaluating the collapse resistance of the ground floor. The structural details presented below are based on surveys and analyses of the existing (*as-is*) condition carried out during the development of the retrofitting project.

The building consists of multi-span frames in both main directions. Typical beams feature a cross-section of 30 × 60 cm^2 with variable longitudinal reinforcement and stirrup spacing. The columns have a consistent cross-section of 30 × 60 cm^2 and uniform reinforcement along the height. Dimensions of both columns and beams reduce along the height of the building. The floors are clay-concrete slabs (i.e., a structural floor system that uses hollow clay blocks within the concrete to create a lighter ribbed slab).

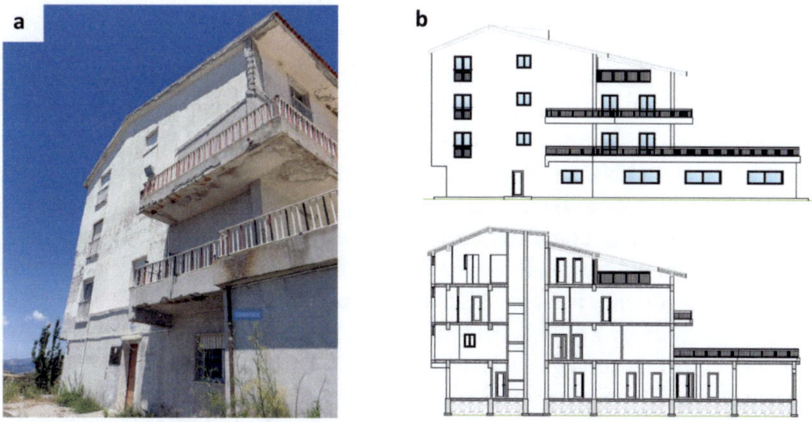

Fig. 1. Building overview: **a** photo of the as-is building and **b** front view and longitudinal section

3.2 Material Properties

The characterization of material properties for both concrete and steel elements was conducted through destructive testing (according to UNI EN 12390-3:2003 and UNI EN 12504-1:2021 for concrete, and UNI EN 10002-1:2004 for steel rebars) on samples extracted from the structure. The number of tests was defined to reach the highest conventional level of knowledge (i.e., level 3) according to NTC18 and its guidelines (C.S.LL.PP., 2019). The results of the tests are shown in Fig. 2 in terms of concrete compressive strength (Fig. 2a) and yield and ultimate strength of steel (Fig. 2b). The structural average compressive strength of concrete is 17.4 MPa, while the tensile tests performed on steel rebars reveals an average yield strength of 533 MPa, an average ultimate strength of 800 MPa, and an ultimate strain exceeding 10 %. By disaggregating the material testing results of samples taken at the ground floor, for the analysed floor, the reference values for concrete are equal to 13 MPa and 16 MPa for columns and beams, respectively.

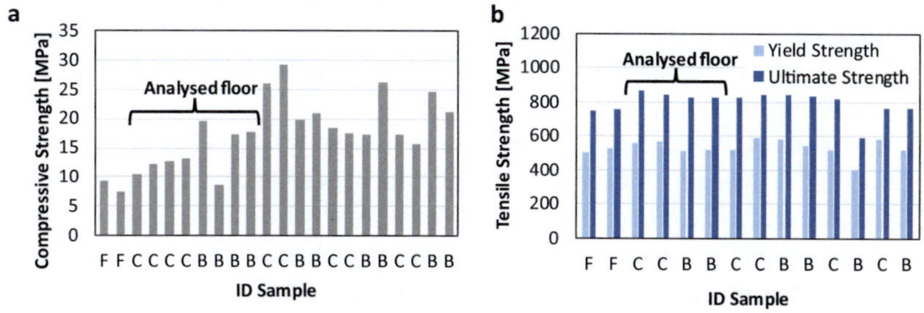

Fig. 2. Building material testing results: **a** concrete compressive strength and **b** yield and ultimate strengths of steel (F: foundation; C: column; and B: beam)

3.3 Climate Data

This study relies on site-specific environmental data assessed according to the Representative Concentration Pathway (RCP) referred as RCP8.5 that represent the pessimistic scenario of no CO_2 reduction (see, IPCC, 2007). The data here adopted are those provided by Lucanto and Nava (2024), and the yearly average temperature and humidity at the site of interest, from the current year (i.e., 2025) to 2080, are shown in Fig. 3a and b, respectively. To define a scenario over a longer time frame, it was assumed that temperature and humidity remain constant at the forecasted levels of 2080 (i.e., no further climate change) until the year 2175 as can be gathered from Fig. 3. This 150-year duration (from 2025 to 2175) is chosen because, according to the adopted corrosion models and modelling strategies (described in Sect. 2), the stirrups in the columns will experience complete cross-sectional loss due to the progression of corrosion in such a time interval and thus it represents a limit condition for the structural behaviour.

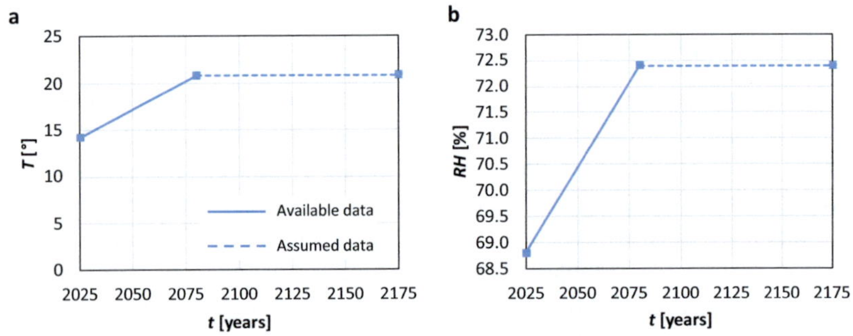

Fig. 3. Yearly average **a** temperature and **b** relative humidity due to climate change at the site (Lucanto and Nava, 2024)

The study assumes that corrosion initiates in 2025, i.e., $t_i = 2025$. On the other hand, under the absence of data and considering the distance of the building from the coast of about 4 km, the f_{Cl} factor that accounts for the chloride concentration (in Eq. 4) is assumed equal to 1. The latter results from the assumption that the surface chloride concentration is equal to the critical threshold (e.g., Guo et al., 2015b). Nonetheless, it is recognized that the C_s parameter is also dependent on several other local factors such as wind patterns and precipitation (which are both also affected by climate change), distance of structure from ocean, and surface orientation, see Stewart et al. (2011) and Shekhar et al. (2018). The unavailability of the climate data needed hinders the quantification of this effect. As result, according to the model described by Eq. 1, the $i_{corr}(t)$ function has been determined and is shown in Fig. 4 considering and neglecting climate change (the latter assumes that the average yearly temperature and humidity remain constant over time).

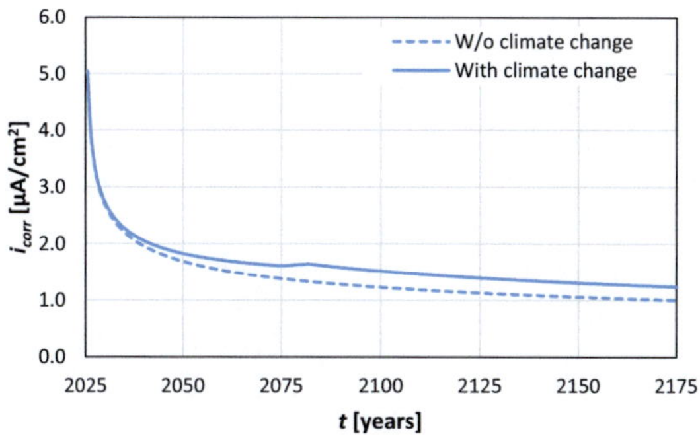

Fig. 4. Corrosion current density profile over time since its initiation ($t_i = 2025$)

4 Overview of the Applied Element Method

The AEM is a numerical approach that recently has been proven to predict damage propagation up to failure, while also capturing deterioration and self-collisions and debris distribution of collapsed structures. It addresses several limitations posed by FE-based methods for the explicit collapse modelling of heavily damaged structures, including problems at regions where mesh sizes abruptly change (Grunwald et al., 2018). On the other hand, standard discrete element approaches (Malomo and Pulatsu, 2024) are much more computationally expensive.

While interested readers are referred to Meguro and Tagel-Din (2000, 2001, 2002), among others, for a broad discussion about the AEM theory, a brief overview is provided here for the sake of clarity. In the framework of the AEM, a given solid member is modelled as a virtual assembly of small 6-DOFs rigid units connected to each other by means of zero-thickness springs uniformly distributed on elements edges (see Fig. 5). The springs, in which the material properties are lumped, are characterized by both normal $k_n = E \cdot d \cdot s/L$, and shear $k_{sx-sy} = G \cdot d \cdot s/L$ stiffnesses, with E and G Young's and shear moduli, and describe the stresses and deformations of a certain unit volume δV (i.e. the 3D region defined by $d \cdot s \cdot L$, where d is the distance between two consequent springs, while s and L are thickness and length of the element respectively). As results, RC structures can be easily modelled by explicitly coupling the mechanical contribution of concrete elements, whose behaviours from the elastic to inelastic branches are assigned to interface springs which response is dependent on the adopted constitutive law. Particularly, the embedded steel bars can be represented by equivalent "reinforcement springs" which can be placed in the same position of their real counterparts.

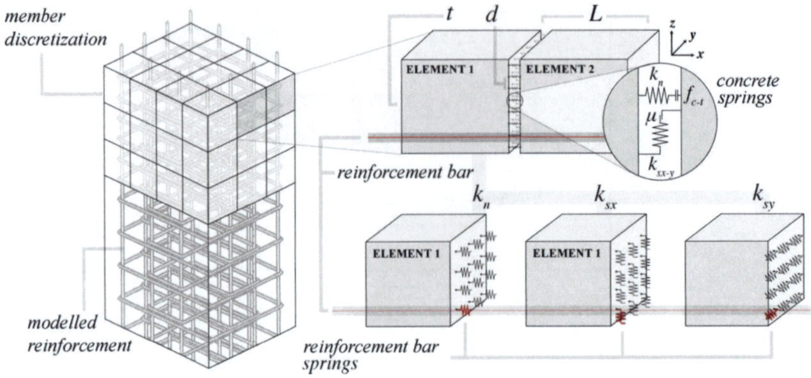

Fig. 5. Schematic of modelling RC components in the framework of Applied Element Method (adapted from Malomo et al., 2020)

For modelling the response of concrete under uniaxial cyclic tension-compression loading, the Maekawa and Okamura (1983) model is adopted. Young's modulus of concrete, fracture and compressive strain parameters are thus used to define an elastic bound, beyond which k_n is assumed as a minimum value (1 % of initial k_n in this work) to avoid stiffness matrix singularities. Tension cracking is accounted by neglecting the contribution of tensile strength right after reaching the peak stress ($f_{c\text{-}t}$). When considering shear-compression stress-states, instead, a simplified Mohr-Coulomb failure envelope is typically employed. The hysteretic model proposed by Menegotto and Pinto (1973) is instead assigned to normally-loaded reinforcement springs. Using the assumptions above, while failure phenomena such as buckling of reinforcement and spalling of concrete are neglected, shear transfer among RC components and shear softening behaviour can be accounted for (which also depend on the contact friction coefficient μ).

5 Numerical Modelling

The structural model, shown in Fig. 6, is developed within the *Extreme Loading for Structures* software (ASI, 2024), which uses the AEM. In this model, the primary components of the structure are represented by 3D elements. These elements capture the behaviour of the structural members under extreme conditions, including shear. The use of 3D elements capable of accounting for their axial, shear, flexural and torsional response yields to a global model that enable the simulation of most of the mechanism that may develop at both element and structure level.

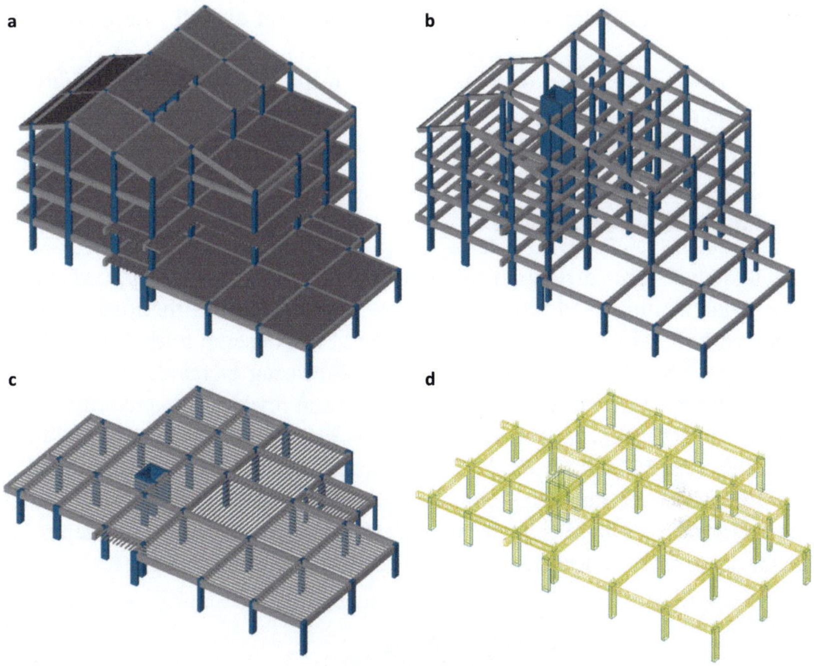

Fig. 6. AEM-based numerical models: solid view of the global model (**a**) with and (**b**) without floor slabs (RC joists are turned off to allow the view of RC beams structure), and ground floor model (**c**) in its solid view and (**d**) the reinforcement layout for columns and beams (top slab is turned off to allow the view of RC joists' directions)

The mesh size has been defined to capture the fully nonlinear behaviour of the structure according to what described in Scattarreggia et al. (2022b). A finer mesh is adopted where plastic hinges are expected to develop, while a coarse mesh is adopted in the portions where an almost elastic behaviour or lower stress-strain are expected. The chosen mesh size allows for a detailed representation of the structural elements, facilitating the capture of critical nonlinear phenomena such as plastic deformation, cracking, and local failure, which significantly affect the structure's performance.

The geometry of columns and beams is explicitly modelled, as well as that of the RC walls. The diaphragms are modelled by assuming the RC joists (0.08×0.21 m^2) (Fig. 6c) behave elastically with their 50 mm thick slab being modelled assuming a grid 0.3×0.3 m^2 of 6 mm-diameters rebars. It incorporates the reinforcement layout based on the as-is details, ensuring that the specific arrangement and quantity of reinforcement used in the actual structure are reflected (see Fig. 6d). Soil-structure interaction is neglected, to focus on the behaviour of the superstructure under the applied extreme loads. To account for the axial load and rotational restraints on top of the columns of the ground floor, the other floors are also modelled. The dead load of non-structural elements is modelled in terms of their weight. For example, partition walls are represented as a uniformly distributed load of approximately 1.6 kN/m^2, floor finishing loads are 1.5 kN/m^2, and infill walls are applied as a linear load of about 7.3 kN/m. The self-weight of the hollow clay blocks used in the floor slabs has been accounted for as well, being not explicitly modelled.

Another key feature of the model is the uniform application of corrosion throughout the entire ground floor, which marks the areas most affected by material degradation. It is applied to both longitudinal and transverse reinforcements.

5.1 Pushdown Analyses Results

Pushdown analyses are carried out by incrementally applying vertical loads on top of the floor slabs, as illustrated in Fig. 7 (the red colour indicates the loaded regions of the floor), with the aim of capturing the structural response up to the onset of failure, defined as the loss of resistance of primary load-bearing elements to vertical loads. Cantilever balconies are not loaded. For the structural response characterization of the entire building, the following scenarios are analysed: (a) the current as-is condition; (b) the year 2080 (i.e. 55 years from 2025) considering the effects of climate change; (c) the year 2080 without considering climate change. Moreover, a hypothetical scenario is considered for the year 2175 (i.e. 150 years from 2025), assuming the climate condition will be the same as those forecasted in 2080 considering climate change.

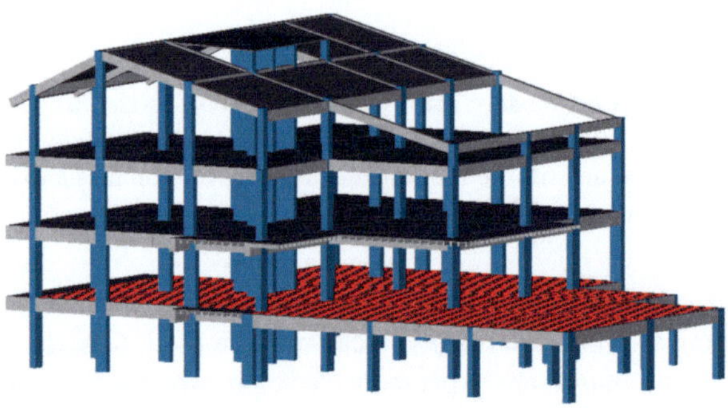

Fig. 7. Schematic representation of the load pattern adopted for the pushdown analyses, with red areas indicating the regions where the load is applied

The analysis is conducted for all the scenarios described above, and the results, in terms of collapse mechanism, are presented in Fig. 8, 9 and 10 In particular, the ultimate vertical load in the as-is condition is approximately 21 kN/m^2. Failure initiates in the external columns due to a combined axial, flexural, and shear mechanism, as highlighted in Fig. 8. The failure propagates with the failure of a few sectors of the analysed floor.

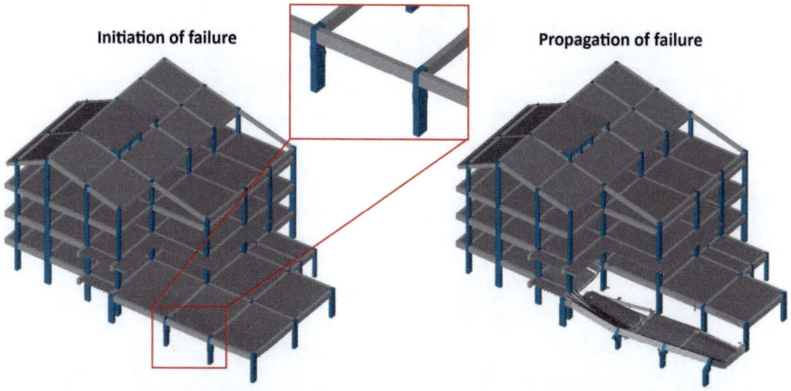

Fig. 8. Collapse mechanism of the structure in the as-is condition under incrementally applied vertical loads on top of the ground floor (no magnification factors applied to the deformed shape)

The ultimate vertical load decreases to approximately 14 kN/m^2 in 2080, with no significant difference observed whether or not the effects of climate change are included (Fig. 9). In this regard, the comparison between the failure mechanisms without (Fig. 9a) and with (Fig. 9b) climate change effects confirms the negligible influence of climate variations. However, the collapse mechanism at the onset of failure differs notably from that observed in the as-is condition. In this case, failure affects a different portion of the building and is primarily governed by a shear-flexural mechanism in the beams. As in the as-is scenario, the failure subsequently propagates through a few sectors of the analysed floor.

For the scenario in the year 2175 (assuming the same climate parameters as in the year 2080, i.e., no further climate change), the maximum vertical load capacity is reduced to approximately 6 kN/m^2. Failure initiates in the same portion observed for the as-is condition; however, in contrast to the as-is case where failure is governed by the columns, the collapse mechanism is now driven by the reduced flexural capacity of the beams (Fig. 10).

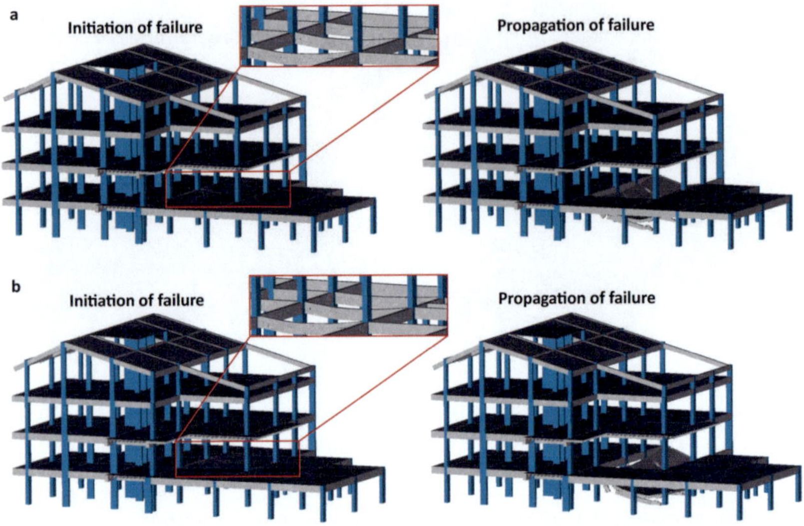

Fig. 9. Collapse mechanisms of the structure in the year 2080 (after 55 years of corrosion progression) under incrementally applied vertical loads on top of the ground floor: (**a**) without and (**b**) with the effects of climate change (no magnification factors applied to the deformed shape)

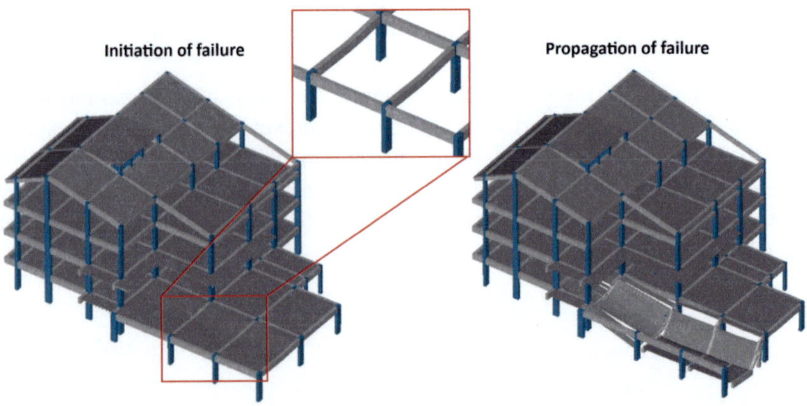

Fig. 10. Collapse mechanism of the structure in the year 2175 (after 150 years of corrosion progression, assuming climate conditions remain the same as those forecasted for the year 2080), under incrementally applied vertical loads on top of the ground floor (no magnification factors applied to the deformed shape)

To provide an overall view of all the analysed scenarios, Fig. 11 reports the maximum vertical load computed for each case. The vertical load at the onset of failure decreases from approximately 21 kN/m^2 to about 14 kN/m^2 in 2080, with no significant influence from climate change. It further reduces to around 6 kN/m^2 in 2175, assuming climate conditions remain consistent with those projected in 2080, including the effects of climate change. The results indicate that the progression of corrosion over time can significantly reduce the building's capacity to resist vertical loads. In contrast, the additional impact due to climate change projections appears to be marginal. These findings suggest that, under the same environmental conditions and in the absence of maintenance interventions aimed at slowing the deterioration process, buildings of this type may require future usage restrictions in order to meet the safety requirements set by the current NTC18 code. Nevertheless, it is important to note that these results are based on a specific corrosion progression model from literature. Not all potential impacts of climate change as well as of corrosion-related phenomena have been considered in the present study.

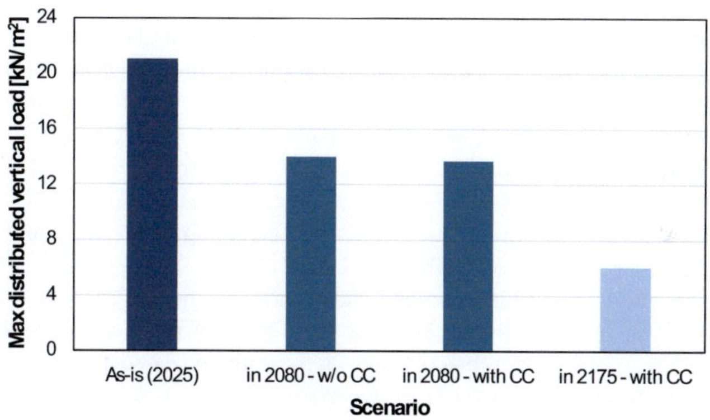

Fig. 11. Distributed vertical load at the onset of failure for the building in the as-is condition, in the year 2080 (i.e., after 55 years of corrosion progression) with and without climate change effects, and in the year 2175 (i.e., after 150 years of corrosion progression assuming climate conditions remain the same as those forecasted for the year 2080)

6 Conclusions

The study is developed in the context of the Tech4You PNRR research project. It discusses a preliminary deterministic modelling of climate change effect on existing structures. The case-study building is a reinforced concrete frame building located in southern Italy and highlights as corrosion may be accelerated by climate change (i.e., higher temperatures and variations in humidity) and produces an increase in structural vulnerability over time. In particular, the pushdown analyses conducted at the ground floor have provided insights into the ultimate vertical load capacity and collapse mechanisms of the building over time. The study focused on two primary corrosion-related effects: the reduction in the cross-sectional area of the steel reinforcement and the degradation of steel ductility. Over a 55-year period (from 2025 to 2080), the analysis revealed an approximately 35 % reduction in the vertical collapse load, accompanied by a negligible decrease due to the additional impacts of climate change. A scenario in 150 years (i.e., in the year 2175), assuming the climate conditions as in 55 years (no further climate change), corresponds to a reduction of about 70 % in the vertical collapse load. The results demonstrated that the structural aging may reduce the structure's capacity and alter the collapse mechanism, highlighting the long-term structural risks posed by the progression of corrosion. Although, climate change may exacerbate the effects of degradation, its influence is relatively minor compared to the dominant role of structural aging without consideration of humidity and temperature increase. While the findings are valid for the specific case study under the assumptions made, it is important to note that not all possible aging effects, such as concrete degradation and bond deterioration, are considered in this analysis. Moreover, the considered effects are applied in accordance with literature-based models and in a deterministic manner. Future work may overcome these limitations. Moreover, they may also include a study of the seismic response of the structure due to its location in a high seismic area of Italy.

Acknowledgments. The presented work has been supported by the project NGEU PNRR–MUR—Call for tender for the presentation of proposals for the Strengthening of research structures and creation of R&D "innovation ecosystems", set up of "territorial leaders in R&D" — to be funded under the National Recovery and Resilience Plan (NRRP), Mission 4, Component 2 Investment 1.4, funded from the European Union—NextGenerationEU. PROPOSAL: Tech4You—Technologies for climate change adaptation and quality of life improvement. NRRP THEMATIC AREA: CLIMATE, ENERGY, SUSTAINABLE MOBILITY TERRITORY OF REFERENCE: Calabrian and Lucanian territorial areas. Spoke 4—Technologies for resilient and accessible cultural and natural heritage: Unibas Affiliates: Unical, Unirc, CNR—Goal 4.7: Protection and enhancement of natural and cultural heritage and identity of the inner territories. Pilot Project 4.7.1—Open platform "phigital space" (physical and digital) of the type "user profiling" for the advanced and dynamic codesign of interventions on the built and ex novo. The authors gratefully acknowledge Prof. Ricardo Monteiro (IUSS Pavia) for providing access to the *Extreme Loading for Structures* software used in this work.

Competing Interests The authors have no conflicts of interest to declare that are relevant to the content of this chapter.

References

Apostolopoulos, C.A., Papadakis, V.G.: Consequences of steel corrosion on the ductility properties of reinforcement bar. Constr. Build. Mater. **22**, 2316–2324 (2008). https://doi.org/10.1016/j.conbuildmat.2007.10.006

ASI – Applied Science International. Extreme Loading for Structures (ELS) (2024)

Balafas, I., Burgoyne, C.J.: Environmental effects on cover cracking due to corrosion. Cem. Concr. Res. **40**, 1429–1440 (2010). https://doi.org/10.1016/j.cemconres.2010.05.003

Bao, C., Lv, D., Wang, H., et al.: Progressive collapse resistance mechanism of RC frame structure considering reinforcement corrosion. Model Simulat Eng. **4451011** (2023). https://doi.org/10.1155/2023/4451011

Bao, C., Wang, H., Long, H., et al.: Study on progressive collapse mechanism of RC frame structures corroded by chloride. Case Stud. Constr. Mater. **22**, e04472 (2025). https://doi.org/10.1016/j.cscm.2025.e04472

Bastidas-Arteaga, E., Schoefs, F., Stewart, M.G., Wang, X.: Influence of global warming on durability of corroding RC structures: a probabilistic approach. Eng. Struct. **51**, 259–266 (2013). https://doi.org/10.1016/j.engstruct.2013.01.002

Biondini, F., Vergani, M.: Deteriorating beam finite element for nonlinear analysis of concrete structures under corrosion. Struct. Infrastruct. Eng. **11**(4), 519–532 (2015). https://doi.org/10.1080/15732479.2014.951863

Brunesi, E., Nascimbene, R., Parisi, F., Augenti, N.: Progressive collapse fragility of reinforced concrete framed structures through incremental dynamic analysis. Eng. Struct. **104**, 65–79 (2015). https://doi.org/10.1016/j.engstruct.2015.09.024

Coronelli, D., Gambarova, P.: Structural assessment of corroded reinforced concrete beams: modeling guidelines. J. Struct. Eng. **130**(8), 1214–1224 (2004)

C.S.LL.PP.: Decreto Ministeriale 17 gennaio 2018: Norme tecniche per le costruzioni, Gazzetta Ufficiale della Repubblica Italiana, n. 42, 20 febbraio, Suppl. Ordinario n. 8. Ist. Polig. e Zecca dello Stato S.p.a., Rome (in Italian) (2018)

C.S.LL.PP.: CIRCOLARE 21 gennaio 2019, n. 7. Istruzioni per l'applicazione dell'"Aggiornamento delle Norme tecniche per le costruzioni di cui al decreto ministeriale 17 gennaio 2018", Gazzetta Ufficiale della Repubblica Italiana, n. 35, 11 febbraio 2019, Suppl. Ordinario n. 8. Istituto Poligrafico e Zecca dello Stato S.p.a., Rome, Italy (in Italian) (2019)

DuraCrete: General Guidelines for Durability Design and Redesign (Report Number BE95–1347/R17). Prepared for DuraCrete: Probabilistic Performance-Based Durability Design of Concrete Structures, The European Union—Brite EuRam III (2000)

Dimova, S., Polo López, C.S., Sousa, M.L., Rianna, G., Bastidas-Arteaga, E., Nogal, M., Gervásio, H., Martorana, E., Reder, A., Athanasopoulou, A.: Impact of Climate Change on the Corrosion of the European Reinforced Concrete Building Stock. Publications Office of the European Union (2024). https://doi.org/10.2760/016004

Grunwald, C., Khalil, A.A., Schaufelberger, B., Ricciardi, E.M., Pellecchia, C., de Iuliis, E., Riedel, W.: Reliability of collapse simulation—comparing finite and applied element methods at different levels. Eng. Struct. **176**, 265–278 (2018). https://doi.org/10.1016/j.engstruct.2018.08.068

Guo, A., Li, H., Ba, X., Guan, X., Li, H.: Experimental investigation on the cyclic performance of reinforced concrete piers with chloride-induced corrosion in marine environment. Eng. Struct. **105**, 1–11 (2015a). https://doi.org/10.1016/j.engstruct.2015.09.031

Guo, Y., Trejo, D., Yim, S.: New model for estimating the time-variant seismic performance of corroding RC bridge columns. J. Struct. Eng. **141**(6), 1–12 (2015b), https://doi.org/10.1061/(ASCE)ST.1943-541X.000114

Hanjari, K.Z., Kettil, P., Lundgren, K.: Analysis of mechanical behavior of corroded reinforced concrete structures. ACI Struct. J. **108**(5), 532–541 (2011). https://doi.org/10.14359/51683210

Intergovernmental Panel on Climate Change (IPCC): Climate Change 2007: The Physical Science Basis. Contribution of Working Group I to the Fourth Assessment Report of the Intergovernmental Panel on Climate Change. Cambridge University Press (2007)

Li, Y., Lu, X., Guan, H., Ren, P.: Numerical investigation of progressive collapse resistance of reinforced concrete frames subject to column removals from different stories. Adv. Struct. Eng. **19**(2), 314–326 (2016). https://doi.org/10.1177/1369433215624515

Liu, T., Weyers, R.W.: Modeling the dynamic corrosion process in chloride contaminated concrete structures. Cem. Concr. Res. **28**, 365–379 (1998). https://doi.org/10.1016/S0008-8846(98)00259-2

Lucanto, D., Nava, C.: Climate Data Projection for the Site of Bova. Abita Lab, University of Reggio Calabria (2024) https://www.abitalab-unirc.com/

Maekawa, K., Okamura, H.: The deformational behavior and constitutive equation of concrete using elasto-plastic and fracture model. J Faculty Eng. **37**, 253–328 (1983)

Malomo, D., Pulatsu, B.: Discontinuum models for the structural and seismic assessment of unreinforced masonry structures: a critical appraisal. Structures. **62**, 106108 (2024). https://doi.org/10.1016/j.istruc.2024.106108

Malomo, D., Scattarreggia, N., Orgnoni, A., Pinho, R., Moratti, M., Calvi, G.M.: Numerical study on the collapse of the Morandi bridge. J. Perform. Constr. Facil. **34**(4), 04020044 (2020). https://doi.org/10.1061/(ASCE)CF.1943-5509.0001428

Masson-Delmotte, V., Zhai, P., Pirani, A., Connors, S.L., Péan, C., Berger, S., Caud, N., Chen, Y., Goldfarb, L., Gomis, M.I., et al.: Summary for policymakers. In Climate Change 2021: The Physical Science Basis. Contribution of Working Group I to the Sixth Assessment Report of the Intergovernmental Panel on Climate Change (pp. 3–32). Cambridge University Press (2021)

Meguro, K., Tagel-Din, H.: Applied element method for structural analysis: theory and application for linear materials. Struct. Eng. Earthquake Eng. **17** (2000)

Meguro, K., Tagel-Din, H.: Applied element simulation of RC structures under cyclic loading. J. Struct. Eng. **127**(11), 1295–1305 (2001). https://doi.org/10.1061/(ASCE)0733-9445(2001)127:11(1295)

Meguro, K., Tagel-Din, H.: Applied element method used for large displacement structural analysis. J. Nat. Dis. Sci. **24**(1), 25–34 (2002)

Menegotto, M., Pinto, P.E.: Method of Analysis for Cyclically Loaded RC Plane Frames Including Changes in Geometry and Non-elastic Behavior of Elements under Combined Normal Force and Bending. In Proceedings of the IABSE Symposium on Resistance and Ultimate Deformability of Structures Acted on by Well Defined Repeated Loads, Zurich, 15–22 (1973)

Mishra, V., Sadhu, A.: Towards the effect of climate change in structural loads of urban infrastructure: a review. Sustain. Cities Soc. **89**, 104352 (2023). https://doi.org/10.1016/j.scs.2022.104352

Mortagi, M., Ghosh, J.: Concurrent modelling of carbonation and chloride-induced deterioration and uncertainty treatment in aging bridge fragility assessment. Struct. Infrastruct. Eng. **18**, 197–218 (2020). https://doi.org/10.1080/15732479.2020.1838560

Nasr, A., Honfi, D., Larsson Ivanov, O.: Probabilistic analysis of climate change impact on chloride-induced deterioration of reinforced concrete considering Nordic climate. J. Infrastruct. Preserv. Resil. **3**, 1–16 (2022). https://doi.org/10.1186/S43065-022-00053-6

Raupach, M.: Models for the propagation phase of reinforcement corrosion—an overview. Mater. Corros. **57**(8), 605–613 (2006). https://doi.org/10.1002/maco.200603991

Scattarreggia, N., Qiao, T., Malomo, D.: Earthquake response modeling of corroded reinforced concrete hollow-section Piers via simplified Fiber-based FE analysis. Sustainability **13**(16), 9342 (2021). https://doi.org/10.3390/su13169342

Scattarreggia, N., Salomone, R., Moratti, M., Malomo, D., Pinho, R., Calvi, G.M.: Collapse analysis of the multi-span reinforced concrete arch bridge of Caprigliola, Italy. Eng. Struct. **251**, 113375 (2022a). https://doi.org/10.1016/j.engstruct.2021.113375

Scattarreggia, N., Galik, W., Calvi, P.M., Moratti, M., Orgnoni, A., Pinho, R.: Analytical and numerical analysis of the torsional response of the multi-cell deck of a collapsed cable-stayed bridge. Eng. Struct. **265**, 114412 (2022b). https://doi.org/10.1016/j.engstruct.2022.114412

Shekhar, S., Ghosh, J., Padgett, J.E.: Seismic life-cycle cost analysis of ageing highway bridges under chloride exposure conditions: modelling and recommendations. Struct. Infrastruct. Eng. **14**, 941–966 (2018). https://doi.org/10.1080/15732479.2018.1437639

Shetty, A., Venkataramana, K., Narayan, K.S.B.: Experimental and numerical investigation on flexural bond strength behavior of corroded NBS RC beam. Int. J. Adv. Struct. Eng. **7**, 223–231 (2015)

Stewart, M.G., Wang, X., Nguyen, M.N.: Climate change impact and risks of concrete infrastructure deterioration. Eng. Struct. **33**(4), 1326–1337 (2011). https://doi.org/10.1016/j.engstruct.2011.01.010

Stewart, M.G., Wang, X., Nguyen, M.N.: Climate change adaptation for corrosion control of concrete infrastructure. Struct. Saf. **35**, 29–39 (2012). https://doi.org/10.1016/j.strusafe.2011.10.002

UNI EN 10002-1:2004 – Materiali metallici – Prova di trazione – Parte 1: Metodo di prova a temperatura ambiente. Rome, Italy (in Italian)

UNI EN 12390-3:2003 – Prova sul calcestruzzo indurito – Parte 3: Resistenza alla compressione dei provini. Rome, Italy (in Italian)

UNI EN 12504-1:2021 – Prove sul Calcestruzzo nelle Strutture – Parte 1: Carote—Prelievo, Esame e Prova di Compressione. Rome, Italy (in Italian)

Vu, K.A.T., Stewart, M.G.: Structural reliability of concrete bridges including improved chloride-induced corrosion models. Struct. Saf. **22**, 313–333 (2000)

Wang, X., Liu, X.: Modeling bond strength of corroded reinforcement without stirrups. Cem. Concr. Res. **34**(8), 1331–1339 (2004). https://doi.org/10.1016/j.cemconres.2003.12.028

Wang, Q., Wei, J., Dong, R., Xu, G.: Numerical analysis of carbonation process for concrete structures. J. Wuhan Univ. Tech. **36**(5), 91–96 (2014)

Wang, X., Nguyen, M., Stewart, M. G., Syme, M., Leitch, A.: Analysis of Climate Change Impacts on the Deterioration of Concrete Infrastructure—Synthesis Report. CSIRO, Canberra, 54 p. (2010). http://www.csiro.au/resources/concrete-durabilityreport-synthesis.html

Xu, Q., Zhen, X., Zhang, Y., Han, M., Zhang, W.: Numerical simulation study of progressive collapse of reinforced concrete frames with masonry infill walls under blast loading. Model. Simul. Eng. **2022**, 1781415 (2022). https://doi.org/10.1155/2022/1781415

Inerter-Based Base Isolation System for Masonry Structures: Prototype Design and Realization

Giuseppe Failla(✉), Andrea Francesco Russillo, and Gioacchino Alotta

Department of Civil, Energy, Environmental and Materials Engineering (DICEAM), University of Reggio Calabria, Via Zehender, 89124 Reggio Calabria, Italy
{giuseppe.failla,andreaf.russillo,gioacchino.alotta}@unirc.it

Abstract. This paper describes the detailed design and realization of a hybrid base isolation system for seismic protection of masonry structures. The system combines standard base isolators with a tuned inerter damper, consisting of a grounded rack-and-pinion inerter connected in series to a set of linearly-elastic springs and a viscous dashpot in parallel. On adopting a fixed-point tuning procedure to estimate optimal stiffness and viscous damping coefficient of the springs/dashpot, for a given inertance of the inerter, numerical analyses demonstrate the effectiveness of the proposed hybrid base isolation system.

1 Introduction

Protecting the monumental heritage against earthquake-induced vibrations is of paramount importance in areas at seismic risk. Among various potential strategies, base isolation (Clemente and Martelli 2019; De Luca and Guidi 2019) is certainly very effective and particularly suitable for monumental heritage assets (Clemente et al. 2012, 2016; Osgooei et al. 2015; Habieb et al. 2019a, b, 2023 Losanno et al. 2021, 2022; Basar et al. 2021; Zhang et al. 2021; Faiella et al. 2023), being deployed at the foundation level without requiring invasive or unaesthetic modifications to the structure above.

Although base isolation is certainly a well-established technique for seismic protection, several studies highlighted some drawbacks such as, for example, the high displacement demand at the isolation floor, which requires resorting to large-size isolators capable of withstanding the design displacements, as well as allowing an adequate separation distance between buildings to prevent mutual collisions or structural pounding (De Domenico and Ricciardi 2018). Furthermore, base isolation may not be effective in presence of long-period ground motions that might severely increase the lateral displacements of the isolators and cause long-lasting vibrations due to the resonance effects (De Domenico and Ricciardi 2018; Naumann Masoom et al. 2023).

In view of these shortcomings, research studies on base isolation is still ongoing. A potential improvement to base isolation is obtainable via coupling the base isolation system with a tuned mass device targeting the first natural frequency of the isolated structure; in this context, a typical example of tuned mass device is a tuned mass damper

C. Nava et al. (Eds.): *Climatic and Structural Safety in Multi-Hazard Regime of Cultural and Natural Heritage*, GREEN, pp. 108–123, 2026.
https://doi.org/10.1007/978-3-032-11926-1_7

(TMD), i.e., a secondary mass connected to the structure by a linearly-elastic spring and a viscous dashpot acting in parallel at the isolation floor of the structure, where the maximum relative displacements occur; this strategy was proposed by some studies on steel and reinforced concrete structures (Tsai 1995; Taniguchi et al. 2008; Xiang et al. 2014).

Although this strategy does reduce the displacement demand on the base isolation system, its effectiveness is strictly related to the amount of the TMD mass that can be used: for example, the displacement demand can be reduced by 15% to 25% provided a TMD is installed having a mass ratio of 5% or more of the building total mass (De Domenico and Ricciardi 2018), which may become hardly feasible requiring a very large space and inevitable costs. Another significant limitation is related to the TMD stroke, which may be unacceptably large, thus requiring a large space/clearance in the building to accommodate the displacement demand of this secondary mass/subsystem, sometimes even exceeding the maximum admissible displacement of the base isolation system.

As numerical evidence demonstrates that the larger the tuned mass is, the larger the benefits are in reducing the displacements of the base isolated system, recent studies investigated the so-called inerter (Smith 2002) as alternative to the classical tuned mass for coupling with the base isolation system. Indeed, the inerter is a two-terminal mechanical device whose inertia provides a force proportional to the relative translational acceleration at the terminals (Smith 2002). It is a very promising concept in mechanics, because it is capable of providing a very large inertia with a relatively small mass, which is possible by exploiting, e.g., rack-and-pinion or ball-screw mechanisms to transform, by means of gearing, the translational kinetic energy associated with the relative motion of the device terminals into rotational kinetic energy of a lightweight flywheel (Papageorgiou et al. 2009; Marian and Giaralis 2017; Ge and Wang 2018). An impressive number of recent studies demonstrate the great potential of the inerter in a wide range of applications (Marian and Giaralis 2014; Hu and Chen 2015; Alotta and Failla 2021; Russillo et al. 2022; Alotta et al. 2023; Sciutteri et al. 2024)

However, research on inerters is at the early stage and extensive effort is needed for experimental validation in structural engineering applications. This is even true for potential applications in conjunction with a base isolation system for monumental masonry structures, not yet explored.

Motivated by the interest in developing innovative and effective strategies for the protection of monumental heritage assets, this study proposes a hybrid base isolation system combining classical base isolators with a tuned-inerter-damper (TID), for seismic protection of masonry structures. The TID consists of a rack-and-pinion inerter in series with a set of linearly-elastic springs and a viscous dashpot in parallel. The TID is installed at the level of the isolators, between the isolation floor of the masonry structures and the ground, to exploit the relative motion between the two for activating the inerter and leverage its large inertia effects. The inerter is designed to be easily re-tunable; re-tuning is important to make the proposed control system adaptable to potential variations in the structural properties of the masonry structure to be protected. Indeed, non-negligible variations in mechanical parameters such as stiffness and strength may be envisaged in masonry structures as a result of climate-change related phenomena: e.g., chemical degradation due to the increase of CO_2 in the atmosphere and/or high thermal stresses during increasingly frequent heat waves, as recommended by the "National Climate Change Adaptation Plan" released by the Italian Ministry of Environment and Energy Security in December 2022 (Italian Ministry of Environment and Energy Security 2022). This study describes the detailed design and realization of the proposed TID and illustrates a suitable tuning procedure of the TID for the TID-base isolation system of a benchmark two-floor masonry structure.

The paper is structured as follows. The proposed TID is described in Sect. 2. The tuning procedure of the TID for the TID-base isolation system is illustrated in Sect. 3. An Appendix is included, with a detailed description of all components of the TID.

2 Proposed Tuned Inerter Damper (TID)

The proposed TID consists of a rack-and-pinion inerter in series with a set of linearly-elastic springs and a viscous dashpot in parallel. The idea is to install the TID at the level of the isolators, between the isolation floor of the masonry structure and the ground, to exploit the relative motion between the two for activating the inerter and exploit its large inertia effects. Therefore, the rack-and-pinion inerter is grounded, while the springs and the dashpot are connected to the isolation floor, as depicted in Fig. 1.

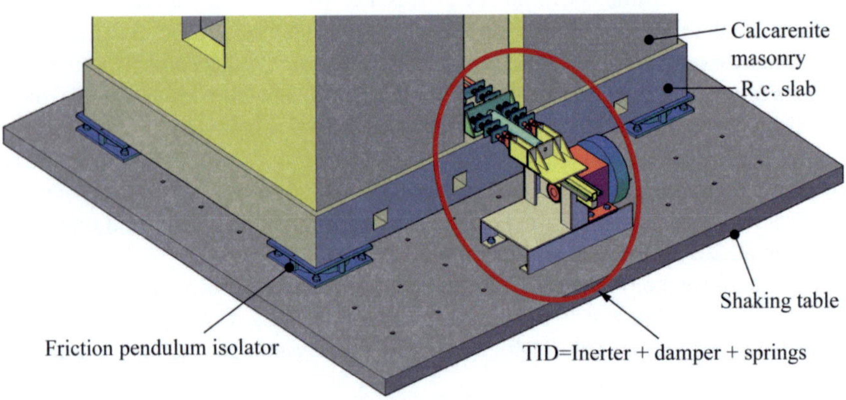

Fig. 1. Proposed TID-base isolation system for masonry structure.

Among the most popular inerter mechanisms (Papageorgiou et al. 2009; Marian and Giaralis 2017; Ge and Wang 2018), the rack-and-pinion inerter is selected. It is a two-terminal device where gearing transforms the translational kinetic energy associated with the relative motion of the two terminals into rotational kinetic energy of a lightweight flywheel. As a result, the rack-and-pinion inerter can provide large inertia effects with a relatively small mass. Specifically, its inertia effects are measured by the so-called inertance (kg), which depends on the number of gears, the gearing ratio and on the mass moment of inertia of the flywheel. The main components of the proposed rack-and-pinion inerter are depicted in Fig. 2 and Fig. 3: a flywheel, a gearbox and a rack, all supported by a grounded support structure. The rack slides on two parallel rails and is connected in series to the set of linearly-elastic springs and the viscous dashpot in parallel by an end stiffened plate. As a result of the relative motion between structure and ground, the translational motion of the rack on the rails is converted into rotational motion of the flywheel via the gearbox. Since the nominal inertance of the device depends on the flywheel mass moment of inertia attached to the output gearbox shaft, changing mass/diameter of the flywheel allows a straightforward re-tuning of the device. Furthermore, notice that the physical mass of the inerter that may contribute to the motion of the device relative to the ground is rather small, as pertains only to the rack and to the end stiffened plate above the rack. On the other hand, the flywheel is mounted to the lower part of the device, which is almost non-deformable and rigidly fixed to the ground. Therefore, it may be concluded that the device meets the requisite of an inerter, as its mass contributing to motion relative to the ground is small and independent of the nominal inertance (Pietrosanti et al. 2021).

The springs are linearly-elastic compression springs, the dashpot is a fluid viscous damper, they are all connected to a support attached to the structure. Being compression springs, two sets of springs are used, placed on two opposite sides of the support. For the purpose of estimating the optimal design parameters of the TID, it is assumed that the fluid viscous damper behaves linearly.

All components of the TID are made of steel. The TID is designed for shaking-table experimental tests on a benchmark two-floor masonry structure, base-isolated by the proposed TID-base isolation system, see Fig. 1.

The TID has been realized by Metalmeccanica Falcone, a steel manufacturer based in Reggio Calabria, Italy. A detailed list of all the components is provided in the Appendix, along with some pictures.

In the following Section, the tuning procedure providing the optimal design parameters of the TID is illustrated in detail.

3 Tuning Procedure and Numerical Results

The tuning procedure relies on the numerical scheme presented in ref. (Krenk and Høgsberg 2016; Krenk 2019), which is readily applicable for the problem at hand considering that both the TID and the base-isolated masonry structure can reasonably be treated as single-degree-of-freedom (SDOF) systems, see Fig. 4.

Indeed, for a preliminary design, the base-isolated masonry structure can be represented as a rigid mass oscillating in the horizontal plane with the base isolators providing elastic restoring forces and viscous damping forces; the inerter can be modelled as a linear massless mechanical element resisting relative acceleration at its two terminals through its inertance and the whole TID can be modelled as a linear SDOF system, whose degree of freedom is given by the displacement of the connection point between the inerter and the springs/dashpot. On these assumptions, the equations of motion of the coupled system (base-isolated masonry structure and TID) read

$$\begin{cases} m_0\ddot{x}(t) + c_0\dot{x}(t) + k_0 x(t) + c_d(\dot{x}(t) - \dot{x}_d(t)) + k_d(x(t) - x_d(t)) = -m_0\ddot{x}_g(t) \\ b\big(\ddot{x}_d(t) - \ddot{x}_g(t)\big) + c_d(\dot{x}_d(t) - \dot{x}(t)) + k_d(x_d(t) - x(t)) = 0 \end{cases} \quad (1a,b)$$

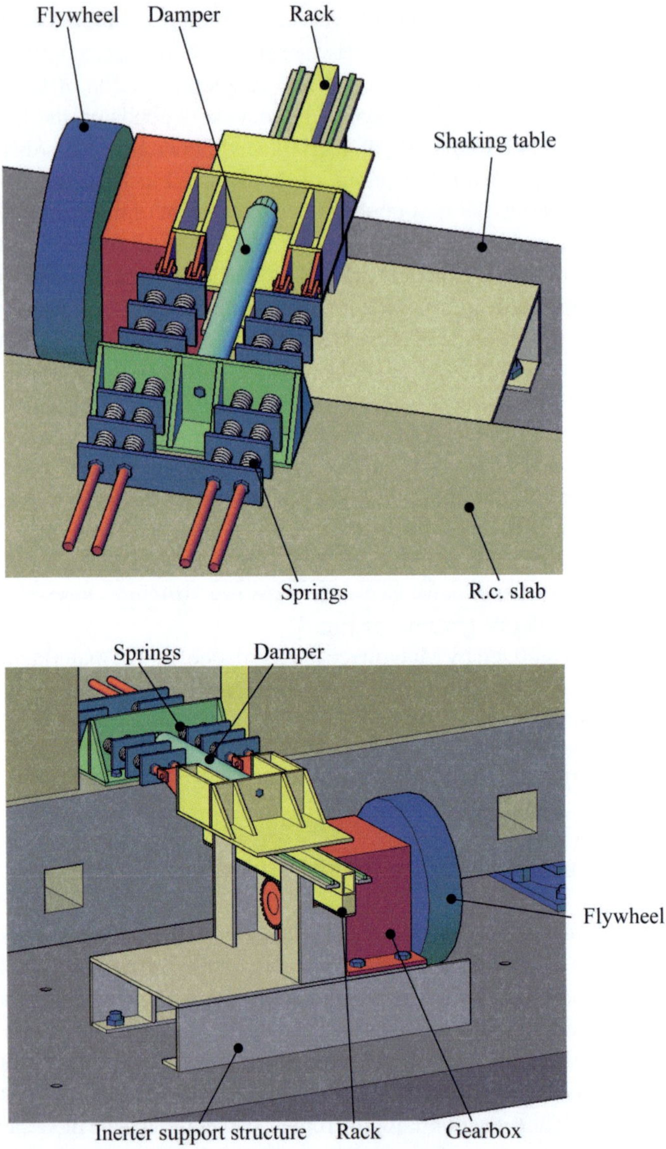

Fig. 2. Main components of TID.

where m_0 is the mass of the SDOF modeling the base-isolated masonry structure, k_0 and c_0 are the stiffness and viscous damping coefficient of the base isolators, while b, k_d and c_d are the inertance, stiffness and viscous damping coefficient of the TID. In Eq. (1), the force developed by the inerter is given by

$$F_b = b\big(\ddot{x}_d(t) - \ddot{x}_g(t)\big) \tag{2}$$

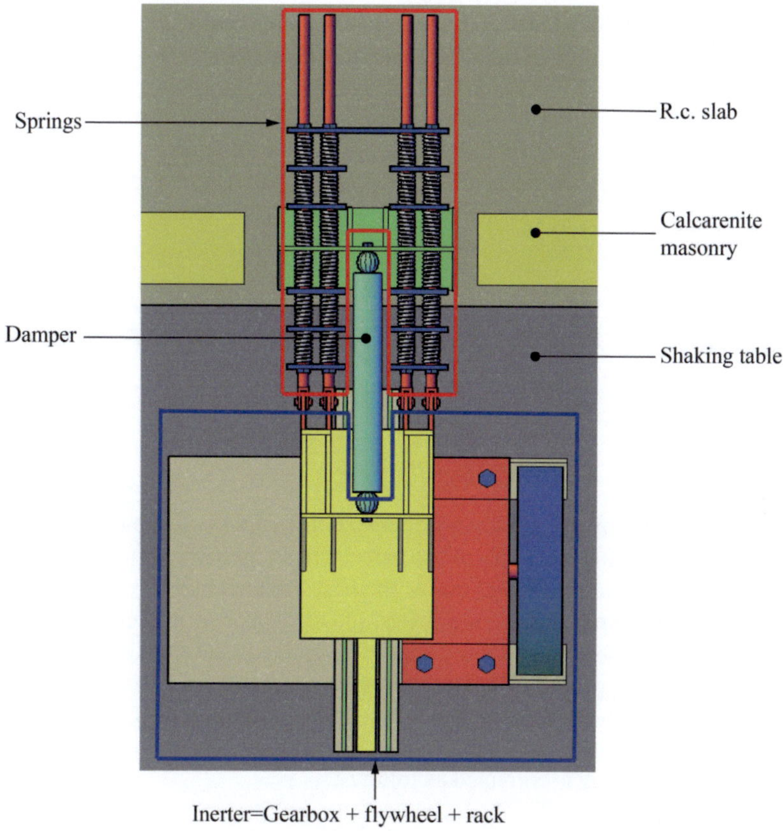

Fig. 3. Main components of TID.

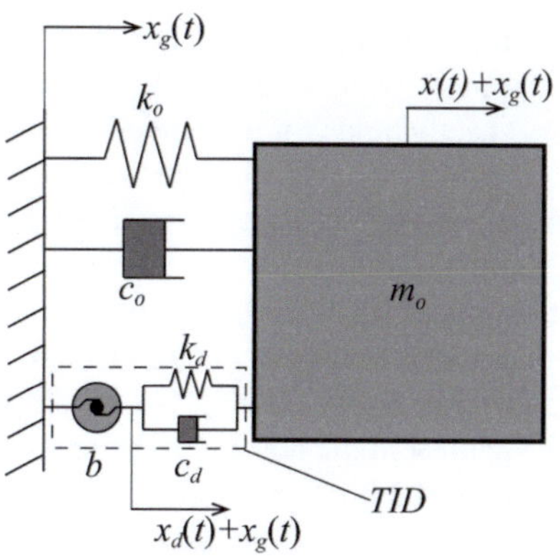

Fig. 4. Reference model for tuning procedure of proposed TID-base isolation system.

The optimal design parameters of the TID are sought by selecting the inertance b based on engineering judgement and using the well-known fixed-point theory for tuned mass devices (Krenk 2019; Krenk and Høgsberg 2016). In particular, the fixed-point theory provides the following optimal tuning frequency and damping ratio of the TID:

$$\omega_d = \frac{\omega_0}{1+\mu};\ \zeta_d = \sqrt{\frac{\mu}{2(1+\mu)}} \tag{3a,b}$$

where μ is the mass ratio

$$\mu = \frac{b}{m_0} \tag{4}$$

and ω_0 is the natural frequency of the SDOF modelling the base-isolated masonry structure

$$\omega_0 = \sqrt{\frac{k_0}{m_0}} \tag{5}$$

It is known that aim of the fixed-point theory is to obtain two separate peaks in the frequency response function (FRF) of the base-isolated masonry structure with TID, in the vicinity of the natural frequency ω_0 of the base-isolated masonry structure without TID; obviously, the two peaks are lower than the resonance peak occurring in the FRF of the base-isolated masonry structure without TID, thus ensuring vibration mitigation. That said, in this study the optimal tuning frequency (3a) is considered as a first tentative value to be slightly corrected, via a trial-and-error procedure, in order to obtain two peaks of approximately the same height; specifically, in the trial-and-error procedure the optimal tuning frequency is corrected as follows:

$$\omega_d = \alpha \frac{\omega_0}{1+\mu} \tag{6}$$

where α is a dimensionless parameter close to 1. The trial-and-error procedure is required because, if the optimal tuning frequency (3a) was used as obtained by the fixed-point theory, the two peaks in the FRF would feature slightly different heights, as expected since the fixed-point theory strictly holds for undamped structures while, in contrast, the base-isolated masonry structure is damped by the base isolators. In the trial-and-error procedure, the discrepancy between the two peaks is corrected by: (1) exploring different potential values of the optimal tuning frequency on assuming ω_d in Eq. (3a) as the first tentative value; (2) calculating the FRF of the base-isolated masonry structure with TID, for each potential value of optimal tuning frequency and, finally, (3) choosing the optimal tuning frequency (6) as that yielding the FRF with two peaks of approximately the same height in the vicinity of the natural frequency ω_0 of the base-isolated masonry structure without TID. In the trial-and-error procedure, all the FRFs are calculated by taking ζ_d in Eq. (3) as optimal damping ratio of the TID. As anticipated above, the optimal tuning frequency (6) and the corresponding first tentative value (3a) are close to each other (correction value α close to 1), meaning that the trial-and-error procedure does not lead to an optimal tuning frequency far from that predicted by the fixed-point theory. Therefore, the fixed-point theory can still be considered as a valuable tuning procedure that, upon slight corrections introduced by the trial-and-error procedure, proves suitable for optimizing the parameters of the TID without introducing any further theoretical assumptions.

To assess the effectiveness of the tuning procedure, a two-floor masonry structure with mass equal to 29 t is considered. Stiffness and viscous damping coefficient assumed for the base isolators are $k_0 = 4 \times 29000$ N/m and $c_0 = 4 \times 2664$ Ns/m (the experimental set up requires four isolators at the four corners of the masonry structure, see Fig. 1), in agreement with typical values in pertinent literature. For three values of inertance b = 4, 8, 12 t. Table 1 reports the optimal stiffness k_d and optimal viscous damping c_d of the TID, as obtained from optimal frequency ω_d in Eq.(3a) (slighlty corrected by the trial-and-error procedure) and damping ratio ζ_d in Eq.(3b) using the following relations:

Table 1. Optimal stiffness and viscous damping coefficient of TID

b (t)	k_d (N/m)	α	c_d (N s/m)
4	11000	0.945	3270
8	16900	0.928	7670
12	20300	0.920	11950

$$k_d = \omega_d^2 b; \qquad c_d = 2\zeta_d \omega_d b \tag{6a,b}$$

Fig. 5 shows the frequency response function (FRF) amplitude for the displacement relative to ground of the base-isolated masonry structure, considering two configurations: (a) the classical base-isolation system without TID and (b) the proposed TID-base isolation system, for the three selected values of the TID inertance b. The FRF is calculated from Eq.(1), assuming $x(t) = X(\omega)e^{i\omega t}$ and $x_d(t) = X_d(\omega)e^{i\omega t}$ with $X(\omega)$ and $X_d(\omega)e^{i\omega t}$ complex functions of the frequency ω, under a unit harmonic ground acceleration $\ddot{x}_g(t) = 1e^{i\omega t}$; in particular, Fig. 5 shows the amplitude $|X(\omega)|$, with $|X(0)| = m_0/k_0$ = 0.25 m the amplitude of the relative displacement under a unit pseudo-static ground acceleration. It is apparent that the TID causes a considerable reduction of the FRF amplitude, meaning that the TID greatly reduces the displacement demand on the base isolators, which is a critical issue in the design of any base isolation system. This is relevant conclusion of this study, unveiling the potential of the proposed TID-base isolation system. The reduction becomes more significant as the inertance increases, consistently with the classical behavior of tuned mass devices whose effectiveness increases with the mass of the device. Unlike standard tuned mass devices, however, the proposed TID develops large inertia effects with a relatively small mass, with obvious benefits in terms of costs and practical realization. Indeed, the inertance b = 12 t is that provided by gearbox and flywheel of the proposed TID (see Appendix) and, if needed, smaller/larger values of the inertance can be obtained by simply changing mass/diameter of the flywheel. Finally, it is noteworthy that the FRF exhibits two peaks of the same height in the vicinity of the natural frequency ω_0 of the base-isolated masonry structure without TID, as expected from the adopted tuning procedure.

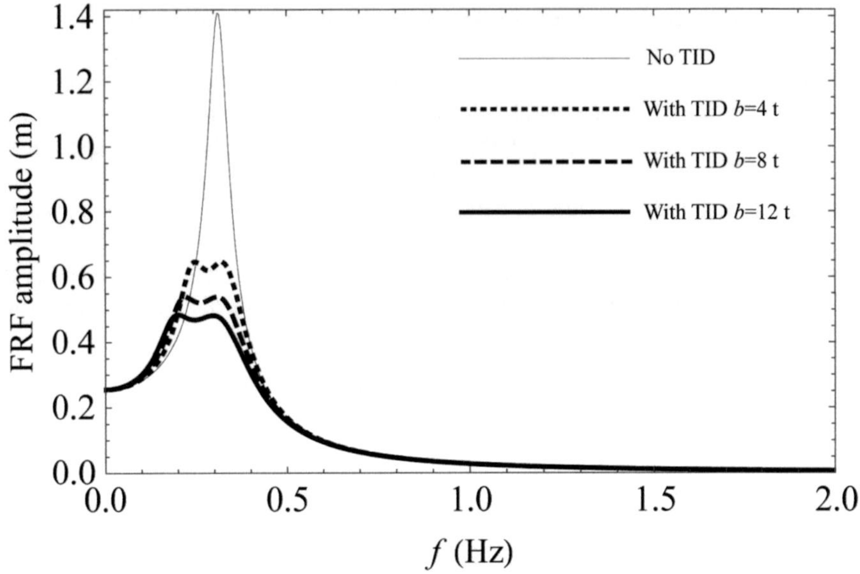

Fig. 5. Frequency response function (FRF) amplitude for displacement relative to ground of base-isolated masonry structure without and with TID.

Based on the results in Fig. 5, it can be concluded that the proposed TID-base isolation system has a strong potential for base isolation of masonry structures. Like the performances of any tuned mass device with parameters derived from the fixed-point theory or alternative methods implemented in the frequency domain, the performances of the TID-base isolation system may vary under non-stationary earthquake ground motion, due to its inherent transient nature. In general, reductions of the system response are expected under earthquake records with main frequency content falling in the range where the frequency response of the system is significantly reduced, e.g., see Fig. 5.

4 Concluding Remarks

This paper has described the detailed design and realization of a hybrid base isolation system for seismic protection of masonry structures, combining standard base isolators with a TID. The TID consists of a grounded rack-and-pinion inerter connected in series to a set of linearly-elastic springs and a viscous dashpot in parallel. A fixed-point tuning procedure has been adopted to estimate optimal stiffness and viscous damping coefficient of the springs/dashpot, upon selecting the inertance of the inerter based on engineering judgement. Frequency response analyses demonstrate the potential of the proposed TID-base isolation system versus a classical one without TID. The TID has been realized for experimental testing and technical details on all its components are included.

5 Appendix

Here follows a detailed list of all components of the TID along with some images of the TID, as realized by Metalmeccanica Falcone, Reggio Calabria, Italy (Figs. 6, 7, 8, 9, 10, 11, 12, 13 and 14).

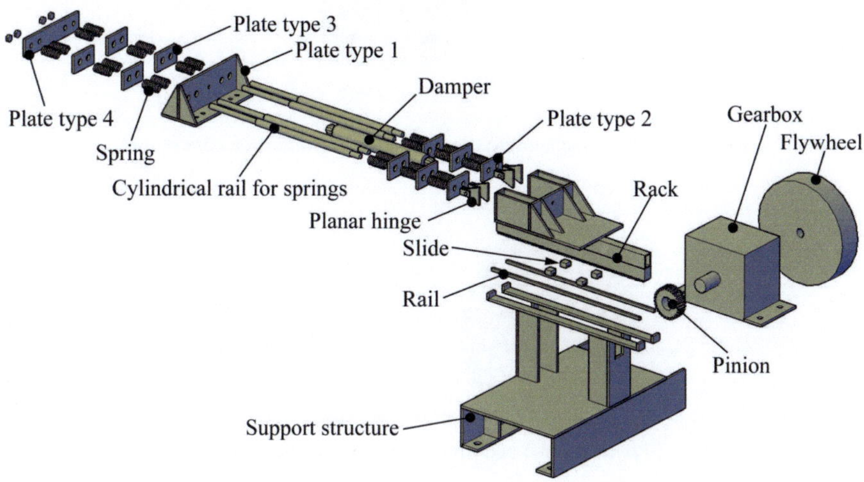

Fig. 6. Overview of all components of TID.

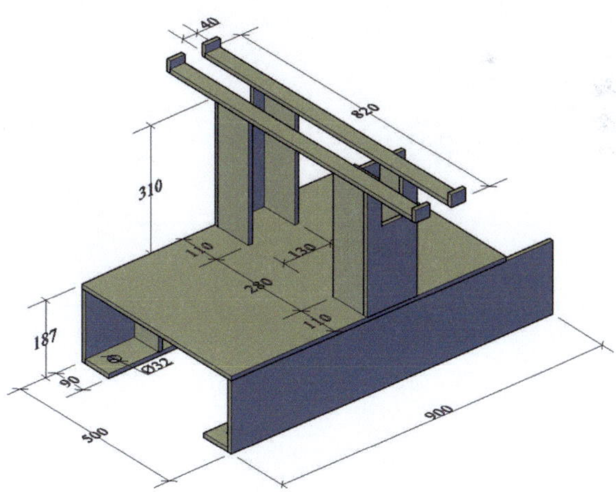

Fig. 7. Support structure. Dimensions are in mm

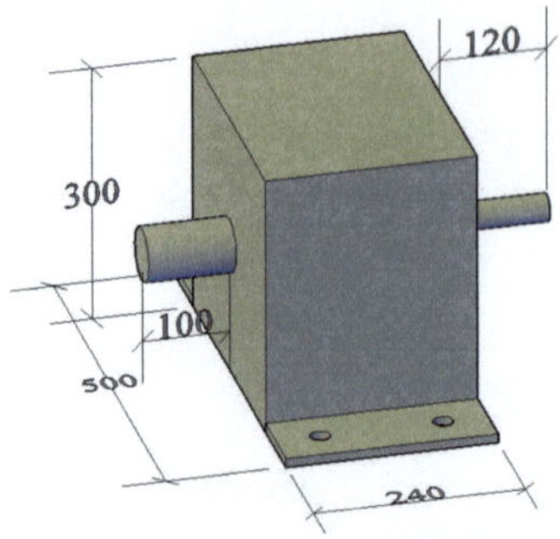

Fig. 8. Gearbox (schematic representation). Dimensions are in mm

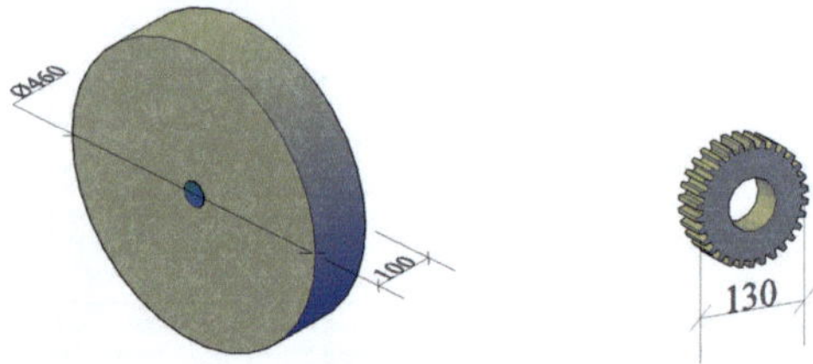

Fig. 9. Flywheel (left) and pinion (right). Dimensions are in mm

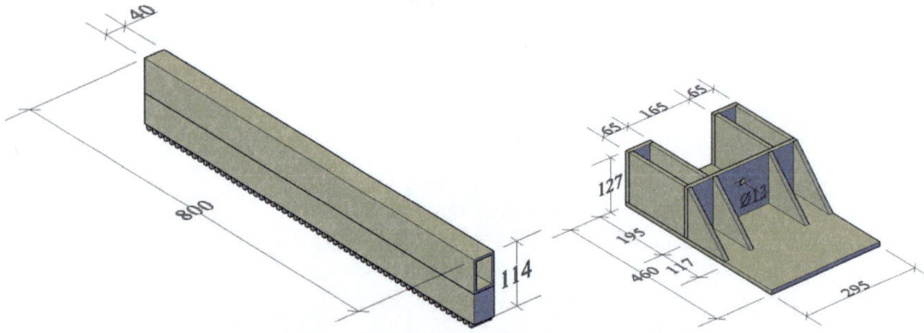

Fig. 10. Rack (left) and end stiffened plate connecting rack to springs/dashpot (right). Dimensions are in mm

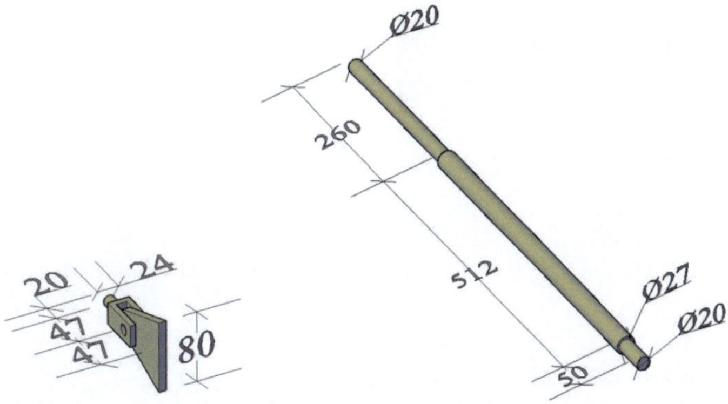

Fig. 11. Plane hinge (left) and cylindrical rail for springs (right). Dimensions are in mm

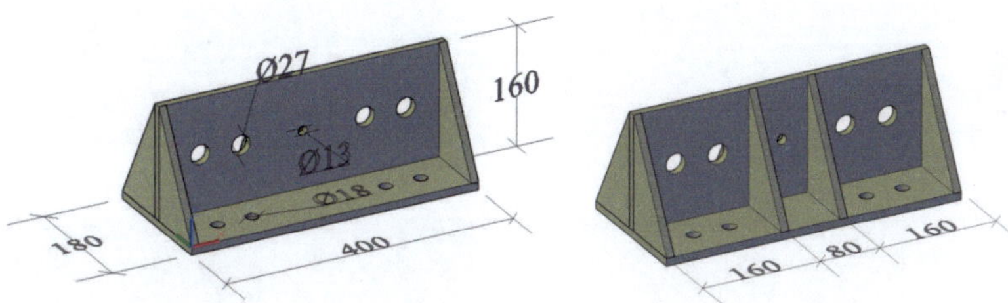

Fig. 12. Plate type 1 (opposite views). Dimensions are in mm

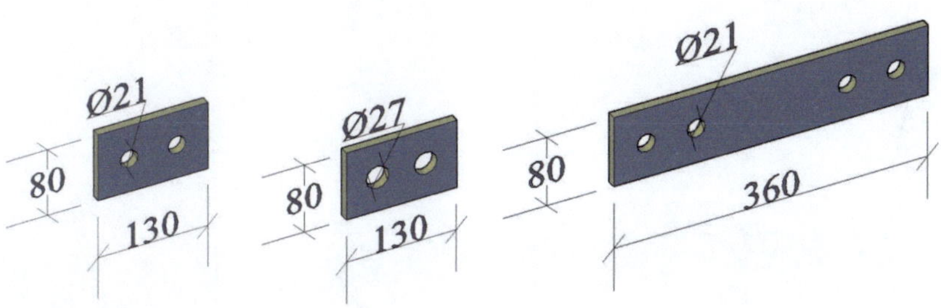

Fig. 13. Plates type 2 (left), type 3 (center), type 4 (right). Dimensions are in mm

Fig. 14. Gearbox (top-left), set of compression springs (top-right), support structure (bottom-left), rack (bottom-right) as realized by Metalmeccanica Falcone, Reggio Calabria, Italy.

References

Alotta, G., Failla, G.: Improved inerter-based vibration absorbers. Int. J. Mech. Sci. **192**, 106087 (2021). https://doi.org/10.1016/j.ijmecsci.2020.106087

Alotta, G., Biondo, C., Giaralis, A., Failla, G.: Seismic protection of land-based wind turbine towers using the tuned inerter damper. Structures. **51**, 640–656 (2023). https://doi.org/10.1016/j.istruc.2023.03.004

Basar, T., Deb, S.K., Das, P.J., Sarmah, M.: Seismic response control of low-rise unreinforced masonry building test model using low-cost and sustainable un-bonded scrap tyre isolator (U-STI). Soil Dyn. Earthq. Eng. **142**, 106561 (2021). https://doi.org/10.1016/j.soildyn.2020.106561

Clemente, P., Martelli, A.: Seismically isolated buildings in Italy: state-of-the-art review and applications. Soil Dyn. Earthq. Eng. **119**, 471–487 (2019). https://doi.org/10.1016/j.soildyn.2017.12.029

Clemente, P., Bontempi, F., Boccamazzo, A.: Base isolated masonry buildings. In: Proceedings of the 15th World Conference on Earthquake Engineering (15WCEE), Portugal, Lisbon, 24–28 September (2012)

Clemente P, Bontempi F, Boccamazzo A (2016) Seismic isolation in masonry buildings: Technological and economic issues. In: Modena C, da Porto F, Valluzzi MR (eds) Brick and Block Masonry: Trends, Innovation and Challenges. 16th International Brick and Block Masonry Conference (IB2MAC), Padua, Italy, June 2016. Taylor & Francis Group, London, UK, 2207–2215

De Domenico, D., Ricciardi, G.: An enhanced base isolation system equipped with optimal tuned mass damper inerter (TMDI). Earthq. Eng. Struct. Dyn. **47**(5), 1169–1192 (2018). https://doi.org/10.1002/eqe.3011

De Luca, A., Guidi, L.G.: State of art in the worldwide evolution of base isolation design. Soil Dyn. Earthq. Eng. **125**, 105722 (2019). https://doi.org/10.1016/j.soildyn.2019.105722

Faiella, D., Argenziano, M., Esposito, F., Brandonisio, G., Fraldi, M., Mele, E.: Effectiveness of isolated vertical extension of masonry buildings as nonconventional TMD. Soil Dyn. Earthq. Eng. **165**, 107675 (2023). https://doi.org/10.1016/j.soildyn.2022.107675

Ge, Z., Wang, W.: Modeling, testing, and characteristic analysis of a planetary flywheel inerter. Shock Vib. **2018**, 2631539 (2018). https://doi.org/10.1155/2018/2631539

Habieb, A.B., Valente, M., Milani, G.: Effectiveness of different base isolation systems for seismic protection: numerical insights into an existing masonry bell tower. Soil Dyn. Earthq. Eng. **125**, 105752 (2019a). https://doi.org/10.1016/j.soildyn.2019.105752

Habieb, A.B., Valente, M., Milani, G.: Hybrid seismic base isolation of a historical masonry church using unbonded fiber reinforced elastomeric isolators and shape memory alloy wires. Eng. Struct. **196**, 109281 (2019b). https://doi.org/10.1016/j.engstruct.2019.109281

Habieb, A.B., Valente, M., Milani, G.: Base seismic isolation of a historical masonry church using fiber reinforced elastomeric isolators. Soil Dyn. Earthq. Eng. **120**, 127–145 (2019). https://doi.org/10.1016/j.soildyn.2019.01.022

Hu, Y., Chen, M.Z.: Performance evaluation for inerter-based dynamic vibration absorbers. Int. J. Mech. Sci. **99**, 297–307 (2015). https://doi.org/10.1016/j.ijmecsci.2015.06.003

Italian Ministry of Environment and Energy Security: Piano Nazionale di Adattamento ai Cambiamenti Climatici (2022). https://www.mase.gov.it/sites/default/files/archivio/allegati/clima/PNACC_versione_dicembre2022.pdf. Accessed 14 June 2025

Krenk, S.: Resonant inerter based vibration absorbers on flexible structures. J. Franklin Inst. **356**(14), 7704–7730 (2019). https://doi.org/10.1016/j.jfranklin.2018.11.038

Krenk, S., Høgsberg, J.: Tuned resonant mass or inerter-based absorbers: Unified calibration with quasi-dynamic flexibility and inertia correction. Proc. R. Soc. A. **472**(2185), 20150718 (2016). https://doi.org/10.1098/rspa.2015.0718

Losanno, D., Ravichandran, N., Parisi, F., Calabrese, A., Serino, G.: Seismic performance of a Low-Cost base isolation system for unreinforced brick Masonry buildings in developing countries. Soil Dyn. Earthq. Eng. **141**, 106501 (2021). https://doi.org/10.1016/j.soildyn.2020.106501

Losanno, D., Ravichandran, N., Parisi, F.: Seismic fragility of base-isolated single-storey unreinforced masonry buildings equipped with classical and recycled rubber bearings in Himalayan regions. J. Build. Eng. **45**, 103648 (2022). https://doi.org/10.1016/j.jobe.2021.103648

Marian, L., Giaralis, A.: Optimal design of a novel tuned mass-damper-inerter (TMDI) passive vibration control configuration for stochastically support-excited structural systems. Probab. Eng. Mech. **38**, 156–164 (2014). https://doi.org/10.1016/j.probengmech.2014.03.007

Marian, L., Giaralis, A.: The tuned mass-damper-inerter for harmonic vibrations suppression, attached mass reduction, and energy harvesting. Smart Struct. Syst. **19**(6), 665–678 (2017) https://doi.org/10.12989/sss.2017.19.6.665

Nauman Masoom, M., Q, u.A.K., Badar, I., Khushnood, R.A., Najam, F.A., Naseer, A.: Development of a new base isolation system using the concept of metamaterials. Eng. Struct. **286**, 116151 (2023). https://doi.org/10.1016/j.engstruct.2023.116151

Osgooei, P.M., Van Engelen, N.C., Konstantinidis, D., Tait, M.J.: Experimental and finite element study on the lateral response of modified rectangular fiber-reinforced elastomeric isolators (MR-FREIs). Eng. Struct. **85**, 293–303 (2015). https://doi.org/10.1016/j.engstruct.2014.11.037

Papageorgiou, C., Houghton, N.E., Smith, M.C.: Experimental testing and analysis of inerter devices. J. Dyn. Sys. Meas. Control. **131**(1), 011001 (2009). https://doi.org/10.1115/1.3023120

Pietrosanti, D., De Angelis, M., Giaralis, A.: Experimental seismic performance assessment and numerical modelling of nonlinear inerter vibration absorber (IVA)-equipped base isolated structures tested on shaking table. Earthq. Eng. Struct. Dyn. **50**(10), 2732–2753 (2021). https://doi.org/10.1002/eqe.3469

Russillo, A.F., Failla, G., Alotta, G.: Ultra-wide low-frequency band gap in locally-resonant plates with tunable inerter-based resonators. Appl. Math. Model. **106**, 682–695 (2022). https://doi.org/10.1016/j.apm.2022.02.015

Sciutteri, S., Russillo, A.F., Santoro, R., Ricciardi, G., Failla, G.: An inerter-based concept of locally resonant fluid-conveying pipe. Eur. J. Mech. A/Solids. **106**, 105316 (2024). https://doi.org/10.1016/j.euromechsol.2024.105316

Smith, M.C.: Synthesis of mechanical networks: the inerter. IEEE Trans. Autom. Control. **47**(10), 1648–1662 (2002). https://doi.org/10.1109/TAC.2002.803532

Taniguchi, T., Der Kiureghian, A., Melkumyan, M.: Effect of tuned mass damper on displacement demand of base-isolated structures. Eng. Struct. **30**, 3478–3488 (2008). https://doi.org/10.1016/j.engstruct.2008.05.027

Tsai, H.C.: The effect of tuned-mass dampers on the seismic response of base-isolated structures. Int. J. Solids Struct. **32**, 1195–1210 (1995). https://doi.org/10.1016/0020-7683(94)00150-U

Xiang, P., Nishitani, A.: Optimum design for more effective tuned mass damper system and its application to base-isolated buildings. Struct. Control Health Monit. **21**, 98–114 (2014). https://doi.org/10.1002/stc.1556

Zhang, C., Ali, A., Sun, L.: Investigation on low-cost friction-based isolation systems for masonry building structures: experimental and numerical studies. Eng. Struct. **243**, 112645 (2021). https://doi.org/10.1016/j.engstruct.2021.112645

Locally Resonant Metamaterial Foundation for Masonry Structures: Prototype Design and Realization

Giuseppe Failla(✉), Luigi Cartone, Gioacchino Alotta, and Andrea Francesco Russillo

Department of Civil, Energy, Environmental and Materials Engineering (DICEAM), University of Reggio Calabria, Reggio Calabria, Italy
{giuseppe.failla,luigi.cartone,gioacchino.alotta, andreaf.russillo}@unirc.it

Abstract. In the framework of ongoing research on seismic metamaterials, this study presents a novel locally resonant metamaterial foundation for seismic base isolation of masonry structures. The foundation is conceived as an assembly of steel cells of circular shape, vertically stacked and mutually connected by circular steel laminated elastomeric bearings. The steel cell consists of a hollow steel cylinder, connected by a rubber ring to an annular steel ring, which acts as resonant unit oscillating in the horizontal plane. Elastic wave propagation analysis, performed on detailed 3D finite element model and on a simplified mass-spring model, demonstrates its effectiveness.

1 Introduction

Base isolation is well-established for seismic protection of structures (Naeim and Kelly 1999; Clemente and Martelli 2019). Yet, many studies are actively seeking for potential advancements of this technique and, in this framework, a novel strategy attracting attention is the concept of periodic foundation, that is, a foundation made of unit cells arranged periodically in space. The working mechanism of periodic foundations hinges upon the capability of infinite periodic media to forbid the propagation of elastic waves within frequency ranges called bandgaps. Correspondingly, the transmittance of finite periodic foundations features strong attenuation zones preventing elastic waves from propagating from the soil to the structure, which match the bandgaps predicted by wave dispersion analysis. In other words, periodic foundations can act as mechanical filters between soil and structure, inhibiting earthquake-induced structural vibrations at frequencies falling within the theoretical bandgaps predicted by wave dispersion analysis. Based on the characteristics of the unit cell, bandgaps may arise from Bragg scattering and/or local resonance; while the first are caused by multiple scattering that elastic waves experience when propagating through the foundation, the second stem from coupling between elastic waves propagating through the foundation and vibrations of periodically arranged resonant units, in the vicinities of the respective resonant frequencies. In locally resonant periodic foundations, therefore, the opening frequencies of the bandgaps occur approximately at the tuning frequencies of the resonant units, which drives the selection of their mass/stiffness parameters.

C. Nava et al. (Eds.): *Climatic and Structural Safety in Multi-Hazard Regime of Cultural and Natural Heritage*, GREEN, pp. 124–135, 2026.
https://doi.org/10.1007/978-3-032-11926-1_8

Locally resonant periodic foundations are commonly referred to as locally resonant metamaterial (LRM) foundation, with the term metamaterial signifying, in a broad sense, an artificially engineered material with exotic properties not found in nature, obtained by appropriate architecture either at macro-scale or at micro-/nano-scales. LRM foundations as well as LRM structural components in general (beams/plates) are hot topics in current research, with an impressive number of studies in the last two decades (Chen et al. 2011; Xiao et al. 2012, 2013; Baravelli and Ruzzene 2013; Peng and Pai 2014; Qian and Shi 2016; Al Ba'ba'a et al. 2017; Miranda et al. 2019; Wang et al. 2019; Failla et al. 2020; Russillo and Failla 2020; Russillo et al. 2021, 2022, 2024a, b Burlon and Failla 2022; Russillo and Failla 2022; Burlon et al. 2023; Santoro et al. 2023; Failla et al. 2024a, b, c; Sciutteri et al. 2024; Zhou et al. 2012).

Several types of LRM foundations were proposed, with fundamental differences in the design of the unit cell. Jia and Shi (2010) and Cheng and Shi (2013) proposed a unit cell consisting of a steel circular cylinder, coated by rubber and embedded in concrete; Zheng et al. (2024) devised a super-cell including four types of cells, each consisting of a steel scatter, a rubber coating layer and a concrete matrix. Further, Casablanca et al. (2018) introduced a LRM foundation consisting of reinforced concrete plates hosting small resonators made by steel cylinders and rubber, vertically stacked and interconnected by ultralow damping layers of steel and Teflon, while Naumann Masoom et al. (2023) designed a LRM foundation made of high-density concrete panels embedding steel-silicone resonators, placed on top of each other and mutually connected by frictionless layers. For shear waves caused by horizontal ground motion, parametric analyses revealed band gaps of different opening frequency/size; typically, band gaps were obtained in frequency ranges from 2.49 Hz to 3.72 Hz (Jia and Shi 2010), from 7.22 Hz to 10.16 Hz (Cheng and Shi 2013); multiple band gaps were also obtained in the frequency ranges 9.3–11.3 Hz, 26.8–29.7 Hz, 34.5–38.8 Hz, and 43.9–53.3 Hz (Zheng et al. 2024), 5–14 Hz (Casablanca et al. 2018), 0.3–7.35 Hz (Naumann Masoom et al. 2023). Numerical analyses were conducted on 3D multi-story frame structures (Jia and Shi 2010; Zheng et al. 2024) and on a nuclear power plant (Cheng and Shi 2013). Moreover, scale experimental tests were made on the foundation alone, i.e. without considering any base isolated structure on top of the foundation (Naumann Masoom et al. 2023). While the works in ref. (Jia and Shi 2010; Cheng and Shi 2013; Casablanca et al. 2018; Naumann Masoom et al. 2023; Zheng et al. 2024) focused on shear waves caused by horizontal ground motion, other studies were concerned with waves caused by horizontal/vertical ground motions, e.g., Cheng and Shi (2018) proposed a LRM foundation including two different unit cells resonating in two directions (one horizontal, one vertical), made by a reinforced concrete matrix with a solid inclusion (reinforced concrete or steel) coated by a soft silicone rubber, which provides complete and directional band gaps in the low frequency range below 10 Hz for seismic protection of nuclear power plants. Furthermore, Shi and Huang (2013) and Colombi et al. (2020) devised LRM foundations featuring unit cells resonating in three directions, e.g., a unit cell consisting of a cubic concrete cube with cubic rubber-coated steel inclusions (Shi and Huang 2013) or consisting of a concrete cube-shaped mass supported by rubber elastic connectors and enclosed in a concrete hollow box (Colombi et al. 2020). Shi and Huang (2013) obtained a complete band gap between 8.79 Hz and 15.72 Hz, while Colombi et al. (2020) demonstrated by numerical tests that the LRM foundation, as applied at the base of a frame structure, provides attenuation zone approximately in the frequency range 5–8 Hz. Some works also presented the results of experimental testing, e.g., Yan et al. (2014) performed experimental tests on a LRM foundation with unit cell resonating in two directions, made by a steel circular cylinder with rubber coating and embedded in concrete, as installed at the base of a small-scale frame structure and subjected to ground motions in one horizontal direction and in the vertical direction; Li et al. (2023, 2024) proposed a LRM foundation with unit cell made of a rubber layer and a concrete layer with resonators acting in three directions, each made up of silicone/silica gel and metallic mass, and tested it numerically and experimentally at small scale, under vertical ground motion and horizontal ground motion in one direction; as in ref. (Naumann Masoom et al. 2023), the LRM foundation alone was tested, without any base isolated structure. In general, numerical/experimental analyses were conducted assuming that the LRM extend over the whole base of the structure to be protected.

The recent studies described above demonstrate that LRM foundations are of strong interest for base isolation. However, advancing research in this field requires tackling some challenges. When dealing with LRM foundations and LRM structural components in general, a main and well-known issue is to obtain bandgaps of considerable size at low frequency because, for a given resonator-to-cell mass ratio, the size generally reduces as the tuning frequency decreases. Moreover, not only ensuring an adequate size at low frequency but also enhancing the attenuation within the bandgaps is a fundamental research target in this field.

Focusing on seismic protection of masonry structures, this paper proposes an innovative concept of LRM foundation, made by an assembly of steel cells of circular shape, vertically stacked and mutually connected by circular steel-laminated elastomeric bearings. The cell consists of a steel hollow cylinder welded to top and bottom steel plates; local resonance is obtained by a steel annular ring, connected to the steel hollow cylinder via a rubber ring, which acts as mass-spring resonant unit oscillating in the horizontal plane. For elastic wave propagation analysis, a detailed 3D finite element model is implemented in ABAQUS and a corresponding mass-spring model is developed in Matlab. Numerical results show that the proposed LRM foundation can attenuate the effects of earthquake ground motion by inducing relevant bandgaps with strong attenuation in the low frequency response of the structure. Another relevant feature of the proposed LRM foundation is that the resonant units are external, readily accessible and not embedded as in alternative LRM foundations; this allows the resonant units to be easily re-tuned by changing the rubber ring and/or the annular steel ring. Re-tuning is important to make the proposed LRM foundation adaptable to potential variations in the structural properties of the masonry structure to be protected. Indeed, non-negligible variations in mechanical parameters such as stiffness and strength may be envisaged in masonry structures as a result of climate-change related phenomena: e.g., chemical degradation due to the increase of CO_2 in the atmosphere and/or high thermal stresses during increasingly frequent heat waves, as recommended by the "National Climate Change Adaptation Plan" released by the Italian Ministry of Environment and Energy Security in December 2022 (Italian Ministry of Environment and Energy Security 2022).

2 Proposed Locally Resonant Metamaterial (LRM) Foundation

The proposed LRM foundation is described in Fig. 1. Each cell has circular shape and is made of a steel hollow cylinder welded to top and bottom steel plates; it hosts a steel annular ring, connected to the steel hollow cylinder via a rubber ring, which acts as mass-spring resonant unit oscillating in the horizontal plane. Teflon annular rings are interposed between the steel annular ring and bottom/top steel plates to avoid friction. The cells are vertically stacked and mutually connected by circular steel-laminated elastomeric bearings; the latter are connected to the steel plates of two consecutive cells by high-strength, double-sided thin adhesive layers. Each steel-laminated elastomeric bearing is composed of four steel layers and five rubber layers.

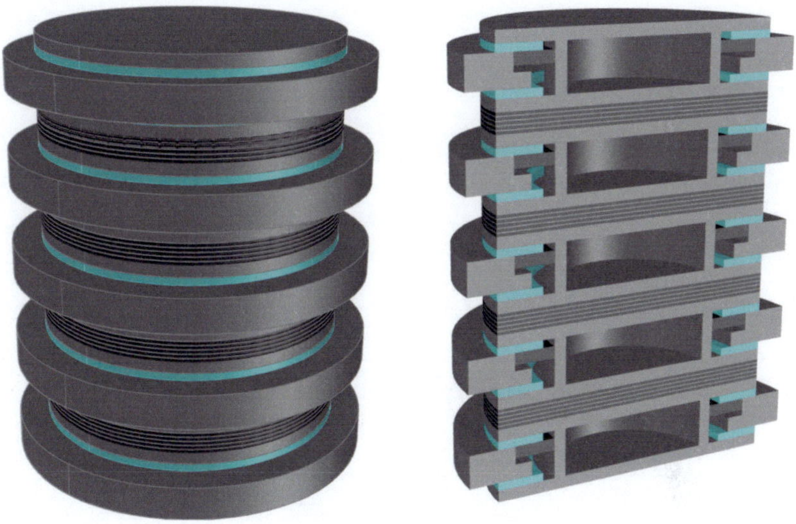

Fig. 1. Proposed LRM foundation.

It is assumed that the foundation is made by five cells overall. The geometry is detailed in Fig. 2 and the material parameters are reported in Table 1. The geometry is chosen based on engineering judgement and in agreement with alternative concepts of LRM foundations or periodic foundations in the literature (Xiang et al. 2012; Nauman Masoom et al. 2023); indeed, the total height is similar to those of the LRM foundation proposed in ref. (Nauman Masoom et al. 2023) and of the two-layer periodic foundation in ref. (Xiang et al. 2012). Material parameters are typical of applications in this field. For example, rubber parameters are of the same order of magnitude as those in ref. (Cheng et al. 2020) and as those in ref. (Strauss et al. 2014) assumed for the steel-laminated bearings.

The LRM foundation is designed to be tested experimentally as base isolation system for a 2:3 scale masonry structure. Four LRM foundations will be placed at the four corners of a benchmark, rectangular-shaped masonry structure and experimental tests will be performed on a shaking table.

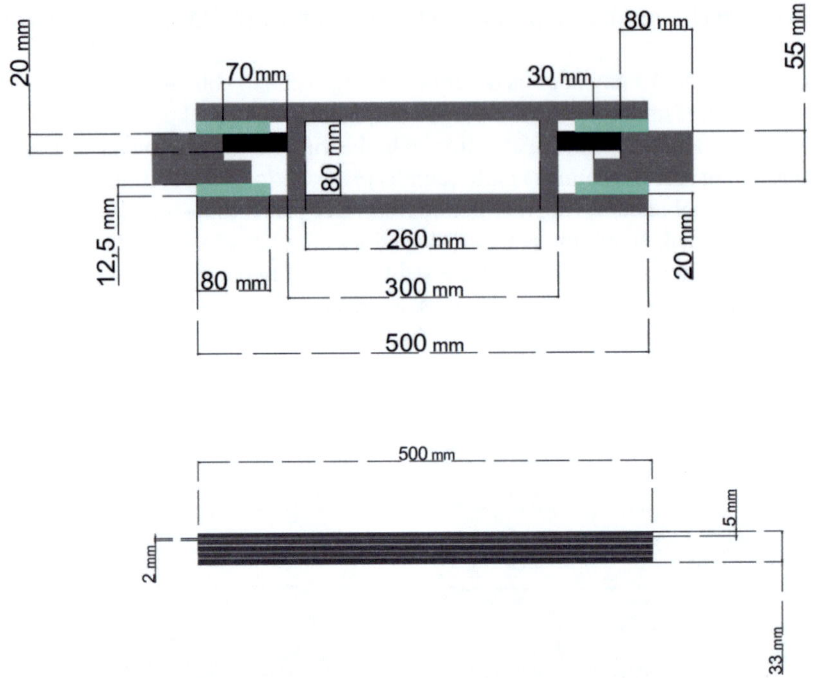

Fig. 2. Unit cell and steel-laminated elastomeric bearings of proposed LRM foundation.

Table 1. Material parameters of LRM foundation

	Young modulus (GPa)	Poisson coefficient	Mass density (kg/m^3)
Steel	210	0.3	7800
Rubber	9×10^{-4}	0.46	1350

The LRM foundation is currently under realization by Phononic Vibes, a spin-off company of the Polytechnic of Milan.

3 Numerical Models

The proposed LRM foundation in Fig. 1, with parameters in Table 1 and Table 2, is modelled in ABAQUS using hexahedral elements, as shown in Fig. 3. All components are modelled in detail, except for the Teflon annular rings, interposed between the steel annular ring and bottom/top steel plates to avoid friction, which are reverted to vertical kinematic constraints between the steel plates and the steel annular rings, for simplicity. Meshes of the steel cell and the steel-laminated elastomeric bearings are created separately and then assembled by the tie-mesh option of ABAQUS. This choice is made to obtain accurate and non-distorted meshes in the thin steel and rubber layers composing the steel-laminated elastomeric bearings. Overall, 78512 finite elements model the LRM foundation.

Table 2. Mechanical parameters of double-sided thin adhesive layers in LRM foundation

Adhesion to BA steel	1600 cN/20mm (EN 1939)
Static shear	1 mm/3 h (EN 1943)
Dynamic shear	450 N/4 cm^2 (Tested after 15 min, 300 mm/min)

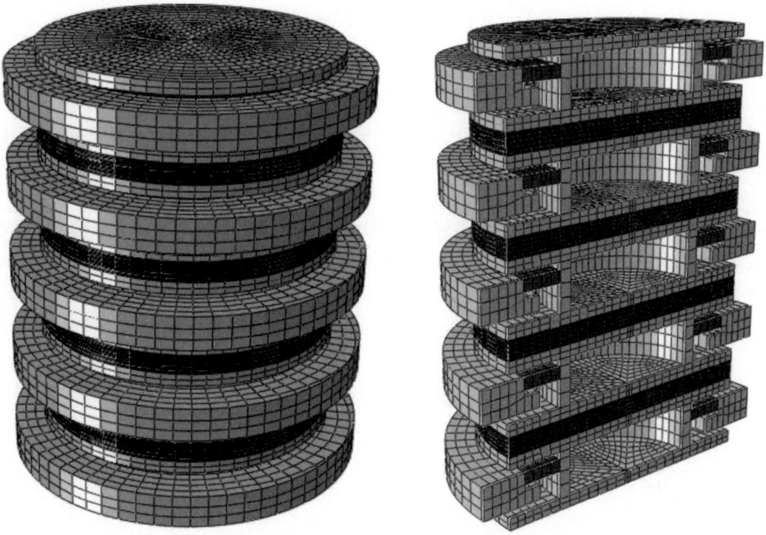

Fig. 3. Finite element model of proposed LRM foundation.

For a preliminary elastic wave propagation analysis, the LRM foundation can be modelled as a mass-spring chain with horizontal displacements in one direction. Mass M of the cell indicates the mass of the hollow cylinder and the masses of top/bottom steel plates welded to the hollow cylinder, mass m denotes the mass of the steel annular ring. The masses of the other components are well smaller and can be neglected. Under horizontal motion, the steel-laminated elastomeric bearings can be modelled as (massless) linear springs of equivalent stiffness

$$K = GA/t_r \tag{1}$$

where A is the area of the cross section, $t_r = nt_i$ with n and t_i number and thickness of the rubber layers, G is the shear modulus of the rubber; Eq. (1) is widely used in literature (Strauss et al. 2014). Moreover, the rubber rings are modelled as (massless) linear springs of stiffness (Xiao et al. 2013)

$$k = \frac{\pi\left(5 + 3.29S^2\right)Gl}{\ln(r_2/r_1)} \tag{2}$$

where $S = l/[(r_1 + r_2)\, ln\, (r_2/r_1)]$ with l, r_1 and r_2 the thickness, inner and outer radii of the rubber ring, G the shear modulus of the rubber. Consistently with typical assumptions in literature, damping in the components is neglected for elastic wave dispersion analysis.

The mass-spring chain is a classical diatomic chain (Hussein et al. 2014). The dispersion diagram can readily be built considering the mass-spring chain as an infinite chain and applying Bloch-Floquet theory to the motion equation of the unit cell:

$$K(u_{n+1} - u_n) - K(u_n - u_{n-1}) + k(z_n - u_n) = M\ddot{u}_n \tag{3a}$$

$$k(u_n - z_n) = m\ddot{z}_n \tag{3b}$$

with u_n the displacement of the mass M of the cell (mass of the hollow cylinder and masses of top/bottom steel plates welded to the hollow cylinder), z_n the displacement of the mass m (mass of the steel annular ring), K and k the stiffnesses in Eq. (1) and Eq. (2). The dispersion diagram of the infinite mass-spring chain is shown in Fig. 4. A bandgap is observed, opening at 10.1 Hz and closing at 13.85 Hz, where elastic wave propagation through the mass-spring chain is forbidden. This frequency range is typical of earthquake-induced elastic waves and substantiates the potential of the LRM for seismic protection. Notice that, as expected, the opening frequency of the bandgap predicted by the mass-spring model is equal to the tuning frequency of the resonant units

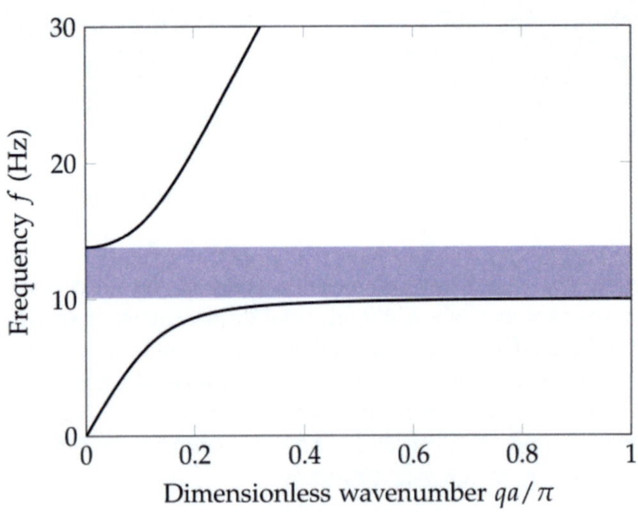

Fig. 4. Dispersion diagram of mass-spring model of LRM foundation.

$$f = \frac{1}{2\pi}\sqrt{\frac{k}{m}} = 10.1 \text{ Hz} \tag{4}$$

with m the mass of the steel annular ring and k the stiffness provided by the rubber annular ring, given by Eq. (2).

4 Elastic Wave Propagation Analysis

Figure 5 shows the transmittance of the finite element model of the LRM foundation under a unit harmonic ground displacement, calculated as the ratio of the absolute displacement amplitude measured at the center of the LRM foundation top plate to the ground displacement unit amplitude; both displacements are taken in the same direction, arbitrarily chosen due to the polar symmetry of the LRM foundation. A significant attenuation zone is observed starting from 9.4 Hz, which is approximately equal to the tuning frequency $f = 10.1$ Hz of the resonant units, calculated in Eq. (4) from the mass-spring model. It is noteworthy that the opening frequency of the attenuation zone agrees with that of several LRM foundations proposed in the literature (Cheng and Shi 2013; Shi and Huang 2013; Casablanca et al. 2018; Colombi et al. 2020; Zheng et al. 2024). Moreover, Fig. 5 includes the transmittance calculated from the mass-spring model of the LRM foundation. The transmittances obtained from the two models are in good agreement, especially in the frequency range of the bandgap; that is, the mass-spring model can correctly capture the bandgap formation. This is a remarkable result, meaning that the mass-spring model can be used at the early stages of design to select/check opening frequency and size of the bandgap, without recurring to a detailed finite element model of the whole LRM foundation.

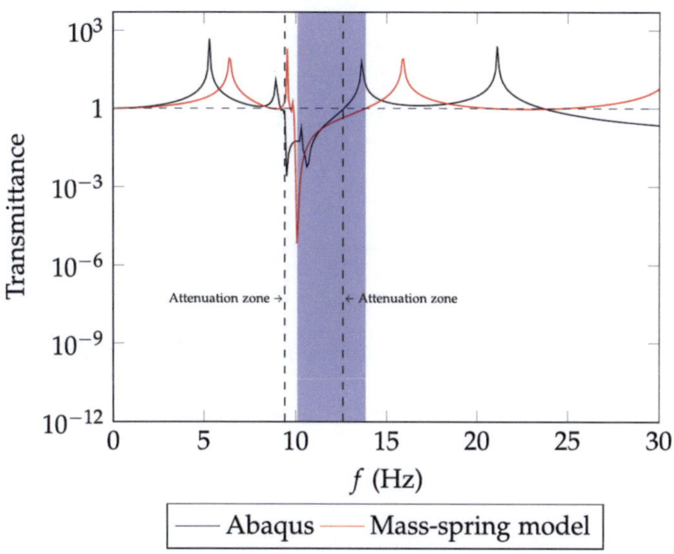

Fig. 5. Transmittance of LRM foundation without damping.

For a further insight, Fig. 6 reports the transmittance obtained from the finite element model of the LRM foundation, when damping is considered in the rubber components, i.e., in the rubber rings of the resonant units and in the rubber layers of the steel-laminated elastomeric bearings. Specifically, the "Composite Modal Damping" option in ABAQUS is used, as it allows pertinent damping factors to be assigned to each material. The material-dependent damping factors are combined into a modal damping ratio for each mode, obtained as weighted averages of the mass matrix associated with each material. Attributing an appropriate damping ratio to each material is important in the problem under study, because damping effects play an important role in LRM structural components: in particular, it is known that high level of damping in the resonant units may be detrimental and reduce the attenuation within the bandgaps. In agreement with current literature (Pianese et al. 2024), high damping factors are attributed to the rubber components, 10% or 15%. Damping in steel is neglected, for conservativeness. Fig. 5 shows that damping in the rubber components has beneficial effects on the resonance peaks outside the bandgaps and leaves almost unchanged the very low values of transmittance attained within the bandgaps. Moreover, no substantial differences are seen in the transmittance within the bandgaps for either 10% or 15% damping factor. This is important for design purposes, in view of the uncertainties inherent in the estimation of actual damping in materials.

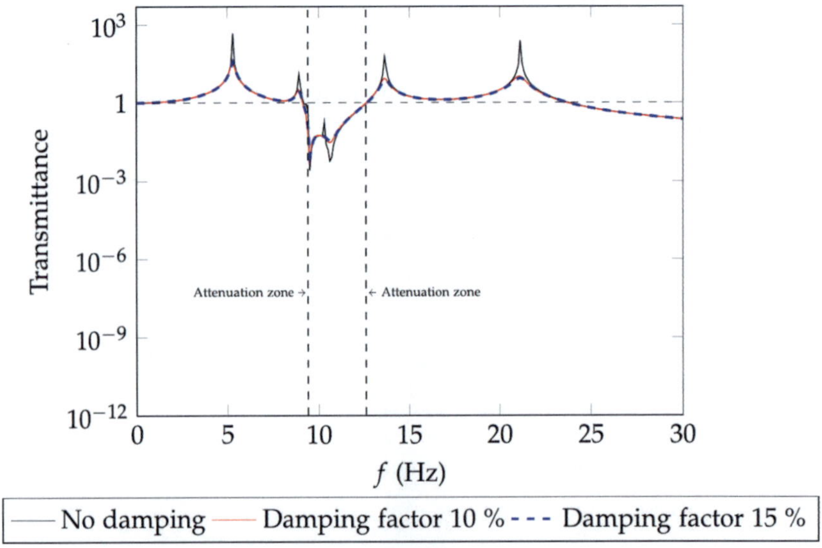

Fig. 6. Transmittance of LRM foundation with damping.

5 Concluding Remarks

This study has presented a novel LRM foundation for seismic base isolation of masonry structures, conceived as an assembly of steel cells of circular shape, vertically stacked and mutually connected by circular steel laminated elastomeric bearings. The steel cell consists of a hollow steel cylinder, connected by a rubber ring to an annular steel ring, which acts as resonant unit oscillating in the horizontal plane. Numerical analyses performed on a detailed 3D finite element model and a simplified mass-spring model provide insight into the working mechanism of the foundation and demonstrate its effectiveness for seismic protection.

References

Al Ba'ba'a, H., Nouh, M., Singh, T.: Formation of local resonance band gaps in finite acoustic metamaterials: A closed-form transfer function model. J. Sound Vib. **410**, 429–446 (2017). https://doi.org/10.1016/j.jsv.2017.08.009

Baravelli, E., Ruzzene, M.: Internally resonating lattices for bandgap generation and low-frequency vibration control. J. Sound Vib. **332**, 6562–6579 (2013). https://doi.org/10.1016/j.jsv.2013.08.014

Clemente, P., Martelli, A.: Seismically isolated buildings in Italy: State-of-the-art review and applications. Soil Dyn. Earthq. Eng. **119**, 471–487 (2019). https://doi.org/10.1016/j.soildyn.2017.12.029

Burlon, A., Failla, G.: Flexural wave propagation in locally-resonant beams with uncoupled/coupled bending-torsion beam-like resonators. Int. J. Mech. Sci. **215**, 106925 (2022). https://doi.org/10.1016/j.ijmecsci.2021.106925

Burlon, A., Failla, G.: On the band gap formation in locally-resonant metamaterial thin-walled beams. Eur. J. Mech. A/Solids. **97**, 104798 (2023). https://doi.org/10.1016/j.euromechsol.2022.104798

Casablanca, O., Ventura, G., Garescì, F., Azzerboni, B., Chiaia, B., Chiappini, M., Finocchio, G.: Seismic isolation of buildings using composite foundations based on metamaterials. J. Appl. Phys. **123**(17), 174903 (2018). https://doi.org/10.1063/1.5018005

Chen, J.S., Sharma, B., Sun, C.T.: Dynamic behaviour of sandwich structure containing spring-mass resonators. Compos. Struct. **93**, 2120–2125 (2011). https://doi.org/10.1016/j.compstruct.2011.02.007

Cheng, Z., Shi, Z.: Novel composite periodic structures with attenuation zones. Eng. Struct. **56**, 1271–1282 (2013). https://doi.org/10.1016/j.engstruct.2013.07.003

Cheng, Z.B., Shi, Z.F.: Composite periodic foundation and its application for seismic isolation. Earthq. Eng. Struct. Dyn. **47**(4), 925–944 (2018). https://doi.org/10.1002/eqe.2999

Cheng, Z., Shi, Z., Palermo, A., Xiang, H., Guo, W., Marzani, A.: Seismic vibrations attenuation via damped layered periodic foundations. Eng. Struct. **211**, 110427 (2020). https://doi.org/10.1016/j.engstruct.2020.110427

Colombi, A., Zaccherini, R., Aguzzi, G., Palermo, A., Chatzi, E.: Mitigation of seismic waves: Metabarriers and metafoundations bench tested. J. Sound Vib. **485**, 115537 (2020). https://doi.org/10.1016/j.jsv.2020.115537

Failla, G., Burlon, A., Russillo, A.F.: A novel metamaterial multiple beam structure with internal local resonance. Acta Mech. **235**, 5885–5903 (2024a). https://doi.org/10.1007/s00707-024-04006-w

Failla, G., Marzani, A., Palermo, A., Russillo, A.F., Colquitt, D.: Current developments in elastic and acoustic metamaterials science. Philos. Trans. R. Soc. A. **382**, 20230369 (2024b). https://doi.org/10.1098/rsta.2023.0369

Failla, G., Marzani, A., Palermo, A., Russillo, A.F., Colquitt, D.: Current developments in elastic and acoustic metamaterials science. Philos. Trans. R. Soc. A. **382**, 20240038 (2024c). https://doi.org/10.1098/rsta.2024.0038

Failla, G., Santoro, R., Burlon, A., Russillo, A.F.: An exact approach to the dynamics of locally-resonant beams. Mech. Res. Commun. **103**, 103460 (2020). https://doi.org/10.1016/j.mechrescom.2019.103460

Hussein, M.I., Leamy, M.J., Ruzzene, M.: Dynamics of phononic materials and structures: Historical origins, recent progress, and future outlook. Appl. Mech. Rev. **66**, 040802 (2014). https://doi.org/10.1115/1.4026911

Italian Ministry of Environment and Energy Security: Piano Nazionale di Adattamento ai Cambiamenti Climatici (2022). https://www.mase.gov.it/sites/default/files/archivio/allegati/clima/PNACC_versione_dicembre2022.pdf. Accessed 14 June 2025.

Jia, G., Shi, Z.: A new seismic isolation system and its feasibility study. Earthq. Eng. Eng. Vib. **9**, 75–82 (2010). https://doi.org/10.1007/s11803-010-8159-8

Li, Y., He, H., Chen, H., Liu, Y., Wang, W., Dong, J., Jin, F., Fan, H.: Composite wave attenuation mechanism of periodic layered metastructure with embedded rubber-mass resonators. Constr. Build. Mater. **384**, 131322 (2023). https://doi.org/10.1016/j.conbuildmat.2023.131322

Li, Y., He, H., Feng, J., Chen, H., Jin, F., Fan, H.: Layered metastructure containing freely-designed local resonators for wave attenuation. Def. Technol. **32**, 23–31 (2024). https://doi.org/10.1016/j.dt.2023.03.026

Miranda Jr., E., Nobrega, E., Ferreira, A., Dos Santos, J.: Flexural wave band gaps in a multi-resonator elastic metamaterial plate using Kirchhoff-Love theory. Mech. Syst. Signal Process. **116**, 480–504 (2019). https://doi.org/10.1016/j.ymssp.2018.06.059

Naeim, F., Kelly, J.M.: Design of Seismic Isolated Structures: From Theory to Practice. Wiley, New York (1999)

Nauman Masoom, M., ul Ain Karim, Q., Badar, I., Khushnood, R.A., Najam, F.A., Naseer, A.: Development of a new base isolation system using the concept of metamaterials. Eng. Struct. **286**, 116151 (2023). https://doi.org/10.1016/j.engstruct.2023.116151

Peng, H., Pai, P.F.: Acoustic metamaterial plates for elastic wave absorption and structural vibration suppression. Int. J. Mech. Sci. **89**, 350–361 (2014). https://doi.org/10.1016/j.ijmecsci.2014.09.018

Pianese, G., Van Engelen, N., Toopchi-Nezhad, H., Milani, G.: An experimental and numerical insight into the unbonded and partially bonded high-damping fiber-reinforced elastomeric isolators. Soil Dyn. Earthq. Eng. **187**, 109016 (2024). https://doi.org/10.1016/j.soildyn.2024.109016

Qian, D., Shi, Z.: Bandgap properties in locally resonant phononic crystal double panel structures with periodically attached spring-mass resonators. Phys. Lett. A. **380**(41), 3319–3325 (2016). https://doi.org/10.1016/j.physleta.2016.07.068

Russillo, A.F., Failla, G.: On the free vibrations of locally-resonant structures. Comput. Struct. **241**, 106356 (2020). https://doi.org/10.1016/j.compstruc.2020.106356

Russillo, A.F., Failla, G.: A novel reduced-order dynamic-stiffness formulation for locally resonant metamaterial plates. Compos. Struct. **280**, 114811 (2022). https://doi.org/10.1016/j.compstruct.2021.114811

Russillo, A.F., Arena, F., Failla, G.: Water-tank metabarrier for seismic Rayleigh wave attenuation. Philos. Trans. R. Soc. A. **382**, 20230363 (2024a). https://doi.org/10.1098/rsta.2023.0363

Russillo, A.F., Failla, G., Alotta, G.: Ultra-wide low-frequency band gap in locally-resonant plates with tunable inerter-based resonators. Appl. Math. Model. **106**, 682–695 (2022). https://doi.org/10.1016/j.apm.2022.02.015

Russillo, A.F., Failla, G., Fraternali, F.: Free and forced vibrations of damped locally-resonant sandwich beams. Eur. J. Mech. A. Solids **86**, 104188 (2021). https://doi.org/10.1016/j.euromechsol.2020.104188

Russillo, A.F., Kouznetsova, V.G., Failla, G., Geers, M.G.D.: A reduced-order computational homogenization framework for locally resonant metamaterial structures. Comput. Mech. **74**, 743–762 (2024b). https://doi.org/10.1007/s00466-024-02453-9

Russillo, A.F., Failla, G.: Seismic metamaterials for Rayleigh wave attenuation: A novel concept of soil-embedded water-tank metabarrier. Int. J. Solids Struct. **324**, 113656 (2026). https://doi.org/10.1016/j.ijsolstr.2025.113656

Santoro, R., Mazzeo, M., Failla, G.: A computational framework for uncertain locally resonant metamaterial structures. Mech. Syst. Signal Process. **190**, 110094 (2023). https://doi.org/10.1016/j.ymssp.2023.110094

Sciutteri, S., Russillo, A.F., Santoro, R., Ricciardi, G., Failla, G.: An inerter-based concept of locally resonant fluid-conveying pipe. Eur. J. Mech. A/Solids. **106**, 105316 (2024). https://doi.org/10.1016/j.euromechsol.2024.105316

Shi, Z., Huang, J.: Feasibility of reducing three-dimensional wave energy by introducing periodic foundations. Soil Dyn. Earthq. Eng. **50**, 204–212 (2013). https://doi.org/10.1016/j.soildyn.2013.03.009

Strauss, A., Apostolidi, E., Zimmermann, T., Gerhaher, U., Dritsos, S.: Experimental investigations of fiber and steel reinforced elastomeric bearings: Shear modulus and damping coefficient. Eng. Struct. **75**, 402–413 (2014). https://doi.org/10.1016/j.engstruct.2014.06.008

Wang, K., Zhou, J., Cai, C., Xu, D., Ouyang, H.: Mathematical modeling and analysis of a meta-plate for very low-frequency band gap. Appl. Math. Model. **73**, 581–597 (2019). https://doi.org/10.1016/j.apm.2019.04.033

Xiang, H.J., Shi, Z.F., Wang, S.J., Mo, Y.L.: Periodic materials-based vibration attenuation in layered foundations: Experimental validation. Smart. Mater. Struct. **21**, 112003 (2012). https://doi.org/10.1088/0964-1726/21/11/112003

Xiao, Y., Wen, J., Wen, X.: Broadband locally resonant beams containing multiple periodic arrays of attached resonators. Phys. Lett. A. **376**(16), 1384–1390 (2012). https://doi.org/10.1016/j.physleta.2012.02.059

Xiao, Y., Wen, J., Yu, D., Wen, X.: Flexural wave propagation in beams with periodically attached vibration absorbers: Band-gap behavior and band formation mechanisms. J. Sound Vib. **332**(4), 867–893 (2013). https://doi.org/10.1016/j.jsv.2012.09.035

Yan, Y., Laskar, A., Cheng, Z., Menq, F., Tang, Y., Mo, Y.L., Shi, Z.: Seismic isolation of two dimensional periodic foundations. J. Appl. Phys. **116**(4), 044908 (2014). https://doi.org/10.1063/1.4891837

Zheng, H., Miao, L., Xiao, P., Lei, K., Wang, Q.: Novel metamaterial foundation with multi low-frequency bandgaps for isolating earthquakes and train vibrations. Structures. **61**, 106070 (2024). https://doi.org/10.1016/j.istruc.2024.106070

Zhou, X., Liu, X., Hu, G.: Elastic metamaterials with local resonances: An overview. Theor. Appl. Mech. Lett. **2**, 041001 (2012). https://doi.org/10.1063/2.1204101

Advanced Methodological Framework for Assessing Climate Change Impacts on Cultural Heritage Toward a Multi-scale Vulnerability Index

Francesca Giglio(✉)

Department of Architecture and Design, Mediterranea University of Reggio Calabria, Reggio Calabria, Italy
francesca.giglio@unirc.it

Abstract. Cultural heritage assets monuments, historical buildings, landscapes are increasingly vulnerable to climate change-induced impacts. Southern European regions, including the Mediterranean basin, are particularly exposed to risks such as heatwaves, hygrothermal oscillations, flooding, wind, and seismicity. In Calabria, local geomorphological and socio-environmental fragilities further exacerbate the condition of minor historic centers. This chapter presents an advanced methodological framework for assessing climate change impacts on cultural heritage, with the goal of constructing a theoretical basis for a replicable and scalable climate vulnerability assessment tool. The framework connects environmental stressors with physical and biological deterioration pathways, extending its focus to the societal implications of heritage degradation. Based on established references, it identifies a coherent set of midpoint and endpoint indices to translate complex climate-heritage interactions into operational knowledge. The approach aims to guide strategic adaptation and co-design through digital and physical Living Labs, involving communities and institutions. While rooted in experimental evidence, the chapter focuses on methodological and modeling aspects, aiming to support innovative and adaptive strategies for the structural and environmental safety of heritage assets.

Keywords: cultural heritage · climate-related hazard · heritage impact · adaptation · risk assessment

1 Introduction

Extreme weather events on cultural heritage in Europe have become increasingly evident in recent years. They are progressing at unprecedented speed and scale. Deep scientific knowledge about future climate projections is required to develop appropriate preservation strategies. Changes in weather and climate conditions aggravate the physical, chemical, and biological mechanisms of degradation, affecting the structure and/or composition of archeological sites, historic buildings or museum collections (Sabbioni

C. Nava et al. (Eds.): *Climatic and Structural Safety in Multi-Hazard Regime of Cultural and Natural Heritage*, GREEN, pp. 136–149, 2026.
https://doi.org/10.1007/978-3-032-11926-1_9

et al., 2012; Sesana et al., 2021; Bertolin 2019, Loli et al., 2020, O'Neill et al., 2022, Grossi et al., 2007).

Changes in preservation conditions due to climate-related decay processes are unavoidable phenomena for both movable and immovable cultural heritage.

Understanding the mechanisms behind these processes, and their real effect on heritage significance, enables a rational use of materials and anticipation of the behaviour. This knowledge supports preventive conservation, heritage management and restoration.

The degradation progress, which depends on external agents of decay, exposure, the intrinsic properties of the material to be studied, and object construction vulnerability, is nowadays exacerbated by both anthropic factors and the impact of climate change. Due to more frequent and severe weather events, greater exposure, material ageing, and past conservation interventions, the need for adapting cultural heritage to both anthropic and climate change-related effects is increasingly urgent (Bertolin, 2019).

On these topics, Cacciotti et al. (2021) showed that the following damages are often reported at different levels:

- At site level: erosion, soil displacement, earth deposition, damage to forest, parks, or individual trees;
- At building level: material degradation, roof and façade damage, and primary and secondary structural damage.
- At object level (i.e. movable heritage): widespread damage to furniture, musical instruments, artworks, books, paper, glass and ceramic objects.

In this context, the paper describes a preliminary study of the mechanisms of climate change impacts and the strategic role of climate scenarios in identifying the most appropriate interventions for different heritage typologies.

The study consists of three parts: 1) a survey at the European and National levels of key climatic factors and impact mechanisms on cultural heritage materials, based on scientific reports (ICOMOS, 2019; Rebollo et al., 2020); 2) a review of modelling approaches for mitigation and adaptation to climate change scenarios, from which methodologies for developing tools and indices for climate vulnerability and risk assessment are systematized; 3).The elaboration of a framework correlating physical and biological impact mechanisms, climate drivers, expected effects on cultural heritage, and community impacts, including a set of midpoint indices as benchmarks to measure vulnerability of buildings, structures, and urban heritage.

2 Climate Drivers and Degradation Mechanisms

In recent years, the increasing frequency and severity of extreme climate events, together with the gradual transformation of global atmospheric conditions, have prompted a deep re-evaluation of the exposure and vulnerability of cultural heritage, particularly its tangible forms. Climate-induced alterations not only accelerate the physical, chemical, and biological deterioration of materials, but also significantly reshape the environmental and territorial contexts in which heritage assets are embedded. These changes create increasingly critical conditions for the preservation, management, and transmission of cultural heritage to future generations.

According to leading international bodies (ICOMOS, 2019; IPCC, 2021), interactions between climate change and cultural materials result in measurable impacts on historic buildings, cultural landscapes, traditional infrastructures, and artistic objects, threatening their integrity, meaning, and utility value. The urgency of integrating climate risk into preventive conservation protocols is now widely recognized across scientific literature and in adaptation strategies at both European and national levels.

To ensure sustainable management of cultural heritage, it is essential to understand how climate will evolve at specific sites and to what extent future changes will influence heritage typologies and materials (Morel et al., 2022; Richards et al., 2022).

UNESCO's 2007 report Climate Change and World Heritage (Augustin, 2007) was the first attempt to provide an overview of threats posed by climate change to cultural heritage. It emphasized that Climate change – alone or in combination with air pollution - can affect materials, artefacts, collections and cultural landscapes.

A subsequent assessment published in 2016 by UNESCO, UNEP and UCS (Markham et al., 2016), further developed monitoring recommendations, actions and policy responses.

More recent frameworks, such as the Climate Vulnerability Index (CVI), highlight the importance of integrating exposure, sensitivity, and adaptive capacity into risk assessments. This section therefore outlines current foundations and methodologies adopted at the European and international level, emphasizing the need for locally adaptable and diagnostically informed indices.

Following IPCC and ARCH models, the main climatic drivers acting on heritage include:

- Temperature rise and variation: thermal stress, freeze–thaw cycles.
- Humidity fluctuations: salt crystallization, surface scaling, bio-colonization.
- Precipitation and runoff: erosion, infiltration, mold.
- Wind and storms: loss of protective layers, structural instability.
- Solar radiation: photochemical decay, surface drying.

As ICOMOS (2019) stresses, responding to climate change requires adjusting to risks, either in reaction to or in anticipation of a changing climate. Understanding its impact on natural and physical systems, human communities and cultural heritage is therefore essential - not only to evaluate risks and adaptive capacity - but also to recognize the positive role heritage can play as a source of resilience for ecosystems, cities, neighborhoods and landscapes.

Predictions of future climate impacts and cultural heritage responses must be developed using recent and current observations as proxies for future change, integrated with the emission scenarios presented in the latest IPCC Assessment Reports (2023). The ability to downscale 20- and 30-year climate scenarios will become an essential competence for heritage managers. In describing climate impacts and their effects on cultural heritage, it is useful to distinguish between rapid-onset and slow-onset events:

- Rapid onset events, which are short-lived, acute and highly damaging (e.g. extreme winds, hurricanes, storm surge, flash floods, landslides, heat waves, wildfires). Their frequency and intensity are expected to increase globally;

- Slow-onset events, which are progressive and potentially permanent (e.g. glacier melt, sea level rise, aridification, desertification, shifts in seasonality and species distribution). For building heritage, longer-term interactions with air pollution are particularly concerning, leading to degradation of limestone and marble façades, soiling of stone surfaces, leaching of stained glass, metal corrosion and salt crystallization in porous materials such as stone, brick, plaster, mosaics, wall paintings.

3 Typological Risk Classification

This section proposes a typological classification of the main hazards threatening cultural heritage, based on their origin and impact mechanisms.

Cultural heritage is increasingly exposed to a wide range of hazards -environmental, anthropogenic, and biological- that often act simultaneously or in sequence. Rarely isolated, these hazards tend to interact, amplifying their impact through both climatic and non-climatic drivers. According to the IPCC (2014), hazards are processes, phenomena, or human activities that may cause loss of life, injury, or damage to property, infrastructure, and ecosystems. The UN General Assembly (2016) further specifies that hazards should be analyzed in terms of their location, intensity, frequency, and probability. The magnitude depends on both exposure and vulnerability and can be intensified by cascading or compound effects.

In the context of cultural heritage, such risks often result in material decay, structural damage, or the loss of historical and aesthetic value. Following the typologies proposed in the H2020 ARCH project (Rebollo et al. 2020), hazards can be grouped into four main categories: climate-related, human-induced, geological, and biological. These categories are not mutually exclusive, as hazards often overlap or trigger one another. Figure 1 illustrates the interdependencies and cascading effect that characterize these hazards, which also vary across geographical and socio-environmental contexts.

Below is a critical analysis of the main hazards affecting European cultural heritage.

3.1 Climate-Related Hazards

This includes hazards driven by short-term meteorological events or long-term climatic variations.

- Extreme temperatures and thermal stress: rising global temperatures affect traditional materials. In Mediterranean areas, heatwaves cause thermal cracking in stone and ceramics, while moisture loss promotes salt efflorescence. Freeze-thaw cycles, c in alpine and continental zones, induce flaking and loss of cohesion. Higher temperatures also enhance biological activity, increasing risk of fungi and insect infestations in wood and plaster.
- Humidity, precipitation, and hydrogeological hazards: irregular rainfall and flooding erode remains, detach mural paintings, and damage mosaics or flooring. Capillary rise transports salts into masonry, causing plaster detachment. Wet-dry cycles degrade lime mortars and adobe, while landslides and subsidence threaten structures on unstable ground.

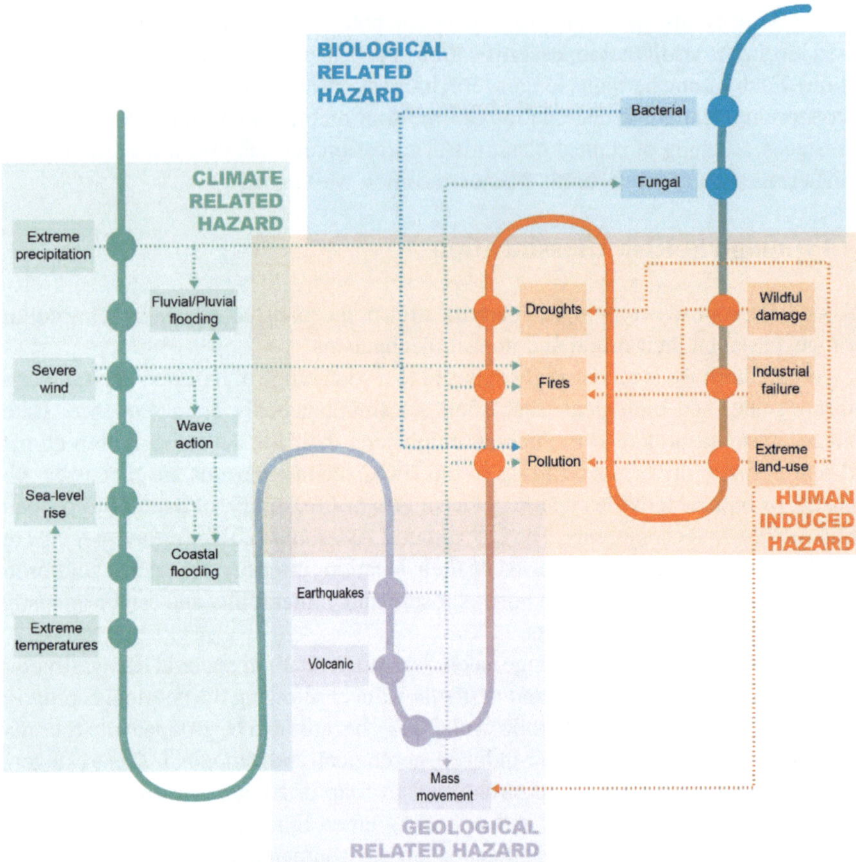

Fig. 1. Schematic categorisation of main hazards affecting European cultural heritage and the inter-connections in between them. The prevalence of the above hazards varies of course depending on geographical variables. 2020 Source: Ri-elaboration of the author from (Giglio et al.2023; Rebollo et al.)

- Wind, salt, sand, and atmospheric pollutants: Wind accelerates abrasion and pollutant transport. Marine salts and airborne particles erode surfaces, while reactive gases (NO_2, SO_2, O_3) form acid rain and corrosive residues, damaging stone, stained glass, and metals.
- Sea level rise and saline intrusion: in coastal areas, sea level rise and storm surges cause erosion and salt infiltration, leading to corrosion, efflorescence, and osmotic damage. This is particularly critical in stilt houses, lagoons, and intertidal zones.

3.2 Geological-Related Hazards

- Earthquakes major structural damage, often followed by cascading effects such as fire, flooding, landslides. Mediterranean countries including Italy, Greece, Turkey, and Spain are particularly vulnerable (Russo et al., 2014; Adnan et al., 2015).

- Mass movements: landslides, rockfalls, avalanches, debris flows, and mudflows triggered by earthquakes, rainfall or snowmelt, threaten heritage structures (Russo et al., 2014).
- Volcanic eruptions: although rare in Europe, they pose significant risks in Italy, Greece, the Canary Islands, and French Antilles. Despite the high density of heritage sites, volcanic risks remain underexplored (Russo et al., 2014).

3.3 Human-Induced Hazards

- Extreme land use: industrial development, mining, agriculture, and fossil fuel extraction damage ecosystems directly (vegetation loss) and indirectly (pollution). These activities increase the risk of landslides and droughts (Turner, 2012).
- Air pollution: pollutant such as NO_2, SO_2, O_3 and formaldehyde accelerate stone decay, chemical alteration, crusting, and discoloration. Acid rain and particulate matter further degrade marbles and stone façades (Rosu et al., 2020; Comite et al., 2020).

3.4 Biological-Related Hazards

Biological agents -from bacteria to plants- alter heritage materials through metabolic activity. These processes intensify in warm, humid climates. Fungi and lichens dissolve minerals and stain surfaces, while plant roots penetrate structures, causing cracks or collapses. The extent of damage depends on organism type, material properties, exposure, and pollutants (Laoupi et al., 2007; Coumou et al., 2012; Nicu, 2017). These biological processes often act synergistically with climatic stressors, exacerbating deterioration.

4 Modelling for Climate Change Mitigation and Adaptation: An Overview on Methodologies and Indices for Assessing Vulnerability and Risk to Cultural Heritage

The increasing exposure of cultural heritage to climate change requires the use of predictive tools and integrated risk assessment methodologies that can inform adaptive decisions at both management and design levels. Although the literature increasingly acknowledges the role of cultural heritage in territorial resilience, there remains a persistent gap in systematically integrating both material and immaterial dimensions of heritage into analytical frameworks for climate vulnerability assessment. Current approaches often emphasize exposure, while overlooking crucial variables such as the adaptive capacity of local systems, the cultural and identity values attributed to heritage, and the historical dimension of conservation knowledge.

Several methodological proposals have sought to address this fragmentation. Carroll and Aarrevaara (2018) introduced a priority intervention index, based on a 1–10 scale, linking local climate factors (e.g. temperature increases, growing seasons length, precipitation, wind speed) to direct visual observations of material damage. Although heuristic, this approach suggests an operational path for allocating resources on a comparative basis.

Similarly, Loli and Bertolin (2018) proposed a multi-risk modelling framework using data from the Climate for Culture project. Their approach integrates physical degradation trajectories (mechanical, chemical, biological) of historic materials with the regulatory protection levels of assets, thereby defining critical thresholds for adaptive interventions. Projections to 2100 indicate increased biological and chemical risks, particularly for exposed structures in southern and central Europe.

Ciantelli et al. (2018) applied high-resolution modelling with the EC-Earth climate system to future scenarios (2039–2068), producing damage functions that reveal accelerating surface recession, biological accumulation, and salt crystallization/solubilization cycles, especially in archaeological sites and porous stone materials subject to capillary moisture.

These approaches informed the experimental strategy adopted in the T4Y case studies, where climate projections, material behaviour, and degradation processes were combined to support the development of a comprehensive assessment tool.

Taken together, these studies converge on several fundamental principles:

- Climate risk assessment should not be limited to exposure alone but must include a dynamic estimate of vulnerability, based on the physical-mechanical properties of materials, the microclimatic context and maintenance conditions.
- Integration of climate data, decay parameters and land use models is essential for defining contextualized and scalable adaptation strategies.
- Continuous monitoring is indispensable, not only to validate predictive models but also to build awareness and legitimacy for adaptation policies.

From an operational perspective, risk matrices derived from the interaction of climate variables (ΔT, ΔP, ΔRH, ΔW) and material types (wood, stone, plaster, metals) provide a forecasting frameworks. Within this framework, accelerated degradation processes are linked to phenomena such as:

- Freeze/thaw cycles
- Salt crystallization
- Corrosion by atmospheric pollutants combined with acid rain
- Biodeterioration in high humidity environments
- Erosion from wind or wind-driven rain (WDR)

Finally, Sesana et al. (2018) emphasizes that adaptation governance must also overcome barriers in perception and institutional capacity. The absence of shared guidelines, limited technical training among managers and the scarcity of georeferenced data on conservation status and climate vulnerability significantly constraint the effectiveness of adaptation strategies. It is therefore necessary to develop decision-support tools that combine quantitative approaches (modelling, monitoring, simulations) with qualitative dimensions (cultural values, local practices, tacit knowledge).

5 Toward a Set of Climate Vulnerability Indices for Cultural Heritage: A Framework Linking Climate Drivers, Impact Mechanisms, and Societal Effects

The data presented in the preceding sections -concerning physical and biological impact mechanisms, climate drivers, projected effects on cultural heritage, and broader societal consequences- have been systematically selected and correlated through a structured methodology aimed at developing a coherent analytical framework (Table 1). This framework identifies the main pathways through which climate-related hazards compromise the material integrity and structural stability heritage assets, as well as their societal and cultural functions.

Building upon foundational contributions (UNESCO, 2007; ICOMOS, 2019) and subsequent analyses (Rebollo et al., 2020; IPCC, 2021), the framework integrates case studies and projections, providing an updated overview of climate- heritage interaction.

Table 1. Analytical framework: Climate Change Impacts on Cultural Heritage (Re-elaboration of F. Giglio and F. Armocida, fonts: ICOMOS, 2019; ICLEI, 2020)

Climate Driver	Impact Mechanism	Hazard / Effect	Heritage Impacts	Urban Heritage & Community Impacts
Temperature Rise	Heatwaves, altered freeze-thaw cycles	Urban Heat Island, material decay	Chemical/mechanical degradation, cracking, salt crystallization	Abandonment of assets, migration, loss of traditional knowledge
Fire	Direct fire damage, post-fire erosion	Wildfire, fire suppression, runoff	Burning, discoloration, collapse, erosion	Loss of sites, increased flooding, shift in materials or layouts
Drought	Soil drying, water scarcity	Desertification, reduced humidity	Cracking, salt spalls, impact on water-based heritage systems	Migration, loss of livelihoods, neglected heritage
Sea Level Rise	Coastal erosion, salinization	Chronic/acute flooding, salt infiltration	Foundation subsidence, salt damage, corrosion, total or partial loss	Inaccessibility, urban transformation, displacement of communities

(*continued*)

Table 1. (*continued*)

Climate Driver	Impact Mechanism	Hazard / Effect	Heritage Impacts	Urban Heritage & Community Impacts
Coastal Flooding	Wave impact, storm surge	Structural failure, salinization	Collapse, salt crystallization, relocation pressures	Historic site loss, damage to roads/ports, erosion of cultural landscapes
Coastal Erosion	Land loss, exposure of structures	Accelerated by sea level rise and storms	Collapse, rust, relocation	Graveyard and landmark loss, landscape degradation
Air Pollution Climate induced hazard	Acidification, deposition, chemical reactions	Combined with temperature/humidity changes	Material erosion, corrosion, façade darkening, glass degradation	Loss of visual integrity, plantings, significant viewsheds
Storms	High winds, surge, debris	Increased frequency/intensity of events	Collapse, water infiltration, physical erosion, damage to utilities	Destruction of historic urban landscapes, access loss
Wind	Direct loading, wind-borne debris	Scouring, horizontal pressure	Surface abrasion, water penetration, cracking	Collapse of weak structures, wind damage to archaeological sites
Extreme Rainfall	Flooding, saturation, landslides	Soil erosion, fungus, instability	Swelling, rot, collapse (especially adobe), spalling	Loss of earthen towns, soil movement, waterlogging of gardens
Pollution	Acid deposition, soot, contaminants	Rain acid interactions, biological growth	Erosion of stone, color change, stained glass degradation	Damage to landscapes, reduced visibility of landmarks
Precipitation & Humidity	Flash floods, capillary rise, saline water	Intense rain, efflorescence	Sediment damage, pollutant accumulation, salt crystallization	Flash flooding of sites, efflorescence, drainage issues

Drawing on this foundation and the extensive literature review, the study advances a proposal for a set functional indices, defined as quantitative indicators linking specific climatic stressor to measurable degradation patterns of materials and structures. These indices are intended to provide reference benchmarks for assessing vulnerability across different scales, from single elements to entire heritage complexes.

A comprehensive assessment must incorporate both structural and non-structural dimensions, acknowledging the cumulative effects of climate change in interaction with non-climatic stressors. Crucially, both rapid onset events (e.g. flood, landslides, storm urges) and slow-onset phenomena (e.g. sea-level rise, desertification) must be considered within governance frameworks that reflect local adaptive capacity.

Equally important is the inclusion of community values, traditional practices, and socio cultural resilience, which can mitigate vulnerabilities often overlooked by purely technical assessments. Long-term monitoring of climate-induced degradation - particularly in historic buildings, interiors and archaeological contexts is essential to track susceptibility and support responsive conservation strategies.

In this perspective, the framework identifies a set of midpoint indices, reflecting average impacts of climate stressors over time and space, and endpoint indices, representing their cumulative effects on heritage assets (Giglio et al. 2024). The following midpoint indices are proposed:

- Deterioration Index (general material degradation rate)
- Carbonation Index (relative to substrate alteration).
- Efflorescence Identification Index (salt crystallization and surface deposits)
- Corrosion Index (specific to metallic elements)

Each index is associated with defined climatic stressors, as summarized in Table 2, and is designed for application across multiple scales – from laboratory experimental analyses (e.g., XRD, SEM) to in situ monitoring and geospatial modelling (Lucanto et al. 2024). This multiscalar flexibility ensures that the indices can function both as a diagnostic tool and as policy-relevant benchmarks for adaptation planning. Offering a structured basis for evaluating climate vulnerability in cultural heritage contexts (Table 2).

Table 2. Identification of a set of midpoint indices as benchmark to measure climate vulnerability on cultural heritage buildings and structures selected from Hazard Climate Impact (ICOMOS, 2019)

Hazard Climate Impact (ICOMOS, 2019)	Functional impact indices to measure climate vulnerability on cultural heritage buildings and structures
Alteration of frost/thaw cycles (temperature level rise)	Deterioration index
	Carbonation index (relative to the reference substrate)
	Efflorescence identification
Heatwaves (temperature level rise)	Deterioration index

(*continued*)

Table 2. (*continued*)

	Carbonation index (relative to the reference substrate)
	Efflorescence identification
Drought and Desertification (temperature level rise)	Deterioration index
	Carbonation index (relative to the reference substrate)
	Efflorescence identification
Chronic or acute flooding (sea level rise)	Deterioration index
	Carbonation index (relative to the reference substrate)
	Efflorescence identification
	Corrosion index of metal elements
Increased intensity and frequency of storms (Combination of climate change effects)	Deterioration index
Increased intensity and frequency of winds (Combination of climate change effects)	Deterioration index
Extreme rainfall (Combination of climate change effects)	Deterioration index
	Carbonation index (relative to the reference substrate)
	Efflorescence identification
	Corrosion index of metal elements
Pollution (Combination of climate change effects)	Deterioration index
	Corrosion index of metal elements
Increase in humidity (Precipitation and humidity)	Carbonation index (relative to the reference substrate)
	Efflorescence identification.

6 Conclusions

Cultural heritage is increasingly exposed to climate-induced risks, requiring a shift in conservation and adaptation strategies. This chapter proposes a methodological framework for assessing such risks, highlighting the relationship between climatic drivers, degradation mechanisms, and material vulnerability. By integrating multi-scale data, ranging from European policy frameworks to site-specific characteristics, the work establishes a theoretical basis for constructing flexible and scalable assessment tools.

The framework links environmental stressors to physical and biological deterioration pathways, while also considering the societal implications of heritage degradation. Based on established references, it identifies a coherent set of midpoint and endpoint indices to translate complex climate-heritage interactions into operational knowledge.

These indices serve as key instruments for monitoring, prioritizing, and guiding adaptive conservation strategies. Midpoint indices (e.g., deterioration, carbonation, efflorescence, corrosion) allow the disaggregation of vulnerability into measurable phenomena, while endpoint indices reflect the cumulative effects of climate stressors over time.

Importantly, the analysis recognizes that climate change impacts are not solely material but also affect identity, memory, and social cohesion. Therefore, vulnerability assessments must integrate technical, cultural, and community-based dimensions.

Future research will focus on operationalizing these indices through field applications in the T4Y pilot area of Bova, aiming is to refine their structure and interdependencies. This approach will support a dynamic decision-support process to enhance long-term resilience in cultural heritage management.

Acknowledgments. The development of the indices was conducted within the Interdepartmental laboratory section MATeRICs, Material Analysis and Technologies for Restoration, Innovation and Climate Exploration (resp. prof Francesca Giglio, dAeD Department), Mediterranea University of Reggio Calabria. The work was carried out in collaboration with Prof. Patrizia Frontera, as well as with RtdB Eng. Angela Malara, Department of Civil, Energy, Environmental and Materials Engineering (DICEAM), Mediterranea University of Reggio Calabria, and PhD's Arch. Francesco Armocida (dAeD Department).

References

Sabbioni, C., Brimblecombe, P., Cassar, M.: The Atlas of Climate Change Impact on European Cultural Heritage: Scientific Analysis and Management Strategies. Anthem Press, London (2012)

Sesana, E., Gagnon, A.S., Ciantelli, C., Cassar, J.A., Hughes, J.J.: Climate change impacts on cultural heritage: a literature review. WIREs Clim. Change **12**(4), e710 (2021). https://doi.org/10.1002/wcc.710

Bertolin, C.: Preservation of cultural heritage and resources threatened by climate change. Geosciences **9**(6), 250 (2019). https://doi.org/10.3390/geosciences9060250

Loli, A., Bertolin, C., Kotova, L.: Service life prediction of building components in the times of climate change. In: IOP Conference Series: Materials Science and Engineering, vol. 949, p. 012048 (2020). https://doi.org/10.1088/1757-899X/949/1/012048

O'Neill, S., Simon, F., Tett, B., Donovan, K.: Extreme rainfall risk and climate change impact assessment for Edinburgh World Heritage sites. Weather Clim. Extremes **38**, 100514 (2022). https://doi.org/10.1016/j.wace.2022.100514

Grossi, C.M., Brimblecombe, P., Harris, I.: Predicting long-term freeze–thaw risks on Europe's built heritage and archaeological sites in a changing climate. Sci. Total. Environ. **377**(2–3), 273–281 (2007). https://doi.org/10.1016/j.scitotenv.2007.02.014

Cacciotti, R., et al.: Climate change-induced disasters and cultural heritage: optimizing management strategies in Central Europe. Clim. Risk Manag. **33**, 100301 (2021). https://doi.org/10.1016/j.crm.2021.100301

ICOMOS Climate Change and Cultural Heritage Working Group. The Future of Our Pasts: Engaging Cultural Heritage in Climate Action. Paris: ICOMOS (2019)

Rebollo, V., Latinos, V.: Arch D 7.2 Good practices in building cultural heritage resilience. ICLEI (2020)

Intergovernmental Panel on Climate Change (IPCC). Climate Change 2021: The Physical Science Basis. Contribution of Working Group I to the Sixth Assessment Report of the Intergovernmental Panel on Climate Change. In: Masson-Delmotte, V., et al. (eds.). Cambridge University Press, Cambridge, United Kingdom and New York, NY, USA (2021)

Morel, H., et al.: Global Research and Action Agenda on Culture, Heritage and Climate Change. Charenton-le-Pont & Paris, France: ICOMOS & ICCROM Climate Heritage Working Group (2022)

Richards, J., Brimblecombe, P.: Moisture as a driver of long-term threats to timber heritage—Part I: changing heritage climatology. Heritage **5**(3), 1929–1946 (2022). https://doi.org/10.3390/heritage5030100

Augustin, C.: Climate change and world heritage: Report on predicting and managing the impacts of climate change on world heritage and strategy to assist States Parties to implement appropriate management responses. UNESCO World Heritage Centre (2007)

Markham, A., Osipova, E., Lafrenz Samuels, K., Caldas, A., Ullstein, B.: World Heritage and tourism in a changing climate. United Nations Environment Programme, Nairobi, Kenya, and United Nations Educational, Scientific and Cultural Organization, Paris, France (2016)

IPCC. Climate Change 2023: Synthesis Report. Contribution of Working Groups I, II and III to the Sixth Assessment Report of the Intergovernmental Panel on Climate Change [Core Writing Team, H. Lee and J. Romero (eds.)]. IPCC, Geneva, Switzerland, pp. 35–115 (2023). https://doi.org/10.59327/IPCC/AR6-9789291691647

IPCC. Annex II: Glossary [Mach, K.J., Planton, S., von Stechow, C. (eds.)]. In: Pachauri, R.K., Meyer, L.A. (eds.) Climate Change 2014: Synthesis Report. Contribution of Working Groups I, II and III to the Fifth Assessment Report of the Intergovernmental Panel on Climate Change. IPCC, Geneva, Switzerland, pp. 117–130 (2014)

United Nations General Assembly. Report of the open-ended intergovernmental expert working group on indicators and terminology relating to disaster risk reduction. United Nations General Assembly, New York, NY, USA (2016)

Giglio, F., Frontera, P., Malara, A., Armocida, F.: Materials and climate change: a set of indices as the benchmark for climate vulnerability and risk assessment for tangible cultural heritage in Europe. Sustainability **16**(5), 2067 (2024). https://doi.org/10.3390/su16052067

Russo, S., Dosio, A., Graversen, R.G., Sillmann, J., Carrao, H., Dunbar, M.B., Vogt, J.V.: Magnitude of extreme heat waves in present climate and their projection in a warming world. J. Geophys. Res. Atmos. **119**(22), 12,500 (2014)

Adnan, A., Ramli, M.Z., Sk Abd Razak, S.M.: Disaster management and mitigation for earthquakes: are we ready. In: Proceedings of the 9th Asia Pacific Structural Engineering and Construction Conference (APSEC 2015), pp. 34–44 (2015)

Turner, S.D.: World Heritage Sites and the extractive industries. Independent study commissioned by IUCN in conjunction with the UNESCO World Heritage Centre, ICMM and Shell (2012). www.icmm.com/document/3787. Accessed 14 June 2025

Rosu, A., Constantin, D.E., Arseni, M., Timofti, M.: Atmospheric measurements in the context of protection and conservation of cultural heritage objects. In: AIP Conference Proceedings, vol. 2218, no. 1, p. 030015. AIP Publishing LLC (2020)

Comite, V., Randazzo, L., La, Russa, M.F., Fermo, P.: A multi-analytical approach for the characterization of black crusts on the facade of an historical cathedral. Microchem. J. 105121 (2020)

Laoupi, A., Tsakiris, G.: Assessing vulnerability in cultural landscapes. In: EWRA Symposium "Water Resources Management: New Approaches and Technologies," Chania, Crete, Greece, 14–16 June 2007 (2007)

Coumou, D., Rahmstorf, S.: A decade of weather extremes. Nat. Clim. Chang. **2**(7), 491–496 (2012). https://doi.org/10.1038/nclimate1452

Nicu, I.C.: Natural hazards—a threat for immovable cultural heritage: a review. Int. J. Conserv. Sci. **8**(3), 391–402 (2017)

Carroll, P., Aarrevaara, E.: Review of potential risk factors of cultural heritage sites and initial modelling for adaptation to climate change. Geosciences **8**(9), 322 (2018). https://doi.org/10.3390/geosciences8090322

Loli, A., Bertolin, C.: Indoor multi-risk scenarios of climate change effects on building materials in Scandinavian countries. Geosciences **8**(10), 347 (2018). https://doi.org/10.3390/geosciences8100347

Ciantelli, C., Palazzi, E., Von Hardenberg, J., Vaccaro, C., Tittarelli, F., Bonazza, A.: How can climate change affect the UNESCO cultural heritage sites in Panama? Geosciences **8**(9), 296 (2018). https://doi.org/10.3390/geosciences8090296

Sesana, E., Gagnon, A., Bertolin, C., Hughes, J.: Adapting cultural heritage to climate change risks: perspectives of cultural heritage experts in Europe. Geosciences **8**(9), 305 (2018). https://doi.org/10.3390/geosciences8090305

Lucanto, D., Nava, C., Mangano, G.: Digital prototyping and regenerative design toward carbon-neutrality and a climate resilient built environment: a multi-scale assessment of environmental multi-risks. Buildings **14**(12), 3934 (2024). https://doi.org/10.3390/buildings14123934

Regenerative Digital Design and Innovative Processes to Define Predictive Models: Workflow on Climate Scenarios

Consuelo Nava(✉)

ABITAlab, dAeD, Mediterranean University of Reggio Calabria, Reggio Calabria, Italy
cnava@unirc.it

Abstract. The *phigital* approach – *physical and digital* – to regenerative design represents the best synthesis between theory and practice in the contemporary era, in cutting-edge research particularly focused on the study of climate vulnerability, for the adaptive and transformative design of the built environment. Predictive and declarative design demonstrates its particular effectiveness through the optimisation of scale, the use of parametric and simulation tools, and digital regenerative design. This paper addresses, from a theoretical and methodological perspective, the themes and insights that have guided the PNRR T4Y applied research activities for Pilot 4.7.1 with actions 8, 9. The text is to be considered an advancement of the author's scientific treatise on emerging technologies for regenerative design (2022–2023), with subsequent elaborations of competitive research trajectories (2023–2025). In the application phase, it proposes a particular examination of concepts related to information processes with "workflow" and the use of "predictive parameter tools", in order to provide useful information for technological and environmental innovation actions, in application of regenerative design and in scenarios of declared climate vulnerability. This methodology was applied in the experimental phase of the research for the case studies of Palizzi (*natural heritage*, coastal area) and Bova (*cultural heritage*, urban and building area). In the Design and Prototyping sections of the ABITAlab university laboratory, part of the Department of Architecture and Design at the Mediterranean University of Reggio Calabria, studies and experiments were conducted, producing both digital and physical prototypes (3D printing), described in a high-information Digital Atlas as a decision support system (Lucanto, Hanida, 2025). They where produced images, navigable models, interoperable data and multi-level information for an *open knowledge user profiling platform* capable of translating structural and environmental impacts on natural and historical heritage, developing solutions for the safety of territories and communities, with projections to 2030, 2050 in urban areas and up to 2085, 2100 in coastal areas.

Keywords: Regenerative Digital Design · Advanced Sustainable Design · Innovative Process · Emerging Technologies · Parametric Tools · Workflow · Climate Scenarios

C. Nava et al. (Eds.): *Climatic and Structural Safety in Multi-Hazard Regime of Cultural and Natural Heritage*, GREEN, pp. 150–173, 2026.
https://doi.org/10.1007/978-3-032-11926-1_10

1 Introduction

The quality that can be achieved through the design and construction of the built environment, in accordance with sustainable strategies, must aim to exceed the benefits and requirements of the project in all its components. The performance attributes with which any technological system enables effective processes of operation and response to expressed needs must find scenarios of 'added quality' in which to express their capabilities, at multiple levels and in multiple relationships. Product technologies are required to have their own qualities depending on their physical and material characteristics, durability, efficiency, configuration, operation, etc. process technologies must be referred to in order to interrogate systems on their capacity for innovation and integration, operability and management, competitiveness and improvement (incremental or radical), adaptation to the evolution of socio-cultural models and their aptitude for creating new meanings within advanced design processes. High performance is also assessed in terms of both short-term and long-term response times. For this reason, its function is to act as a yardstick for interventions, both in transition scenarios and in resilience scenarios (Nava 2019). This landscape-infrastructural strategy shifts previous concepts of hyper-sustainability towards a new challenge, innovating on the concept of 'strong sustainability' (a preventive, pre-visionary and predictive vision of radical sustainability) (Nava 2022). The relationship between innovation capable of addressing design and the discussion around sustainability issues is changing in its operational tools, and "advanced design becomes regenerative", as illustrated below (Fig. 1).

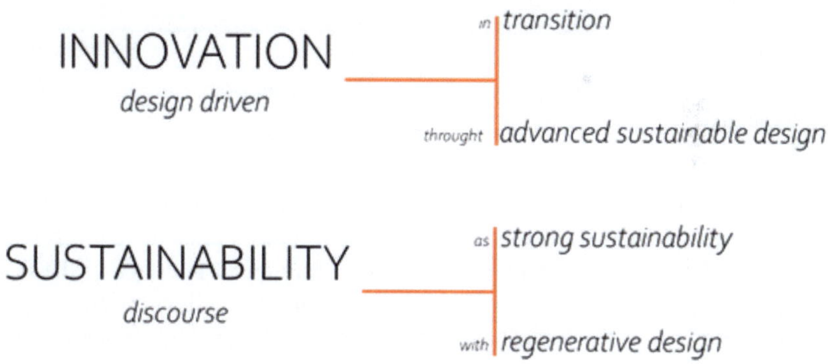

Fig. 1. Relationships between innovation and sustainability in strong sustainability scenarios for regenerative design (C. Nava, 2022)

1.1 Theoretical Background

1.1.1 Transformative Innovation and Climate Risks

This strategy requires a change of approach, looking to the future (future-back) and increasing the opportunities not only for experimentation but also for all options on multiple competing solutions. This involves moving in the opposite direction to incremental innovation, entrusting the future vision with the ability to redefine ecosystems (of innovation action) in order to promote, through a disruptive approach, the launch of initiatives with a strong digital content, with a view to achieving results with an efficient operational method (Fig. 2).

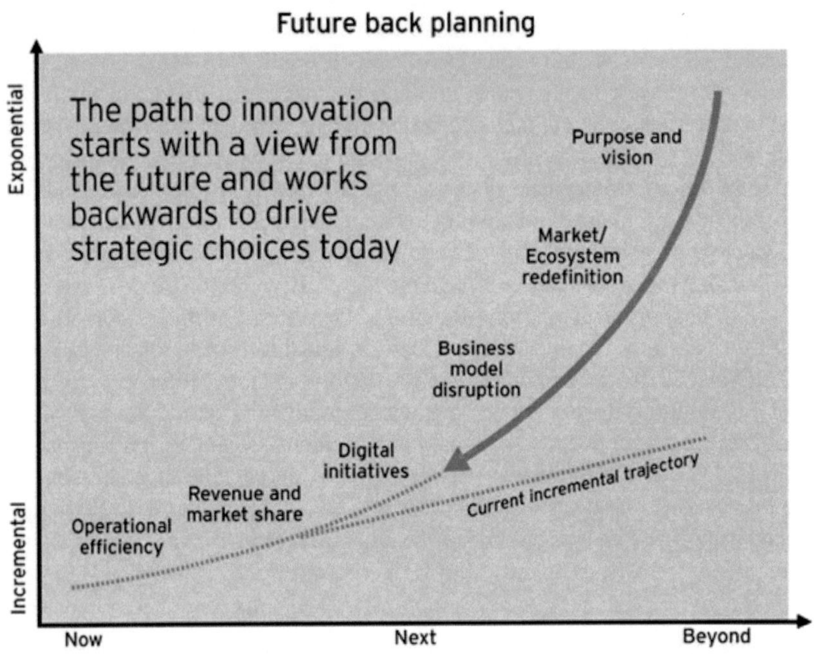

Fig. 2. A vision of future-oriented innovation drives transformation (source: https://www.ey.com/en_uk/consulting/)

The model illustrated, which is recommended for those in the business market who intend to focus on disruptive innovation in order to produce competitive levels of quality in the era of ecological and digital transition, requires an interface in the field of sustainability of built environment transformations, with the ability to pursue 'high performance' objectives. In the context of climate change scenarios, requires the production of levels of quality "beyond performance" in order to aim for a "performance level"; this means working towards scenarios of climate adaptation and/or mitigation and demanding that regenerative design be capable of providing actionable pathways in response to the impact of emerging phenomena. (Fig. 3) This approach is based, also according to the latest IPCC reports (from 2013 to 2022), on the concepts of impact and vulnerability and therefore risk, referring to the "global warming level" (GWL), described as "the increase in the global average air temperature near the surface compared to the pre-industrial period (specifically the period prior to 1750, often approximated by conditions in the period 1850–1900)" (Lionello, Naumann 2022).

The diagram in Fig. 3 is in fact the response generated by the risk categories, which, for example, the IPCC identifies as key risks for Europe, which are susceptible to their level of impact with reference to the increase in global warming (given that the 1.5°

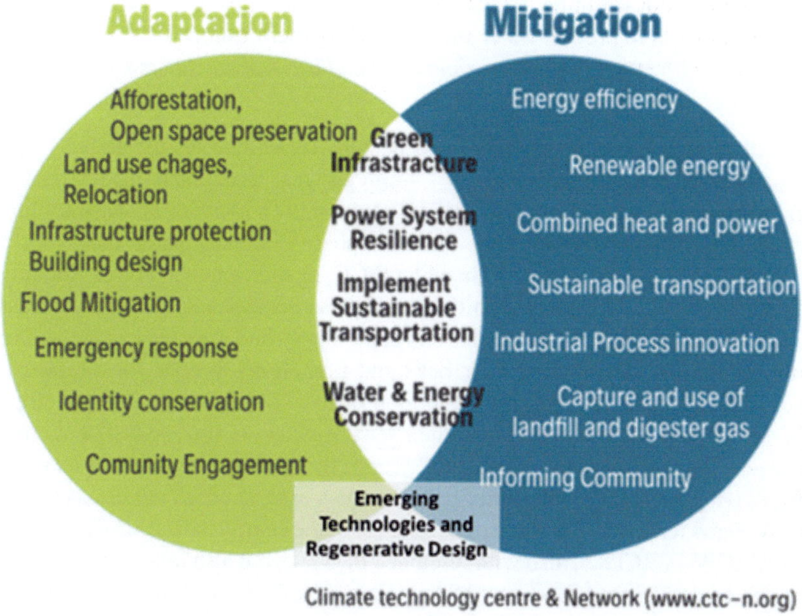

Fig. 3. Schemes on adaptation and mitigation and their areas of application for experimentation (source: www.ctc-n.org)

target of COP 25 has been largely compromised). Even with a 2°C rise in temperature, the risks listed below become more serious and partly irreversible.

The four risk categories to refer to are (Lionello 2022):

- *Risks of heat waves on populations and ecosystems.* The number of deaths and people at risk of heat stress is expected to double or triple for a temperature increase of 3 °C, compared to 1.5 °C. Warming will reduce habitats suitable for current terrestrial and marine ecosystems and irreversibly change their composition, with h us effects increasing in severity above the 2 °C global warming level. Measures to adapt the population to heat stress and mitigate the risks of heat waves require multiple interventions in buildings and urban spaces. These measures must be anticipated in southern Europe, where the risk is greater than in more northern areas.
- *Risks to agricultural production.* Due to a combination of heat and drought, substantial losses in agricultural production are expected in most European areas in the 21st century, which will not be offset by the expected gains in northern Europe.
- *Risks of water scarcity.* In southern Europe, the risk is already high with global warming of 1.5 °C and becomes very high with a 3 °C increase. In these regions, demand for water resources already exceeds supply. This gap is widening due to climate change and socio-economic developments. With a temperature rise of 3 °C, the risk of water scarcity also becomes high in central and western Europe. Even with average warming, adaptation strategies that reduce water demand must be combined with transformations such as diversification of sources and changes in land use. In

the event of high warming, a broad portfolio of measures is required, but this may not be sufficient to prevent water shortages in southern Europe.
- *Risks posed by more frequent and intense flooding*. Due to changes in precipitation and rising sea levels, the risks

The Mediterranean sea, for example, will become a real hotspot for global warming. What caused sea levels to rise by about 1.4 mm per year throughout the 20th century is set to increase and reach about 1 m by 2100. with a 20% increase in sea warming, the ability to absorb carbon dioxide from the atmosphere is already being lost.

It is also essential to understand how the IPCC, in addition to analysing the impacts of climate change, both globally and regionally, on ecosystems, societies, infrastructure, productive sectors, cultures, cities and settlements, and assesses current and future adaptation options, deciphering their risks and limitations, it calls for the promotion of successful practices through the promotion of the SDGs (Sustainable Development Goals) of the UN's 2030 Agenda, in their territorialisation processes at national and local level. The novelty of the latest IPCC report, the sixth, lies not only in its ability to use RCPs (Representative Concentration Pathways), each of which represents a possible pathway characterised by the value of the change in the energy balance in 2100: RCP2.6, RCP4.5, RCP6.0, RCP8.5, also with reference to GWL, but also in its ability to embark on the shared path of the SSP (Shared Socio-Economic Pathways), integrating the RPCs relating to adaptation and mitigation with socio-economic aspects.

"SSPs are based on five narratives describing socio-economic futures characterised by sustainable development (SSP1), regional rivalry (SSP3), inequality (SSP4), fossil fuel-based development (SSP5) and a narrative that lies between those described (SSP2). The combination of SSP-based socio-economic scenarios and RCP-based climate projections provides an integrated framework for climate impact and policy analysis." (CMCC Foundation, IPCC Focal Point for Italy, 2022).

We are looking for an answer in these times of polycrisis, as defined in all its risks in the latest WEF2023 report, those crises (social, environmental, technological, economic, geopolitical) that see design as a translator of the most complex processes useful for the present and future of ecosystems. A "cognitive and scientific challenge", as referred to in M. Losasso's introduction to issue 23 of Techne Journal, entirely dedicated to 'the possible necessary within the polycrisis' and the need to "play a more effective and proactive social and technological role", as mentioned by E. Mussinelli in the editorial of the same issue. This is an issue that remains the responsibility of all citizens, but which certainly calls for planning action imbued with the scientific research of transfer that is characteristic of the effectiveness and application of technologies, as expressed thus far.

1.1.2 The Innovation of Regenerative Design: From Physical Processes to Digital Processes

Further discussion of the role assumed by the project, between emergency and contemporaneity, might seem anachronistic and, at the very least, ill-suited to the speed with which the technologies themselves, their devices, operational capabilities and methodological approaches can apparently take on a decisive and sufficient character. Indeed, immersion in digital capabilities and the absorption of the human element into the products that make them 'devices' seems to exclude the creative processes of design when they are not directed towards the manufacture of those devices. Design can encounter any device, as its unique ability to configure itself in the dimension of the territory, the city, buildings and even public spaces, if the research that produces it intends to consider, in a sort of operational identification process, the project itself as a true enabling and emerging technology.

The complexity of the project means that it takes on the characteristics of physical processes as well as digital ones. Relational organisations, algorithmic writing, data and information exchange, system and resource hierarchies, information and self-learning capabilities, specialisation and coordination of the governing principles of each phase, platforms for client, production and user structures.

Design thus takes on an organisational character, whose field of action becomes the multiple operational fields of action on the objectives and challenges to be pursued. S. Mecca, in his latest text "Il progetto come azione tra ordine e disordine. Alla ricerca dell'Armonia" (2023), outlines a path of "rational organisation" of this complexity, referring to Thompson's classification of three types of technology (concatenation, mediation and intensive) with reference to procedural and methodological issues. He defines intensive technology as "characterised by a varied set of techniques capable of supporting the process of transformation of a specific object, when the choice, combination and ordering of operations can only be determined through feedback interaction with the object itself. Intensive technology has therapeutic characteristics when the object is of a living nature, or design characteristics when the objects are not living, as in construction and research (…). Intensive technology is therefore the technology of non-repetitive processes, whose success depends 'on the availability of all potentially necessary capabilities and, to the same extent, on an adequate combination of capabilities selected according to the needs of the specific case or project".

This is the position taken in this treatise on the operability of technologies in the 'options' of applied experimentation and on the ability of frontier research to take feedback and calibrate itself to each response in an agile manner. The sustainable environmental project has, in fact, always had within its very nature the ability to understand the need for the evolution of its argumentative and procedural structure, bringing with it the instrumental and linguistic necessity of the advancement of architectural technology as an approach and attitude towards contemporaneity (in accordance with the role that technological culture has always played).

However, the temporal and spatial context of sustainability has been defined by complex indicators based on the concept of objectives/visions, design operations and scientific research progress. The evolution of design, classified according to its states of innovation, was traced in 2016 by Fabrizio Ceschin and Idil Gaziulusoy in their article entitled 'Evolution of design for sustainability: From product design to design for system innovations and transitions'. The two scholars propose an evolutionary framework for Design for Sustainability (DfS) and, following an almost chronological outline, provide an overview of the field of DfS, classifying the design approaches developed in recent decades under four levels of innovation: Product, Product-Service System, Spatial-Social System and Socio-Technical System.

The proposed framework summarises the evolution of the field of DfS, showing how it has gradually expanded from a technical and product-centric focus towards large-scale system-level changes in which sustainability is understood as a highly complex socio-technical challenge. The four approaches are classified according to their levels of innovation:

- *Product innovation level*: design approaches that focus on improving existing products or developing completely new products.
- Level of innovation of the product-service system: in this case, the focus goes beyond individual products to integrated combinations of products and services (e.g., the development of new business models).
- *Level of spatial-social* innovation: here, the context of innovation concerns human settlements and the spatial-social conditions of their communities. This aspect can be addressed on different scales, from neighbourhoods to cities.
- *Level of innovation of the socio-technical system*: in this case, design approaches focus on promoting radical changes in the way society's needs are met, such as food and transport/mobility, and therefore on supporting transitions to new socio-technical systems.

Each level of innovation corresponds to a different type of design, which represents the evolution of its definition over time, also with reference to the application areas of research and experimental design.

In terms of product innovation, design takes on the definition of eco-friendly design in terms of environmental impact, emotionally durable design, design for sustainable behaviour, cradle-to-cradle (CTC) design, natural design, and design for improving people's lives. At the level of product-service system innovation, design takes on the definition of service-product system design for efficiency; at the level of social space innovation, design takes on the definition of design for social innovation, systemic design; at the level of socio-technical system innovation, design takes on the definition of design for innovation and system transitions.

The evolutionary framework examines different types of design for sustainability in the space of innovation in relationships and visions (insular and systemic) and, by connecting technologies to communities and the social dimension, embraces the socio-technical challenge (Fig. 4).

The mapping shows that connected design for innovative systems and transitions belongs to the socio-technical evolution framework and can be linked to practices already experienced over the last twenty years. What emerges in this period is that some types of design have undergone evolution, partly influenced by socio-technical design. There is no doubt that some forms of product design connected to nature have given rise to different experiments in the fields of eco-design or green design, biomimetic design, recycling, etc. Product-service, long theorised mainly in academia, has found its applications with the advent of emerging technologies and their transfer to industrial sectors.

In any case, Ceschin and Gaziulusoy's mapping appears interesting, beyond the objectives set by the researchers, precisely because it deciphers those characteristics of 'organisation' and 'processuality' that we have entrusted to the project.

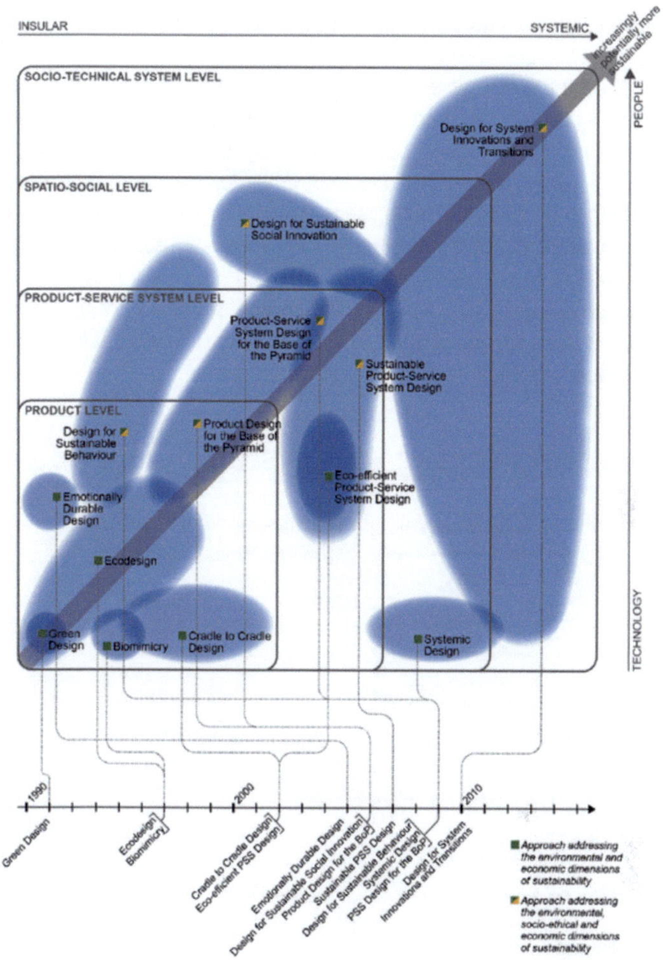

Fig. 4. The evolutionary framework of DfS with existing DfS approaches mapped onto it. The timeline shows the year in which the first key publication of each DfS approach was published (Ceschin, Gaziulusoy 2016)

Regenerative design makes the ecological and social approach 'the distinctive feature' of its transformative and transitional capacity, with a tendency to place the effectiveness of ecosystem services as the theoretical and codification space for a new metric.

Yet it seems that the real challenge towards transition is that of defining new theoretical and applicative frameworks, which are also of interest to technologies as they are understood, working on a regenerative approach based on a socio-ecological (rather than eco-social) approach.

Richard Graves, Bonnie Keeler, Maike Hamann, Elizabeth Kutsckel and Chris Nootenboom published a scientific paper entitled "A Social-Ecological Approach to Architecture and Planning" in the Journal of Architecture and Construction in 2019, discussing this transition in definitional terms and setting out some necessary conditions for its theorisation.

"First, focus on the future potential of systems within the complexity of the changes occurring in the Anthropocene and abandon romantic notions of nature separate from humans that predate development. Second, start by focusing on the social needs of a healthy and prosperous community, and then consider the ecological limits and technological challenges of design. Third, use a measure of ecosystem services to evaluate design that increases the benefits provided by local ecosystems, in pursuit of more livable, sustainable, and equitable communities".

In fact, the proposal is to transform the design process, going beyond the first-generation regenerative approach (that of the last twenty years), which only considers circular 'single-flow' systems (Lyle 1994) (Fig. 5).

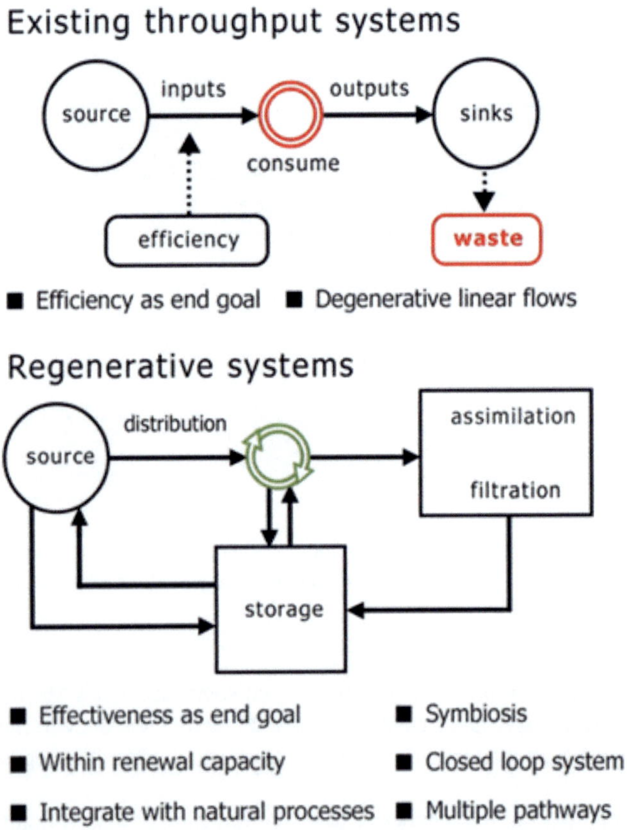

Fig. 5. Adapted from John Tilman Lyle: Regenerative Design for Sustainable Development, 1994

The idea is to go further on the issues of "technological sustainability", which is more directly related to human health. For this latter definition, with a vision that starts from the acceptance of the human and ecological role in the Anthropocene era, the objectives are to integrate social, ecological and technological systems to ensure human well-being, formulating strategies capable of co-evolving towards equity, commitment and capacity building; considering a measurement metric whose indicators are extracted from ecosystem and human services for well-being; with an integral and iterative approach to recognise the complexity of social, ecological and technological systems and their relationships and to promote thinking of complex adaptive systems, with the affirmation of a socio-ecological process [1] (Fig. 6).

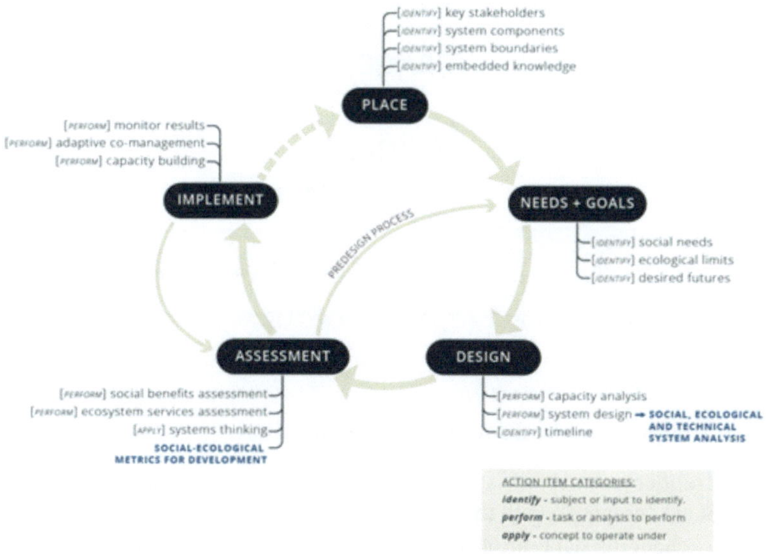

Fig. 6. Socio-ecological process diagram

2 Methodology: Digital Performative Processes

The theoretical path followed so far refers to the scientific positioning assumptions and literature overview that are most consistent with justifying the attempt to construct a method that reconnects the role of enabling and emerging technologies, recognising them in their actions with some differences in the approach to process organisation.

The proposed diagram (Fig. 7) summarises its meaning by identifying a counter-clockwise operational flow for enabling technologies and a clockwise flow for emerging technologies, in defining spaces for innovation in which regenerative and socio-ecological design operates, as previously investigated.

[1] This process has become a benchmark in 15 project experiences for frontier research and communities in transition, in the experiences conducted with ABITALab (dArTe – Mediterranean University of Reggio Calabria) and published in G. Mangano, A. Leuzzo, Co-design and Enabling Technologies, in 2022 by Aracne. With a reinterpretation of the process in seven experiences conducted in inland areas of Calabria and eight experiences conducted in urban areas.

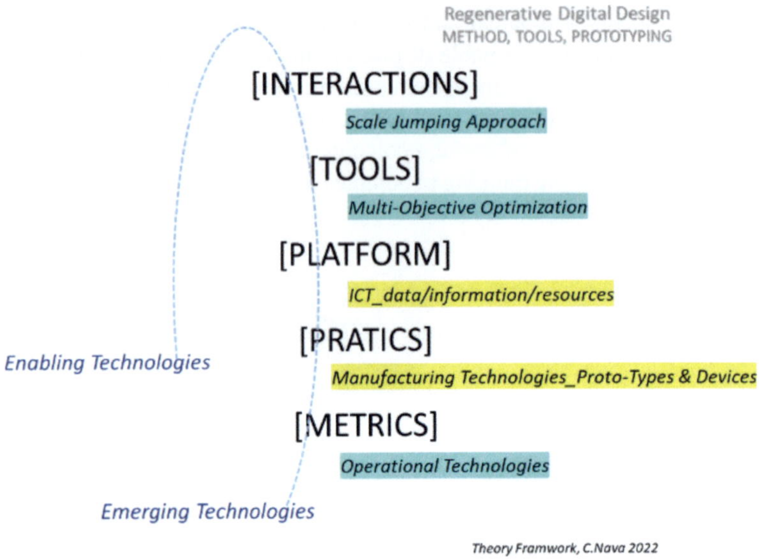

Fig. 7. New operational framework for socio-ecological regenerative design (C. Nava, 2022)

2.1 Scale Jumping and Multi-objective Optimisation

In terms of frontier research, the space for interaction is achieved through the inter-scale approach (Scale Jumping), with the use of tools based on multi-objective optimisation, capable of exchanging data, information and resources on shared and open platforms. Prototypes and devices measure the practices transferred from design experimentation and, by exploiting the operational role of technologies, re-establish a new metric for the organisational action of the regenerative project.

The interscale approach and multi-objective optimisation become the yardstick for the success of practices, supported by digital performance processes.

The extent of interventions in particularly sensitive contexts, such as those relating to natural ecosystems or ecosystems that are fragile due to physical and environmental stress and now also climate change, can always be measured through an inter-scalar process, which must control actions, transformations and impacts, evaluating multiple reference measures. The project and programme that seek to pursue these methods are always 'multidimensional'. Any transformation of the built environment or an ecosystem with a high natural value produces direct and indirect impacts, which can also be measured at different times. The instruction of design processes, through enabling technologies and dedicated devices, has never a hierarchical operational process, but takes information and data and produces them as a design support system, with the logic of an open platform capable of triggering reversible and assessable activities in progress. The size of the object of the intervention, the resources involved and the reference context already belong to three very different configuration scenarios and therefore, being included in any transformation action (including infrastructural), need to enter the processes as receptive or productive systems, with relationships and operations that take into account their specific characteristics and behaviour throughout their life cycle. Multi-scale design control allows for the activation of controlled response conditions towards high performance results already in the design or model/prototype definition phase. It can therefore be said that, according to hyper-sustainability strategies, there may be conditions of scalability of the problems that the project has to face, but these must be matched by an interscalar and integrated response in terms of project definition and performance quality (Nava 2019).

In digital processes, this characteristic becomes even more influential on project operations, as the simultaneous control of multiple factors and the ability to interpolate certain data require reference spaces that can be assumed by systems located in different detailed configurations. The scale of the city, the building and the device is often questioned by data flows that cross all dimensions of the systems, without distinguishing their boundaries, but only working on response indicators for different categories of impact (mesoscale, macroscale, microscale). A new space-time dimension, immaterial and often virtual, transcends the physical and consequently changes the predetermined boundaries and scope of action, in which often even the proactive role of users defines the variability of contextual conditions.

The pressing need to cope with a disruptive demand such as that requiring the expert contribution of digital skills for the management of such radically changed design processes certainly shifts the discussion of innovation and knowledge to the training and educational levels of learning about digital tools and their use. The performative characteristics of these tools certainly cannot replace the grammar of the approaches and methods we have discussed so far, but in characterising the indicators of their new structure, they provide data and information whose structure has changed significantly compared to 20th-century theory. The relationship between design and construction, between configurative research and expressions of the influence of contextual factors on the project, is changing. In terms of sustainable and regenerative design, this condition is even more sought after in the current forecasting, simulation and predictive scenarios to which we have referred the action of strong sustainability in a regime of ecological and digital transition.

Training in this new instrumental skill can once again come from academia and scientific and transfer research, where imagination and creativity can be pushed to the point of realising what has been imagined, thanks to the skills acquired in advanced laboratories, where research, computation and new technologies are measured against material and construction production, for highly advanced solutions (Paris, 2017) (Kozak, Nava 2024).

2.1.1 From Workflow to Predictive Simulations

Digital operational tools are able to report on these paths, from fundamental concepts to their application. The application of multi-objective optimisation algorithms and methods to achieve the desired levels of efficiency and sustainability is of significant importance. The presence of complex scientific issues leads to the use of optimisation methods to solve the problem at hand, even though some of them require solutions with accurate calculations and adequate time frames. Since classic methodologies cannot be used, artificial computational methods must be employed, which often take examples for their algorithmic writing from the structures of certain natural phenomena (e.g., meta-heuristic algorithms referring to intelligent behaviours in nature) (Sadollah, Nasir and Woo Geem Z 2020). For example, in the energy sector, for efficiency and positive building performance, optimisation processes and systems are becoming fundamental for constructing reference scenarios for consumption rather than production, performance and simulation on projective simulation models and the transition to climate and carbon neutrality (2035, 2050, 2085).

Therefore, digital tools that support regenerative design are themselves based on optimisation methods to provide solutions and interpretative models of the data and configurations sought. For this reason, regenerative design is always "integrated" and digital tools can operate from a single problem to multiple scenarios through the use of parametric processes, which are programmable to address customisation and multi-domain issues.

Parametric design for regeneration becomes the most appropriate set of tools for the use of technical optimisation processes in the energy-environmental field, based on models that can be applied to different scales of architectural systems, but also to natural ecosystems, biology, botany, botanical sciences, climatology, comfort and physiology (Naboni 2019).

In terms of solutions, this approach makes it possible to formulate models and solutions based on co-benefits at different scales, providing integrated digital environments in which tools that model climate, energy, nature, the built environment and human behaviour maximise the strategies and benefits of scaling up through the systemic coupling of physical simulation domains that work on data and information flows.

The construction of demand in the "scripted" transfer of the information workflow effectively realises the operational character that we have entrusted to design tools, whereby the coupling of multiple domains in the scale jump and the process of trial and error effectively simulate the creative processes of design in a manner similar to its natural logical-cognitive expression.

In a holistic approach, high-performance solutions are maximised, and therefore positive, and data, information and resources are returned, which once again become the reference structure for identifying "sustainable technologies".

The construction of the 'workflow' becomes a crucial phase for pre-design activities, in identifying the organisation of the process, such as the ability of operational tools to identify its phases, the validation of the outputs of design activities and in "declarative" mode, and therefore the ability to express expected results in terms of defining the actions to be taken in the design and assessment phase (Fig. 8).

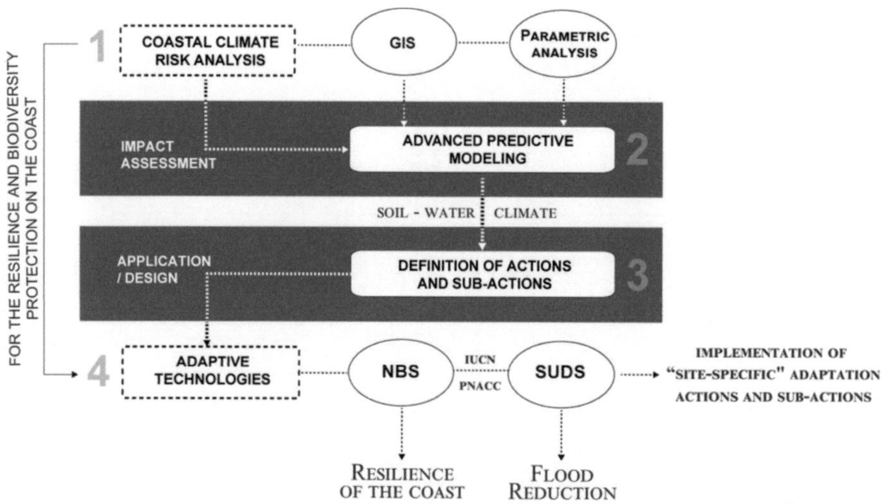

Fig. 8. Adaptive technologies in flooding scenarios through NBS/SUDS. The experimentation of an innovative protocol for resilience and biodiversity protection on the coast of Reggio Calabria (Workflow – D.Laganà, ABITAlab 2024)

Climate vulnerability and resilience analysis, in their procedure for assessing the impacts of climate change, with the selection and classification of multi-risks and hazards, identify the conditions for action through adaptation solutions (Lucanto, Nava, Mangano 2024). The methodology, which can be integrated with the regenerative digital design approach, complies with the provisions of the EU Taxonomy, pursuant to Delegated Regulation (EU) 2023/2486 of 27 June 2023, which supplements Regulation (EU) 2020/852.

Below is a structure of the information, data and resources involved in conducting macro-analysis, starting with climate resilience analysis for all performance levels investigated at the urban cluster and building scale. Studies, types of analysis, simulations and mapping are carried out using digital parametric tools whose outputs can be reported (Table 1).

Table 1. Operational framework for simulations with Regenerative Design tools (E. Catalano, ABITAlab, 2025)

Macro Analysis	Studies	Types of analysis	Parametric tools	OUTPUT
1. Climate resilience analysis	Study of long-term climate change scenarios (temperature increase, variation in rainfall patterns) aimed at assessing the resilience and adaptability of urban and architectural solutions Modelling of the effects of heat waves and extreme weather events (flash floods, hailstorms) on the comfort and safety of spaces	- Urban Heat Island (UHI) analysis - Air Temperature - Humidity - Wind Field/Wind Direction - Heat Stress Index (HSI) - Universal Thermal Climate Index (UTCI) - Mean Radiant Temperature (MRT) Distribution - Cool Island Effect - Simulations of passive cooling strategies at cluster scale (Radiation+Windflow+Evaporation water/Permeable surface) - Analysis of performance development under extreme conditions -Analysis of areas vulnerable to flooding from extreme events	[Rhinoceros+Grasshopper] [Ladybug tools] [ENVI-met] [Forma] [ClimateStudio] [ScalgoLive]	Data Graphs Mapping Project forecast simulations
2. Microclimatic analysis and thermal comfort	Advanced study of urban microclimatic conditions (temperature, humidity, solar radiation, ventilation), aimed at characterising the urban heat island and analysing the thermal gradients induced by impermeable surfaces, building infrastructure and green areas, in order to evaluate design solutions aimed at mitigating local overheating and improving thermo-hygrometric comfort in outdoor spaces Simulations and comparative assessments of different spatial configurations, including the study of impermeable and permeable pavements ( ), the use of materials with high solar reflectance or high thermal emissivity, the optimisation of green surfaces and modelling for the assessment of shading systems to reduce direct solar radiation.	- Predicted Mean Vote index (PMV) - Predicted Dissatisfaction Index (PPD) -Physiological Equivalent Temperature (PET) - Universal Thermal Climate Index (UTCI) -Soil temperature up to 2 metres (Soil Temperature) - Soil moisture content - Solar Reflectance Index -Shadow Analysis -Analysis of impermeable surfaces (asphalt, concrete) vs. draining or vegetated surfaces -Average radiant temperature distribution (MRT) - Cool Island Effect - Analysis of permeable and impermeable surfaces – higher UHI - Direct/diffuse/reflected solar radiation (short-wave radiation) -Long-wave radiation -Albedo analysis	[Rhinoceros+Grasshopper] [Ladybug tools] [ENVI-met] [Shape] [ClimateStudio]	Data Graphs Mapping Project forecast simulations

(*continued*)

Table 1. *(continued)*

Macro Analysis	Studies	Types of analysis	Parametric tools	OUTPUT
3. Advanced analysis of air quality, pollutant dispersion and natural/hybrid ventilation	Analysis of natural ventilation (wind speed and direction) as a key variable in the dispersion of atmospheric pollutants. We assess how the presence of buildings, vegetation or surfaces with different degrees of permeability affects air exchange and the potential formation of areas with high smog concentrations. Mapping and characterisation of air pollutant dispersion using CFD models with scalar transport, assessing the influence of the built environment configuration and ventilation conditions. This analysis aims to identify possible accumulation zones, air exchange paths and the impact on public health. Study to assess the impact of traffic on air quality, aimed at characterising and quantifying the influence of these emission sources on levels of $PM_{2.5}$, PM_{10}, NO_x, O_3 and other pollutants in urban areas, in order to guide effective mitigation strategies and safeguard public health. Comparison of scenarios with different vegetation solutions (trees and green barriers) or optimised urban layouts, aimed at quantifying the effectiveness of these interventions in reducing pollutant concentrations and improving the local microclimate. The analysis focuses in particular on the contribution of vegetation to particulate deposition and the modulation of air flows. Computational fluid dynamics (CFD) modelling allows air circulation to be optimised through the use of r passive ventilation solutions, with the aim of reducing energy consumption and increasing thermo-hygrometric comfort.	-Air quality/Pollutant dispersion -Particulate concentration (PMx PM2.5 PM10), NOx, O3, etc. - Other atmospheric parameters - Effect of vegetation on evapotranspiration - Leaf/canopy temperature - CO2 exchange flux - Wind field/wind direction	[Rhinoceros+Grasshopper] [Ladybug tools] [ENVI-met] [Shape] [ClimateStudio] [OpenFOAM]	Data Graphs Mapping Project forecast simulations

(continued)

Table 1. *(continued)*

Macro Analysis	Studies	Types of analysis	Parametric tools	OUTPUT
4. Energy analyses (Building and Urban Cluster)	Analysis of energy flows at building and cluster level, aimed at quantifying overall consumption (heating, cooling and lighting) and evaluating the effectiveness of different design strategies (shading, materials, natural ventilation) for optimising thermal and energy comfort. Energy simulation of buildings, aimed at comparing and evaluating different construction solutions (building envelope) and systems (high-efficiency heating/cooling systems), with the aim of optimising energy consumption and ensuring optimal thermal and hygrometric comfort conditions. Analysis of the overall energy balance through the study of the balance between heat inputs (solar radiation and internal loads) and heat loss (ly through the building envelope and ventilation), aimed at accurately quantifying actual energy requirements. In-depth analysis of passive solar strategies in order to maximise free heat gains and reduce overall energy consumption. Quantitative analysis of the influence of natural and artificial solar protection systems (trees, internal courtyards, sunshades, tree lines) on the reduction of direct solar radiation and, consequently, on the reduction of cooling requirements.	-Energy balance -Energy simulation of buildings -Effect of shading on energy demand -Passive solar analysis - Sensible/latent heat flux - Sum of incoming components (direct, diffuse, reflected, longwave solar radiation) and outgoing components (longwave, reflection) - Effect of shading on energy demand	[Rhinoceros+Grasshopper] [Ladybug tools] [ENVI-met] [CityEnergy Analyst]	Data Graphs Mapping Project forecast simulations

(continued)

Table 1. *(continued)*

Macro Analysis	Studies	Types of analysis	Parametric tools	OUTPUT
5. Lighting analysis and visual comfort	Assessment of daylight and artificial lighting in public spaces aimed at optimising the visual well-being of users, reducing energy consumption and improving the usability and safety of environments. The analysis focuses on the correct distribution of light sources, the integration of natural and artificial light, and the choice of materials and finishes that promote adequate light diffusion for the different functions and time of day ( ). Glare analysis and identification of possible solutions to reduce visual discomfort (screening, orientation, etc.) aimed at minimising the perception of discomfort and maintaining adequate levels of visual comfort in urban contexts. Luminance mapping (outdoor) to optimise the distribution of natural and artificial light and improve comfort and safety with simulations that highlight any underlit or overexposed areas.	- Outdoor Glare Analysis - Outdoor Luminance Mapping - Combined Daylight+Thermal analysis. Outdoor glare analysis (visual discomfort) + mapping related to outdoor thermal comfort + UTCI + UHI	[DIALux Evo] [Rhinoceros+Grasshopper] [Ladybug tools]	Data Graphs Mapping
6. Analysis of pedestrian flow and use of spaces	Study of pedestrian flows and passage densities aimed at optimising the urban layout to facilitate mobility and ensure safety standards. Areas of high traffic flow are identified, travel times are assessed and critical points are pinpointed, allowing for the evaluation of shaded walkways and rest areas that facilitate social interaction and improve the quality of urban spaces.	-Pedestrian Flow Analysis	[DepthmapX]	Data Mapping Project forecast simulations

3 Discussions – Predictive Tools for Open Knowledge

The parametric simulation tools described (Grasshopper and all application plug-ins such as Lady Bug Tools Environmental, et al.) allow strategies to be developed in the pre-design phase for biomimicry, positive energy buildings, green and blue infrastructure design, urban space design, and the management of natural resources, soil, air, water, solar radiation, etc. Even more so, by using metrics, data and information from relevant databases, they can create simulation models of projected scenarios, such as those needed to simulate configurations, operations and behaviours across multiple contextual options under climate change. This design and operational methodology is able to superimpose all studies and outputs, producing actual digital prototypes that can 'communicate' during the design phase, with specialised and organised flows of information and data (Fig. 9), to be applied in the pre-design phase, together with workflows dedicated to the operational flow process, accompanying all advanced adaptive design and supporting decision-making throughout the regeneration process, even producing new benchmark indicators extracted from the checks, using assessment tools and strategies. This methodology was implemented in the experimental phase of the research for the case studies of Palizzi (*natural heritage,* coastal area) and Bova (*cultural heritage,* urban and building area). In the Design and Prototyping sections of the ABITAlab University Laboratory, part of the Department of Architecture and Design at the Mediterranean University of Reggio Calabria, studies and experiments were conducted, creating both digital and physical prototypes (3D printing), in order to manage all simulations in climate change scenarios, controlling outputs in an interscalar manner and producing images, navigable models, interoperable data and multi-level information. This is an essential phase in the creation of an "open knowledge" user profiling platform capable of translating and disseminating the conditions and scenarios of structural and environmental-climatic impact on natural and historical heritage, with processing designed to ensure the safety of territories and communities, with projections to 2030, 2050 in urban areas and up to 2085, 2100 in coastal areas (Figs. 10, 11).

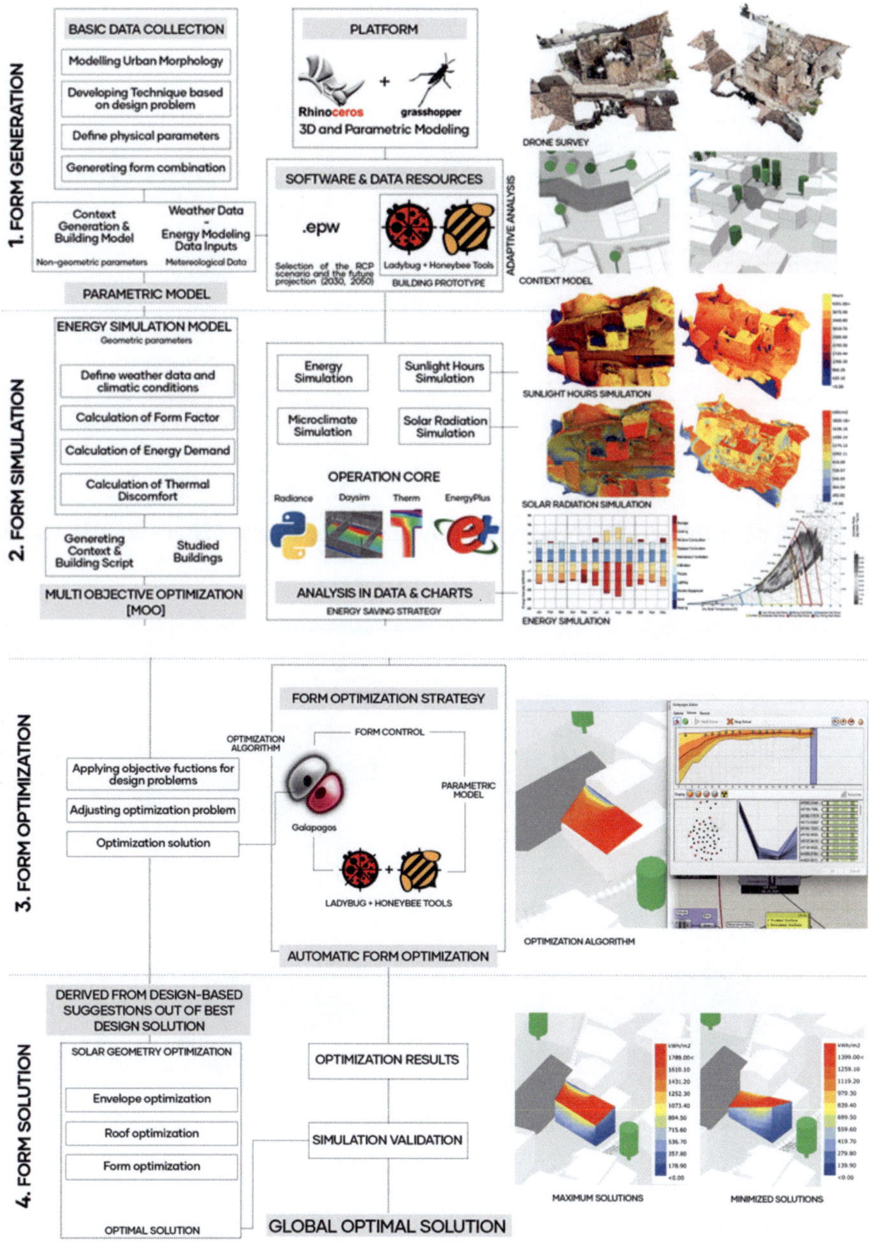

Fig. 9. Energy-environmental pre-design for decarbonisation. Application of a regenerative digital design workflow in the cultural heritage cluster of Bova (RC) (flowchart – E. Catalano, ABITALab, 2024)

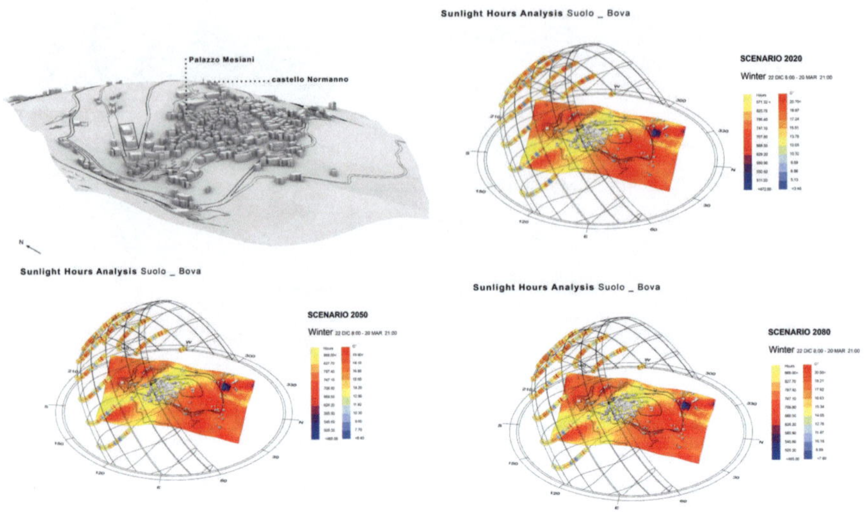

Fig. 10. Climatic Dataset – Soil/Winter_in the cultural heritage cluster of Bova (RC) (D. Laganà, ABITALab, 2024)

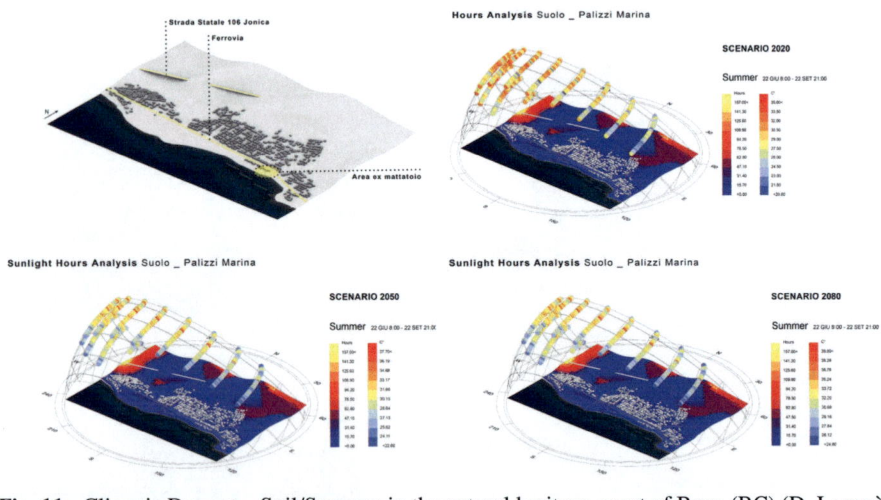

Fig. 11. Climatic Dataset – Soil/Summer in the natural heritage coast of Bova (RC) (D. Laganà, ABITALab, 2024)

4 Conclusions – Beyond *Regenerative Digital Design*

The study proposed in its methodological and applied discussion can express three levels of advancement on the trajectories of digital regenerative design, going beyond the processual nature of the outputs we have presented in this text and preparing its predictive phases for a role of "in and out" information in simulation processes. We consider the possibility of managing the DL (deep learning) of AI's generative capabilities as a further frontier of research, in order to overcome the mathematical and numerical models on which parametric simulation design is based. The discretisation of data and information is a crucial step when using environmental simulation tools (e.g. Lady Bug, ENVI-met), and AI's ability to process large amounts of data offers an interesting opportunity to refine predictive models in the field of climate scenarios (Nava, Melis 2024). **There are three new performance capabilities of a regenerative digital design process** that are truly effective in the field of studies on cultural and natural heritage at risk (Nava 2024), as described below.

First, in parametric design processes, the flexibility to simulate and adjust a wide range of variables allows you to explore latent opportunities within a design space. **The generative capabilities of AI** further enhance this process, allowing designers to explore different permutations and paths that may not be immediately apparent. Through deep learning, AI systems can help recognise patterns of redundancy or untapped potential within data and structures, allowing these elements to be co-opted for new purposes, similar to adaptation in biological systems.

Secondly, through **the construction of workflows that express the complexity** of the methodologies adopted, we propose the construction of a new procedural tool that, by reinterpreting the operational framework adopted for socio-ecological regenerative design, in the circularity of metric, practical, platform, tool and interaction phases (Nava 2022), can work according to the scale-jumping approach, introducing, through DL processes, widespread modelling tools for the production of images, which are inserted into the interactions between the outputs of parametric tools, as a discretisation space for the visualisation of 'scenario options or scenario times', produced with AI.

Third, the regenerative process informed by AI strengthens its phigital profile in an innovative way, because it is capable of guiding the design and discussion of sustainability issues in an "advanced" manner and with "high performance" results in climate adaptation and mitigation scenarios, in order to respond to different states of impact on the physical ecosystem. From a digital point of view, it is capable of producing and exchanging preventive and predictive visions. Regenerative design thus takes on an organisational character to identify options that lead to **new solutions through the use of adaptive technologies.**

Acknowledgement. This study was carried out within the Innovation Ecosystem Tech4You – "Technologies for climate-change adaptation and quality-of-life improvement," Mission 4, Component 2, Investment 1.4 of Italy's National Recovery and Resilience Plan (PNRR), funded by the European Union—NextGeneration EU (project code ECS_00000009, CUP C33C22000290006). The author is pilot project lead, Pilot Project 4.7.1 "Open platform phygital space for advanced, dynamic co-design of built-environment interventions" and scientific responsible for the activities carried out in the Design and Prototyping sections of ABITAlab, Department of Architecture and Design, Mediterranean University of Reggio Calabria.

References

Ceschin, F., Gaziulusoy, I.: Evolution of design for sustainability: from product design to design for system innovations and transitions, 0142-694X Design Studies 47 118e163, Published by Elsevier Ltd (2016)

Change. Contribution of Working Group III to the Sixth Assessment Report of the Intergovernmental Panel on Climate Change [Shukla, P.R., Skea, J., Slade, R., Al Khourdajie, A., van Diemen, R., McCollum, D., Pathak, M., Some, S., Vyas, P., Fradera, R., Belkacemi, M., Hasija, A., Lisboa, G., Luz, S., Malley, J. (eds.)]. Cambridge University Press, Cambridge, UK and New York, NY, USA. 10.1017/9781009157926.001. Graves, R., Keeler, B., Hamann, M., Kutscke1, E., Nootenboom, C. A Social-Ecological Approach to Architecture and Planning. J. Architect Constr. **2**(4), 33–44 (2019)

CMCC: (2017–2023). National Climate Change Adaptation Plan PNACC. Technical and scientific support for the Ministry of the Environment and Protection of Land and Sea (MATTM)

IPCC: IPCC Glossary Search (2022a. Available at: https://apps.ipcc.ch/glossary/

IPCC: Summary for policymakers. In: Climate Change 2022: Mitigation of Climate (2022b)

IPCC: Summary for policymakers. In: Pörtner, H-O., et al., (eds.) Climate Change 2022. Impacts, Adaptation and Vulnerability. Contribution of Working Groups II to the Sixth Assessment Report of the Intergovernmental Panel on Climate Change. Cambridge University Press, Cambridge, United Kingdom and New York, NY, USA (2022c)

Kozak, M., Nava, C.: Application of computer simulation tools in collaborative efforts among scientist, architects, and officials. In Calabrò, F., et al. (eds.) Communities, Institutions and Enterprises Towards Post- Humanism Epistemologies and AI Challenges, vol. 7, NMP 2024, LNNS 1189, pp. 106–114. Springer Nature Switzerland AG (2024). ISBN 978-3-031-74722-9 scopus: 2-s2.0-85211915480

Losasso, M.: Interconnected crises and the complexity of the project, Introduction. TECHNE – J. Technol. Archit. Environ. **23** (2022). https://doi.org/10.36253/techne-12913

Lionello, P., Naumann, G.: The IPCC Report Explained by Italian Experts with the Main Content on Europe, the Mediterranean and Italy.. CMCC for IPCC focal point for Italy (2022)

Lyle, J.T.: Regenerative Design for Sustainable Development. Wiley, New York, NY (1994)

Lucanto, D., Nava, C., Mangano, G.: digital prototyping and regenerative design towards a carbon-neutrality and climate resilient built environment: a multi-scale assessment of environmental multi-risks. Buildings **14**(12), 1–27 (2024). Special Issue- Best Practices of Resilient Buildings (and Districts) and Post-disaster Reconstruction E-ISSN:2075-5309 -open access Q1-. https://doi.org/10.3390/buildings14123934

Mecca, S.: The project as action between order and disorder. In search of harmon, Rubettino ed., Entopan -Harmoni Innovation series, Soveria Mannelli (2022)

Mussinelli, E.: (2022) Editorial TECHNE – J. Techn. Archit. Environ. **23**. 10.36253/techne-12914

Naboni, E., Havinga, L.: Regenerative Design in Digital Practice. A Handbook for the Built Environment, COST Action, Restore WG2. Eurac, Bolzano (2019)

Nava, C.: Reggio Calabria. The technological culture of environmental design. In: Losasso, M., Attaianese, E., (eds.) Research in Environmental Design. The Years 1970–2008, Studies and Projects series, Maggioli ed. Santarcangelo di Romagna (RN) (2022)

Nava, C., Melis, A.: Generative IA and complexity – towards a new paradigm in regenerative digital design, AGATHÓN. Int. J. Archit. Art Des. **16**, 40–49 (2024). https://doi.org/10.19229/2464-9309/1632024

Nava, C.: Human Centered AI in the post optimization regenerative process for climate scenarios. Addressing new workflow for Re-Generative Digital Design. In Calabrò, F., et al. (eds.) Communities, Institutions and Enterprises Towards Post- Humanism Epistemologies and AI Challenges, vol. 7, NMP 2024, LNNS 1189, pp. 76–97. Springer Nature Switzerland AG, pp. 1–22 (2024). https://doi.org/10.1007/978-3-031-74723-6_7. ISBN 978-3-031-74722-9 scopus: 2-s2.0-85211955246

Paris, S.: The transformation of technological culture in design, through new tectonics and digital technologies. International teaching and research scenarios. TECHNE – J. Technol. Architect. Environ. **13**, 194–203 (2017). https://doi.org/10.13128/Techne-19746

Reith, A., Brajković, J.: Scale Jumping: Regenerative Systems Thinking within the Built Environment. A guidebook for regenerative implementation: Interactions, tools, platforms, metrics, practice. COST Action CA16114 RESTORE, Working Group Five: Scale Jumping, printed by Eurac Research (Bolzano, IT) (2021)

Sadollah, A., Nasir, M., Woo Geem, Z.: Sustainability and optimization: from conceptual fundamentals to applications. Sustainability. **12**, 2027 (2020). https://doi.org/10.3390/su12052027

Thompson, J.D.: Organisational Action. Isedi, Turin (1994)

Wuyi, J., Jiawei, X., Yongli, C., Zhiyong, L.: Ecological land adaptive planning in macroscale, mesoscale, and Microscale of Shanghai. Sustainability. **2020**(12), 2142 (2020). https://doi.org/10.3390/su12052142

Case Study Applications and Living Lab Implementation to Bova and Palizzi villages, Italy

The Living Lab Project and Community Engagement in the Pilot Project 4.7.1

Giuseppe Mangano(✉)

Department of Architecture and Design, Mediterranean University of Reggio Calabria, Reggio Calabria, Italy
giuseppe.mangano@unirc.it

Abstract. The paper illustrates the application of the Living Lab methodology in Pilot Project 4.7.1 "Tech4You," aimed at creating a phygital platform for the co-design of cultural and natural heritage interventions in transition and climate change scenarios. After outlining the Living Lab concept and the underlying theoretical framework, the paper describes the integrated methodological process, divided into three phases according to technological readiness levels (TRL), and tested in the case studies of Bova and Palizzi Marina. The results of the laboratory activities, operational mapping, and community engagement events are presented, which allowed for the testing of prototypes and climate scenarios, transferring knowledge for their optimization. The discussion highlights strengths, weaknesses, and future prospects, particularly the transfer of content to the profiled digital platform, as a replicable tool for participatory planning and adaptation to climate change in inland areas.

Keywords: Living lab · Co-design · Community engagement · Phygital platform · User-centered design

1 Introduction

In this era of facing climate change, the LLs Living labs become a tool to improve the response processes of communities in transition, through organizational and management forms of the strategies and solutions that serve the green and digital transition, not forgetting 'the necessary condition of inclusion of the communities' (Nava, Mangano 2023) to the physical-infrastructural processes at all levels, making them capable of measuring themselves with new approaches and visions to trajectories related to environmental issues and climate change; this, in fact, refers to the need not only to operate in a 'transition' regime, but also a 'just transition'.

The co-benefits can be said to refer to the effectiveness of spaces for LLs' actions on the trajectories of interest for the Just Transition Mechanisms (JMT), towards future scenarios 2030, 2050, 2085, in which 'emerging technologies' (Nava, 2023) take on roles in the ecological and digital transition with effectiveness on the socio-economic impacts deriving from activities aimed at 'participation/ownership, lifestyle, low cost

C. Nava et al. (Eds.): *Climatic and Structural Safety in Multi-Hazard Regime of Cultural and Natural Heritage*, GREEN, pp. 177–195, 2026.
https://doi.org/10.1007/978-3-032-11926-1_11

energy bills, social cohesion, education, acceptance and awareness, combating energy poverty, regeneration of the local economy, well-being and health , job creation and local skills' (Caramizaru and Uihlein 2020).

This contribution is part of Pilot Project 4.7.1 PNRR–NGEU Tech4You, "Open platform phygital space (physical and digital) of the type user profiling for the advanced and dynamic co-design of interventions on the built and ex novo," as an operational development of the Living Lab methodology. The latter is intended as a shared physical space to transfer the results of experimental and research actions to the local community, encouraging active user involvement and participatory validation of proposed solutions. Within the framework of PP 4.7.1, the Living Lab assumes the role of a local laboratory for the development and testing of actions 1–7 on structural safety and actions 8–9 on climate change scenarios, in order to verify not only their technical effectiveness but also their comprehensibility and adoption by the community. The final phase of the process involves translating these results into a user-profiled digital interface, capable of adapting content to different user skill levels, enabling dynamic querying and targeted information visualization. The phygital model thus conceived integrates the experiential and participatory dimension of the physical laboratory with the dissemination, scalability, and customization capabilities of the digital platform, laying the foundation for an advanced co-design system that can be replicated in other regional contexts.

The paper's structure follows a path from theoretical framework to discussion of the results. The Introduction presents the context and objectives, situating the work within Pilot Project 4.7.1 and illustrating the role of the Living Lab as a central methodological tool. The Theoretical Background outlines the conceptual and operational framework of Living Labs, supported by literature references and case studies in similar contexts. The Methodology section describes the approach adopted, divided into phases corresponding to the project's technological readiness levels (TRLs) and the phygital co-design methods. The Case Studies analyze the two pilot contexts of Bova and Palizzi, presenting territorial and socio-environmental data and community engagement activities. The Discussions section evaluates the methodology's effectiveness and the critical issues encountered, while the Conclusions & Future Trends section summarizes the contributions made, identifies barriers, and outlines developments toward an adaptive user profile digital platform.

1.1 Theoretical Background

A "Living Lab" is an open innovation ecosystem operating in a territorial context, whose purpose is engaging competence clusters, local authorities, citizens, entrepreneurs, business operators and other possible stakeholders within social innovation environments to create sustainable impact. The main objectives consist in "fostering co-creation among these actors to explore new ideas and concepts, experiment new artifacts and evaluate breakthrough scenarios to promote innovation" (Zanetti, Nollo 2019). Living Labs (LLs) proposes "a research methodology for sensing, prototyping, validating and refining complex solutions in multiple and evolving real-life contexts": this first definition has been coined by William J Mitchell in the 1990s at the MIT Labs in Massachusetts (USA). LLs focus on co-creation, prototyping & testing and scaling-up innovations and services, providing (different types of) joint-value to the involved stakeholders. Thus, know-how, knowledge and participation transfer activities take place in an Open Innovation Environment (aggregation of public bodies, enterprises, universities, research centres and citizens' groups) in a real-life situation, in which the active involvement of end-users enables the co-creation of new services, products and social infrastructures (driven by the needs of end-users) that can be tested in a circumscribed geographical context and within a defined time frame with a view to large-scale application and economic return.

In recent years, Living Labs have emerged as strategic tools to support urban regeneration processes and strengthen environmental safety, especially in contexts linked to cultural and natural heritage. These laboratories, conceived as open and participatory innovation ecosystems, promote co-creation between citizens, institutions, industry, and academia (Quadruple Helix), and are increasingly used to test and develop solutions in real and complex environments (Rosetti & Navarrete Hernandez, 2025). In urban settings, Living Labs have been used to address challenges related to sustainability, risk management, and heritage enhancement through interventions that combine interdisciplinary approaches and co-design methodologies (Florez Ayala et al., 2022; Lupp et al., 2021). Their application in heritage contexts is particularly relevant where conservation must be reconciled with adaptation to climate change and hydro-meteorological risks, through ecosystem-based adaptation strategies (EBA) and nature-based solutions (NBS) (Tiwari et al., 2022).

However, the literature highlights a certain fragmentation in operational definitions of Living Labs and a lack of attention to processual aspects, especially in relation to replicability, impact assessment, and systemic stakeholder engagement (Schuurman et al., 2016; Rosetti & Navarrete Hernandez, 2025). This is particularly critical when working on cultural and natural heritage, where it is essential to adopt participatory governance approaches that respect the plurality of values attributed to heritage (Clark, 2019). The experience of projects such as the Isar-Plan for urban river regeneration (Lupp et al., 2021) or the Coastal City Living Labs (Tiwari et al., 2022) shows how a well-structured co-design process can effectively integrate protection, enhancement, and territorial resilience, offering a model for addressing environmental and social challenges synergistically. It is therefore urgent to define theoretical frameworks and operational models that facilitate the implementation of Living Labs in highly complex areas, such as those related to the regeneration of cultural and natural heritage in urban contexts exposed to environmental risk.

The LLs thus become the physical and digital places, which manage this effectiveness, supporting the co-management of all the operations and activities, made up of citizens, associations, businesses, local and national bodies, study and research centers and universities; a platform of human, technical and digital resources together, in the transition processes, accelerate the methods aimed at founding strategies, selecting solutions, promoting visions.

So, if on the one hand, the socio-economic fragility of the inner areas can represent an impediment to the application of research, on the other its Living Lab responds as a platform of 'innovation and open knowledge', which enables technology transfer processes and sustainability, for innovative models, as reported in the introductory contribution.

The LLs are created to affect the impact of measures on the preservation and conservation of cultural heritage, through organizational and management forms of the strategies and solutions that serve the green and digital transition, not forgetting 'the necessary condition of inclusion of the communities' to the physical-infrastructural processes at all levels, making them capable of measuring themselves with new approaches and visions to issues related to environmental issues and climate change; this, in fact, refers to the need not only to operate in a transition regime, but also a just transition.

2 The Co-design Methodology for Living Lab

From a methodological point of view, LLs are networks composed of heterogeneous actors, resources, and activities that integrate user-centred research and open innovation (Leminen et al., 2012). From the infrastructure perspective, they can be seen as facilities that enable experimentation and co-creation with users in real-life environments (Keyson et al., 2013).

It is a question of experimenting projects with a high rate of innovation, cohesion and promotion of knowledge, in which users are not only final users, but actively participate in the implementation of actions, objectives and new products/services, contributing to the creation of a new circular model, according to an "integrated co-design" approach. The recipients of the research can be identified in the community activated with the Living Lab (citizens, businesses, organizations and local associations) (Nava; Mangano 2023).

It is in this new socio-technical scenario that the Living Labs (LLs) take on a fundamental role for orienting the architectural interventions on built environment and cultural heritage, directing users towards those motivations connected, for example. to the construction of a 'physical space' through digital technologies.

This goal is triggered by a precise methodology that is iterative and based on co-design processes, that consider the phases of the 'interaction between users & providers', the 'user experience value' and the 'partecipatory dimension' as fundamental part of the "user profile approach" (Fig. 1).

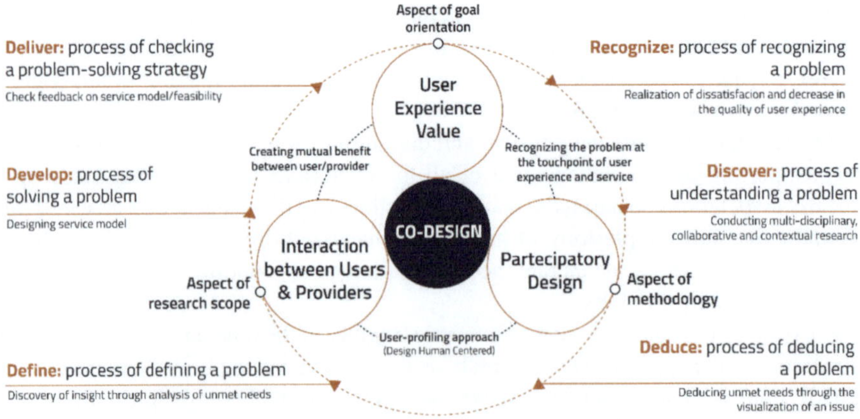

Fig. 1 . Iterative co-design process for the elaboration of a new product-service in Living Lab. Elaboration by G. Mangano (2023)

2.1 The Living Lab Activities Meta-Project in the Grecanica Area

The objective of this phase concerns the definition of the elements that characterize the creative, planning and experimental process of the Living Lab, within the area and the "inner" community identified with a scouting process. Activities will be carried out for prototyping, verification and experimentation for technical and economic feasibility, studies on the level of efficiency of technological choices at the system/component scale and of devices supporting the 'phygital space'.

The LL in support of the project of an 'Open Platform as Phygital Space' will take into account:

- 'territorial' components, i.e. physical capital and natural capital, with the reference community (stakeholders and end users together);
- 'technological' components, i.e. input dataset for predictive models for climate scenarios, energy-environmental data for structures, key enabling tech-nologies in the phase of prototyping and test for the integrated implementation and use-testing of the digital platform itself (Fig. 2).

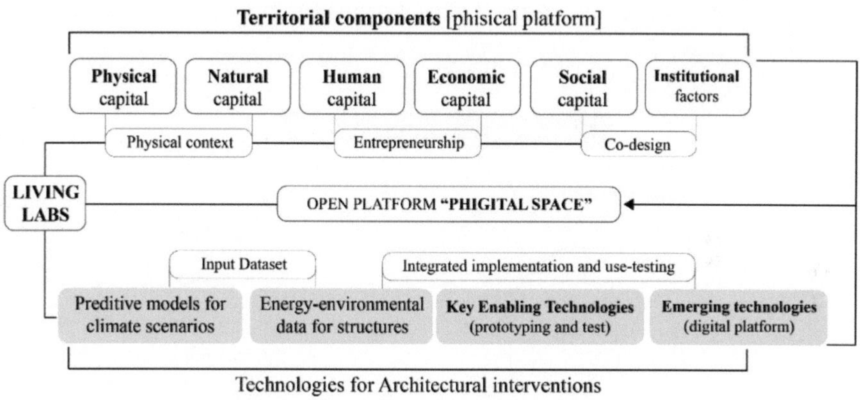

Fig. 2 . The territorial and technological components for the "phygital space". Elaboration by G. Mangano (2023)

Subsequently, the activities will be carried out through a more concrete and operational process within a specific organization defined 'Living Lab', whose collaborative structure, co-design methods, knowledge sharing, and activity program are defined according to a model recognized and documented in the relevant literature.

These aspects become fundamental levers for the development of a LL open innovation ecosystem, by allowing this model to be made more advanced, integrating transdisciplinary and interscalar research, knowledge, technologies and methodologies (building/neighborhood or village, citizen/community, inner areas/urban areas).

The preliminary analysis that guided the selection of the case studies as prototypes was aimed at addressing the challenges aligned with the objectives of the Tech4You project and Spoke 4, as outlined in the pilot project proposal. Particular relevance is given to the identification of two case studies within the Grecanica Area of the Metropolitan City of Reggio Calabria (Ionian side, inner region), where the selected municipalities and their territories demonstrate the potential and feasibility for implementing experimental development actions and establishing the Living Lab framework.

2.1.1 The Operational Methodology of Living Lab

The Ecophygital pilot project's operational methodology is developed according to an integrated, progressive, and participatory approach, divided into three main phases, corresponding to increasing levels of technological readiness (TRL) of the Pilot Project 4.7.1 (Fig. 3). Subsequently, the activities will be carried out through a more concrete and operational process within a specific organization with the Living Lab, whose collaborative structure, co-design methods, knowledge sharing, and activity program are defined according to a recognized model documented in the relevant literature. The activity will be structured on two complementary layers: a physical information layer, activated through the launch of the Living Lab (as a physical platform), facilitating the direct transfer of knowledge to users—whether institutional, professional, or community-based; a digital information layer, which enables the transfer of data to the User Profile Platform. This integrated scenario operationalizes what is defined as 'open knowledge' in the Pilot Project 4.7.1.

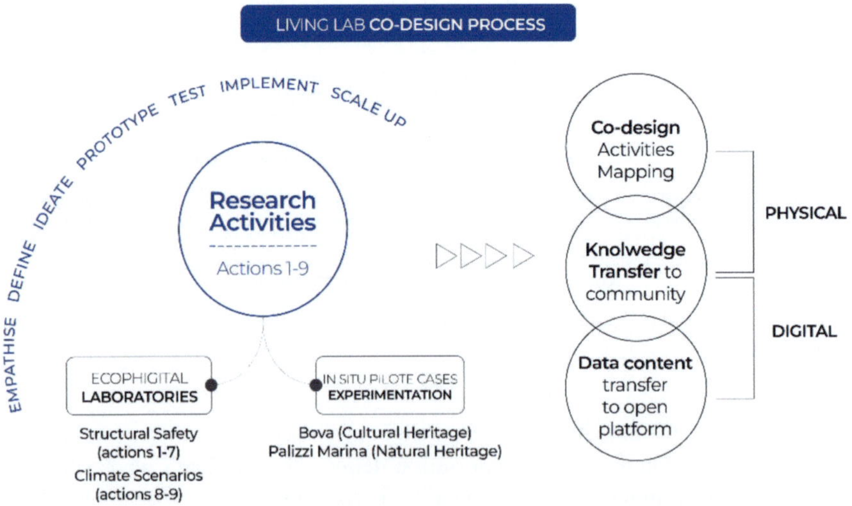

Fig. 3 . Living Lab co-design process for PP 4.7.1. Elaboration by G.Mangano (2025)

The application of its integrated process practices derives from those strategies, solutions and visions, which from the scale can contribute to the global one, with reference to environmental, social, economic, geopolitical and even technological risk factors. The meta-design phase of the LL for the definitions of technological and social innovation spaces it is structured as follow:

- In the first phase Research activities are conducted within the Ecophygital Laboratories (SSL for Structural Safety Laboratories and CSL for Climate Scenarios Laboratories), where technical and scientific solutions are defined, ideated, developed and tested based on existing equipment and expertise, enhanced by the Tech4You project's spending plan, according to the LL co-design process;
- This is followed by the second phase, which focuses on mapping the Living Labs' activities, achieved through a co-design process that systematically involves research laboratories, local governments, economic entities, and citizens (associations, cooperatives, groups), for the construction of a 'human' knowledge platform with context analysis, the involvement of interested users (Fig. 4); this phase includes a desk analysis of the context that will be illustrated in paragraph 3;

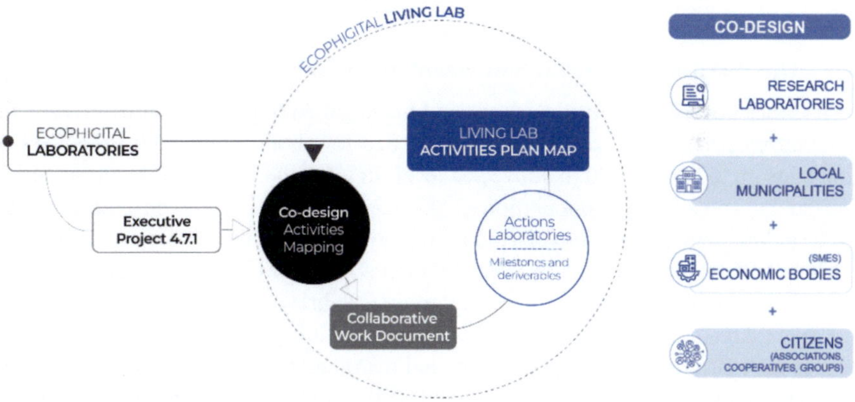

Fig. 4 . Co-Design for Living Lab Activities Map. Elaboration by G. Mangano (2024)

- This process generates a collaborative working document and an operational map of the Living Lab's activities, which defines milestones and deliverables. In this phase, the actions are transferred and tested in two pilot case studies (Bova and Palizzi Marina), representative of cultural and environmental contexts, selected in accordance with Spoke 4. These contexts are analyzed across three climate scenarios (2030, 2050, 2085), based on their ability to generate data and information useful for user profiling. This phase also regars the LL Project programme with reference to transfer activities for "co-design and open Knowledge" (open construction site and platform testing phase);
- The third phase involves the construction of the Ecophygital User Profile Platform, a physical-digital space that integrates content from the previous phases and enables the dynamic co-design of territorial interventions. It is implemented by organizing the data-information space, transferred from the results of the Research Laboratories' activities and the Living Lab, which are used to build the Digital Platform's content archive and therefore the interface structure for user profiling operations.

The whole process is intended to orient the products of research towards a high rate of socio-economic and environmental impacts, including through future valorization in terms of new patents and new competitiveness.

3 The Activation of Ecophygital Living Lab

3.1 The Grecanica Area: Desk Analysis for the Prototype Study Cases

The objective at this stage of the research was to provide a descriptive framework of the study area as an application scenario, making use of the knowledge and studies from the description of the pilot area "Grecanica" in the Regional Strategy for Inner Areas of Calabria, from the territorial, demographic, socio-economic and administrative points of view through the reworking of data from the studies of the bibliographical references.

3.1.1 The Territorial Context: Settlement and Landscape Values

The territory covering the Project Area and Strategic Area municipalities extends in a triangular shape from the coast of the Ionian Sea for about 60 km to the Aspromonte Mountains up to about 1900 m above sea level. In total, the Project Area has an area of 434.8 sq. km and a resident population as of 1.1.2016 of 18,281, with an average density of 42.0 in/km^2 with a minimum of 9.6 in/km^2 for Bova and a maximum of 110.9 in/km2 for Montebello Ionico, Area under consideration is home to 18,821 inhabitants out of the total 551,380 in the Metropolitan City of Reggio Calabria (Fig. 5). The municipalities of the Project Area are all classified as Peripheral and Ultra-peripheral and in the last four decades have suffered from depopulation and marginalization phenomena that in some cases have caused the almost total abandonment of entire inner centers, affected both by hydrogeological instability events that have led to the building of a new settlement far from the places of historical foundation, and by the lack of adequate connections with the metropolitan area of Reggio Calabria and with the coastal centers, where essential health and education services are provided. Although it has strong criticalities, the Grecanica Area also preserves an environmental, landscape and architectural heritage of extraordinary quality, but also a strong cultural identity that derives from the speech of the 'Grecanica' language. In addition, part of the coastal strip and low hills, particularly that between the towns of Palizzi and San Lorenzo, is protected as a Landscape Asset .The municipalities of Bagaladi, Montebello Jonico, San Lorenzo, Bova, Bruzzano Zeffirio, Gallicianò, Amendolea, Ferruzzano, Palizzi, Pietrapennata, Roghudi Vecchio, Staiti) included in the list of Historical Centers and Minor Historical Settlements Susceptible to Protection and Enhancement of the Region of Calabria, in implementation of art. 48 of R.L. n.19/2002. The Grecanica Area possesses important qualities that reside in its unique landscape and historical-cultural features. The Calabria QTRP (Regional Territorial Landscape Framework) traces the elements that distinguish the APTR-UPTR 'Area of the Greeks of Calabria' in the great morphological complexity of the hilly and mountainous reliefs, characterized by strong erosions that in the clayey areas determine the formation of gullies, by a low and sandy coastline, but above all by a hydrographic basin marked by numerous watercourses perpendicular to the coast, such as the River of Amendolea. The high environmental and landscape value due to the presence of as many as four protected areas, geosites, vegetation and valuable tree species has caused some areas to be subject to law preservation. This system of historic centers and villages in the process of depopulation constitutes the precious landscape of the Grecanica area, which if on the one hand provides historical settlement evidence of great relevance on the other hand is embellished by the vast heterogeneity of vegetation and tree species (Mediterranean scrub, reeds, dune beaches) that also characterize the "productive landscape" of the area, rich in terraces and fields intended for the cultivation of bergamot and local agricultural products. In fact, economies from sustainable tourism activities are linked to this natural and cultural heritage of the whole area.

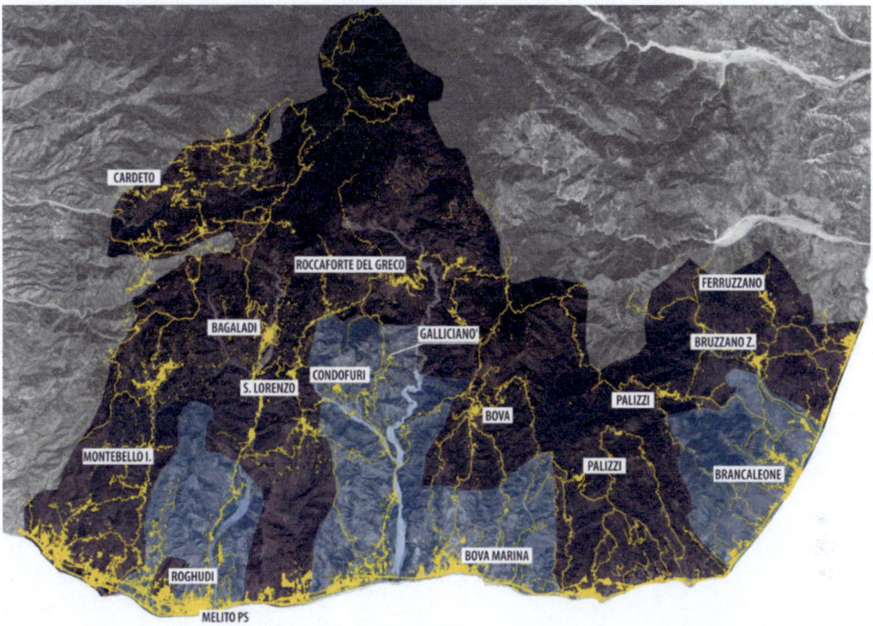

Fig. 5 . The Strategy Area and the Project Area for the Grecanic pilota Area. Source: elaboration by G. Mangano (2019)

The Grecanica Area is the second pilot area for the implementation of the National Strategy of Inner Areas in Calabria (SNAI), occupies a vast territory in the eastern Metropolitan City of Reggio Calabria and includes eleven municipalities in the Project Area (Bagaladi, Bova, Bruzzano Zeffirio, Cardeto, Ferruzzano, Montebello Ionico, Palizzi, Roccaforte del Greco, Roghudi, San Lorenzo, and Staiti) and four municipalities in addition to the Strategy Area (Melito Porto Salvo, Condofuri, Bova Marina, and Brancaleone). The implementation of the SNAI in the Grecanica Area is currently in the implementation phase of the interventions provided for in the Framework Program Agreement, signed in October 2021. The document contains the actions to be implemented in the coming years to halt and reverse the depopulation of the centers and Villages of the Pilot Area Grecanica, through: 1) territorial regeneration; 2) improvement of services for education; 3) improvement of services for health; 4) improvement of services for mobility; 5) development of agri-food chains; and 6) development of sustainable tourism.

3.1.2 1The Prototype Case Studies: Data & Information

The urban area of Palizzi Marina (Prototype 1) is classified as "coastal center predominantly built as a projection to the coast of inner mountain centers after the creation of the railroad" (the so-called "Marine"). The municipality of Bova (Prototype 2) is classified as "historic inner entrenched towns and villages that have preserved the original urban layout and much of the historic-building heritage".

In 2011, the municipality of Palizzi had 1745 houses, of which only 58.1% were occupied and the remaining 41.9% unoccupied, highlighting a significant level of depopulation or underutilization of the building stock (2017 inhabitants, of which 1906 in Palizzi Marina). This figure appears to contradict the significant landscape, natural, and cultural value of the area: Palizzi is crossed by the Fiumara River and includes a coastal and hilly stretch protected as a Landscape Property under Legislative Decree 42/2004. Furthermore, the area is recognized as a "Minor Historic Center" eligible for protection and enhancement by the Calabria Region. The presence of this heritage, combined with its strategic location along important infrastructures such as the State Road 106 and the Ionian railway, suggests potential opportunities for the recovery and revitalization of the urban and residential fabric, including through cultural and environmental enhancement projects (Fig. 6a–d).

a b c d

Fig. 6. a, b, c, d. The experimentation in coastal area in Palizzi Marina – Natural Heritage. Source: ph. G.Mangano (2024).

In the municipality of Bova, classified as an "ultra-peripheral" inner area, the housing situation in 2011 showed a prevalence of vacant homes (59.5%) compared to occupied ones (40.5%), out of a total of only 568 dwellings. This phenomenon reflects a significant depopulation process, aggravated by the low population density (8.52 inhabitants/km^2) and the high percentage of residents living in scattered homes (48.4%) outside the town center. Despite this, Bova represents an important example of a historic village in the Calabrian hinterland, with a well-preserved original urban layout and a rich historic building heritage. Its settlement morphology, linked to its location along rivers and socioeconomic ties almost exclusively with coastal towns, reinforces the idea of a place with a strong yet isolated identity. This cultural heritage, combined with the village's scenic and architectural beauty, represents a key resource for development and redevelopment strategies, essential to counteracting demographic and housing decline (Fig. 7).

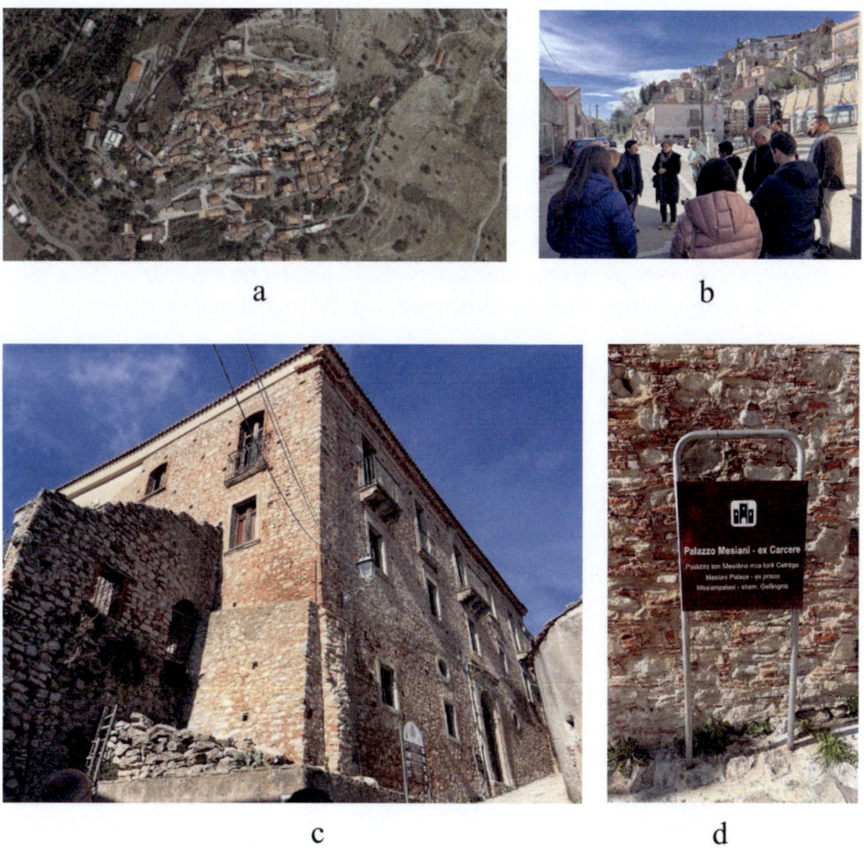

Fig. 7 . **a**, **b**, **c**, **d**. The experimental pilot case of Cultural Heritage in Bova (RC), Palazzo Mesiani. Source: ph. G.Mangano (2024).

The desk analysis constituted the basic investigation phase of the prototype cases from a documentary perspective (demographic data, data on the state of settlement, ongoing policies and strategies, and expeditious surveys of artifacts and natural areas). This activity laid the foundation for organizing the in-situ Laboratory and Living Lab activities, the results of which are presented below.

3.2 The Living Lab Co-Design for Activities Program and Map

The co-design process that led to the generation of the Living Lab's activity map and agenda is divided into three structured and progressive steps, aimed at building an integrated system for planning, accountability, and monitoring project activities.

- Activity Definition (Step 1).

  The starting point of the process is the definition of the activities, organized into two main areas:

  - activities in university laboratories, which include research, development, testing, and validation of deliverables, with reference to the prototypes considered;
  - in-situ laboratory activities, carried out in the participating municipalities (pilot cases), which include knowledge transfer, operational testing, field co-design, and community engagement.

This initial phase is based on the mapping of the project's activities and their chronological order according to the milestones. This is where prototypes and activity categories are considered: workshops on structural safety (SSL) and climate scenarios (SC), in-situ activities, co-design events, and the construction of the digital platform (Fig. 8).

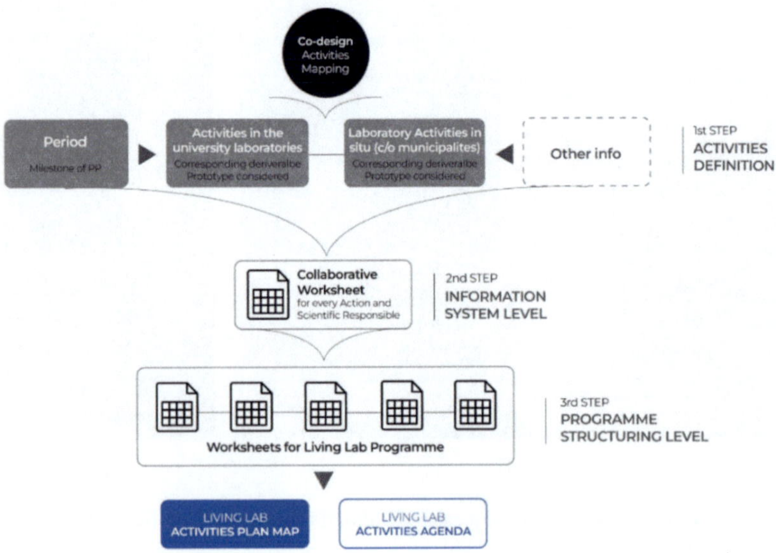

Fig. 8 . Co-design process for LL activities map and agenda. Source: elaboration by G. Mangano (2024)

- Information System Level (Step 2).
  After defining the activities, the process enters the second level, characterized by the creation of collaborative worksheets, one for each action and its corresponding scientific manager. Each worksheet contains:

  - the activities planned in the university and on-site laboratories;
  - the associated deliverables;
  - the implementation period (quarter and milestone);
  - any operational notes.

These documents constitute the information core of the system and enable dialogue between the various managers, providing a shared basis for planning.

- Program Structure (Step 3).
  The third step consists of structuring the overall Living Lab program, starting from the completed worksheets. The data is aggregated and organized, resulting in two main tools:

  - the activity map, which visually represents the temporal and operational organization of the actions, with activities and deliverables tracked across the quarters and pilot cases highlighted with distinct color codes;
  - The Living Lab activity agenda, which presents all actions, their operational content, and objectives in a structured, textual format, facilitating consultation and accountability for all team members.

The entire co-design process ensures that each manager is aware not only of their own activities, but also of the overall project framework and the interactions between the various actions. The map and agenda thus become shared, operational, and dynamic tools for the integrated management of the Living Lab (Fig. 9).

Fig. 9 . First LL meeting with Bova and Palizzi municipalities representatives in oct. 2024. Source: elaboration by G.Mangano (2024)

The map (Fig. 10) also aims to monitor the progress of deliverables and related project activities and to ensure that all action leaders are fully aware of both the specific tasks to be performed and the broader body of research activities within which each intervention is located. This document provides a planning map designed to outline the actions planned within the Living Lab of the PP 4.7.1 Tech4You project, organizing them over time according to the main operational categories: workshops on structural safety (SSL) and climate scenarios (SC), in-situ transfer activities on pilot cases, participatory Living Lab events, and the development of the user-centered digital platform.

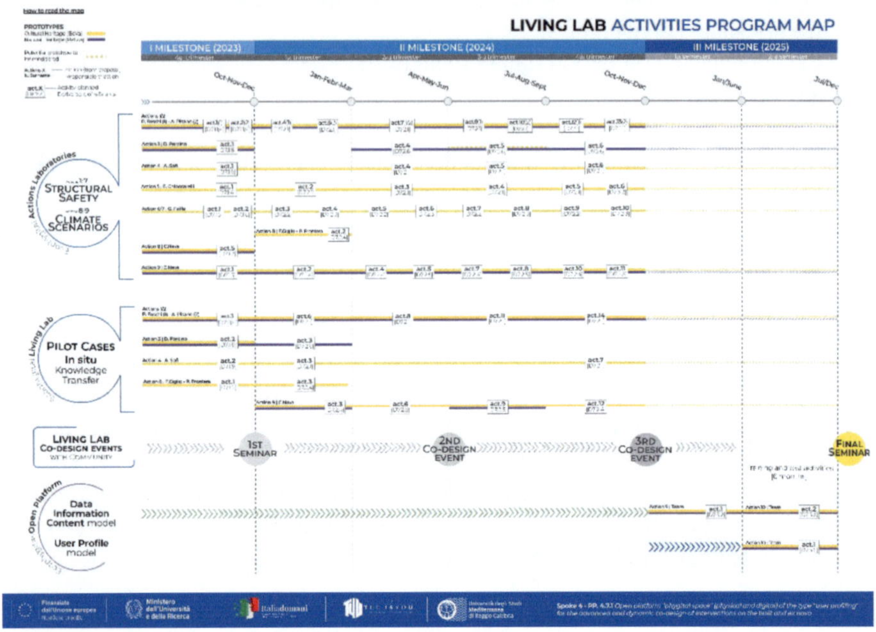

Fig. 10 . LL Activity Map. Source: elaboration by G.Mangano (2024)

The prototype cases included in the map, relating to cultural (Bova) and natural (Palizzi) heritage, are highlighted with color differentiation, allowing the context in which the actions are implemented to be clearly distinguished.

The map is structured along a timeline that covers the entire three-year project period (2023–2025), divided into quarters and annual milestones and allows for a synchronous and progressive interpretation of the activities. The actions are organized by project line and scientific responsibility, each identified by an acronym (act.X) associated with the relevant deliverable (D7.X.X), ensuring the traceability of the expected results. The map includes both experimental and training activities as well as public opportunities for community engagement, fully integrating the dimensions of applied research and participation. It also represents the evolution of the open digital platform "phygital," geared toward dynamic co-design and the management of content and user profiles.

3.2.1 'In Situ' Community Engagement – Living Lab Events

The co-design and community engagement methodology adopted in the Living Lab events of the Tech4You – PP 4.7.1 project was based on a participatory and multidisciplinary approach, aimed at directly involving local stakeholders in the development and testing of prototypes and climate scenarios. During the first event in Bova (April 17, 2024), the methodology was applied through an interactive presentation in which the research team illustrated the project ongoing results using visual tools, concept maps, inviting participants—administrators, associations, and citizens—to express needs, concerns, and suggestions related to the area's cultural and natural heritage. They showed digital and physical prototypes and advanced simulations on climate change scenarios and their effect in terms of structural and environmental safety, both for the built heritage of Bova and for the coastal climate analyses in Palizzi Marina. The meeting fostered the building of a relationship of trust and collaboration between the local community and the project team, laying the foundation for the co-design of the prototypes (Fig. 11).

Fig. 11 . 1st Living Lab event in Bova, 17 April 2024. Source: ph. ABITAlab

In the second event in Palizzi (June 24, 2025), the methodology evolved into an in-situ experiment: participants were directly involved in the field evaluating prototypes and simulating climate scenarios applied to both contexts (cultural and natural heritage). Through hands-on activities and demonstrations, citizens and local representatives were able to test the proposed solutions, providing immediate feedback. It consisted of two distinct sessions: in the morning, an in-situ testing activity was conducted at Palizzi Marina, with the deployment of the 'Eco-Reef' (Lucanto & Hanida, 2025) ecological prototype device for enhancing biodiversity and protecting coastal areas on the seabed. A feedback seminar was held in the afternoon, where the results of the experiment were collectively discussed and integrated into a process of transfer to the Ecophygital digital platform, to enhance and make accessible the shared knowledge (Fig. 12).

Fig. 12 . 2nd Living Lab event in Palizzi Marina, 24 June 2025. Source: ph. ABITAlab

The impacts of engagement for these two LL events can be summarized in the following Table 1:

Table 1 . Number of participants at the LL Events PP 4.7.1.

Type of Participants at LL Events	Number
Professors, Researchers, PhD and PhD candidates	16
Local Administrators	6
Representatives of Associations	12
Local entrepreneurs	2
Citizens	7
Total	43

Both events followed the Living Lab principles: end-user focus, collaboration between public and private stakeholders, and the use of digital technologies to support the co-creation of innovative solutions aimed at resilience and enhancing the local area.

4 Discussion

The application of the Living Lab methodology in PP 4.7.1 has demonstrated high effectiveness in combining technical and scientific research with community engagement. The events held in Bova and Palizzi allowed for in-situ testing of prototypes and climate scenarios, obtaining direct feedback from end users and facilitating a process of adapting solutions to local specificities. The integrated approach, based on a dual physical-digital level, has fostered the creation of a collaborative environment capable of connecting institutional stakeholders, businesses, associations, and citizens. In terms of project progress, the methodology has produced concrete results: mapping of activities, definition of shared operational agendas, testing of prototypes, and the start of the transfer of user profiles to a digital platform. These outcomes have helped strengthen the internal coherence of PP 4.7.1 and its ability to generate replicable impacts. However, some critical issues emerge related to the continuity of community involvement and the need to balance the technical complexity of the content with its usability for a non-specialist audience.

5 Conclusions & Future Trends: Transferring to User Profile Platform

The experience conducted confirms that the Living Lab methodology represents an effective tool for developing and validating innovative interventions in complex territorial contexts, integrating multidisciplinary skills and actively involving local communities. The phygital dimension proposed in PP 4.7.1 has allowed us to broaden the scope of the activities, enhancing the capacity for knowledge transfer and creating a structured archive of content useful for participatory planning.

However, two significant barriers remain. The first concerns the difficulty of activating consistent community involvement in areas subject to depopulation and marginalization, where participation is hindered by socioeconomic and demographic factors. The second is cognitive in nature: the level of technical complexity of the content developed by the research, while necessary for scientific reliability, can be difficult for citizens and administrators to understand, requiring more effective translation and mediation tools.

Looking ahead, the primary objective will be to complete the transfer of the project's action content (1–9) to a responsive and queryable digital interface (action 10), capable of modulating information based on the user's profile. This platform, open to multiple levels of access, will allow access to data and scenarios relating to environmental and structural risks, possible interventions on built and natural heritage, and implementation technologies in the two pilot municipalities of Bova and Palizzi. The system's evolution towards a user-centered and adaptive logic will be a decisive step towards ensuring the model's transferability to other contexts, contributing to the resilience and sustainable development of inner areas.

Acknowledgments. This study was carried out by Giuseppe Mangano, PhD and Researcher of ABITAlab Department of Architecture and Design, Mediterranean University of Reggio Calabria, within the Innovation Ecosystem Tech4You"Technologies for climate-change adaptation and quality-of-life improvement," Mission 4, Component 2, Investment 1.4 of Italy's National Recovery and Resilience Plan (PNRR), funded by the European Union—NextGenerationEU (project code ECS_00000009, CUP C33C22000290006). The authors gratefully acknowledge Actions 8 and 9 of Pilot Project 4.7.1 ("Open platform phygital space for advanced, dynamic co-design of built-environment interventions"), led by Prof. Consuelo Nava, which provided the coordination and organization of the research team, the research theorical framework and the direction of ABITAlab (www.abitalab.unirc.it).

References

ABITAlab Laboratory: Prototipi e dimostratori. ABITAlab - Università Mediterranea di Reggio Calabria. Disponibile online (2025). https://www.abitalab-unirc.com. Accessed 22 July 2025

Aa.Vv. Per un Atlante della Calabria. Territorio, insediamenti storici, architetture e beni culturali, Gangemi Editore, Reggio Calabria, Italy, 2011, pp. 85, 309

Bessembinder, J., Terrado, M., Hewitt, C., Garrett, N., Kotova, L., Buonocore, M., Groenland, R.: Need for a common typology of climate services. Climate Services. **16**, 100135 (2019). https://doi.org/10.1016/j.cliser.2019.100135

Calabria Region; GAL "Grecanica": Strategia Nazionale per le Aree Interne "Area Grecanica". Strategia d'Area (2020)

Caramizaru, A., Uihlein, A.: JRC- Energy Communities: an Overview of Energy and Social Innovation, EUR 30083 EN, Publications Office of the European Union, Luxembourg, ISBN 978–92–76-10713-2 (2020). 10.2760/180576, JRC119433

Clark, K.: Playing with the Past: exploring Values in Cultural Heritage Practice. Routledge, London (2019)

Florez Ayala, I., et al.: Urban living labs: a platform for addressing sustainability transitions in cities. Sustainability. **14**(3), 734 (2022)

Keyson, D., Mahmud, A., Romero, N.: Living lab and research on sustainability: practical approaches on sustainable interaction design. In: Marcus, A. (ed.) Design, User Experience, and Usability: User Experience in Novel Technological Environments. DUXU 2013 Lecture Notes in Computer Science, vol. 8014, pp. 229–234. Springer, Berlin (2013). https://doi.org/10.1007/978-3-319-03647-2_17

Leminen, S., Westerlund, M., Nyström, A.G.: Living labs as open-innovation networks. Technol. Innov. Manag. Rev. **2**, 6–11 (2012) https://doi.org/10.22215/timreview/602

Lucanto, D., Hanida, A.: The digital atlas: structure, information, interoperability of physical and digital models. In: Nava, C., Pisano, A., Giglio, F., Mangano, G. (eds.) Climatic and Structural Safety in Multihazard Regime of Cultural and Natural Heritage. Living Lab and Digital Platform User Profiling. Springer (2025)

Lupp, G., Zingraff-Hamed, A., Huang, J.J., Oen, A., Pauleit, S.: Living labs—a concept for co-designing nature-based solutions. Sustainability. **13**(1), 188 (2021)

Mangano, G.: Which living lab processes for the renewable energy communities? A pilot experimentation in NRRP-T4Y research. In: Sayigh, A., Trombadore, A., Calcagno, G. (eds.) Getting to Zero—beyond Energy Transition Towards Carbon-Neutral Mediterranean Cities. Innovative Renewable Energy. Springer, Cham (2025). https://doi.org/10.1007/978-3-031-82323-7_83

Mangano, G., Leuzzo, A.: Co-design e tecnologie abilitanti: 15 esperienze progettuali per la ricerca di frontiera e le comunità in transizione. Aracne, Rome (2022)

Nava, C.: Tecnologie emergenti per il progetto rigenerativo. Cinque questioni teoriche su innovazione e sostenibilità dalla prassi della ricerca di frontiera. Aracne, Pomezia (2019)

Nava, C., Mangano, G.: Living lab for the design and activation of energy communities in the inner areas. TECHNE—J. Technol. Architect. Environ. **26**, 120–130 (2023) https://doi.org/10.36253/techne-14495

Nava, C.: Tecnologie emergenti per il progetto rigenerativo. 5 questioni teoriche su innovazione e sostenibilità dalla prassi della Ricerca di frontiera. Aracne Editrice (2023). ISBN:979-12-218-0826-1

Nübel, K., Bühler, M.M., Jelinek, T.: Federated digital platforms: value chain integration for sustainable infrastructure planning and delivery. Sustainability. **13**, 8996 (2021). https://doi.org/10.3390/su13168996

Rosetti, I., Navarrete Hernandez, T.: Modelling cultural living labs: a process-based review. J. Innov. Manag. **13**(1), 218–242 (2025)

Schuurman, D., De Marez, L., Ballon, P.: The impact of living lab methodology on open innovation contributions and outcomes. Technol. Innov. Manag. Rev. **6**(1), 7–16 (2016)

Tiwari, A., Rodrigues, L.C., Lucy, F.E., Gharbia, S.: Building climate resilience in coastal city living labs using ecosystem-based adaptation: a systematic review. Sustainability. **14**(17), 10863 (2022)

Zanetti, M., Nollo, G., DeCecco, M.: Livinglab, an innovative approach in healthcare. IEEE Smart Cities Initiative—Trento, white paper (2019)

Photogrammetry-Drone-Based Tool for Structural Vulnerability Assessment in Historic Urban Centers

Giulia Percolla(✉), Mariaceleste Lasorella, Paolo Fuschi, and Aurora Angela Pisano

Department of Architecture and Design – dAeD, University Mediterranean of Reggio Calabria, Reggio Calabria, Italy
{giulia.percolla,mceleste.lasorella,paolo.fuschi, aurora.pisano}@unirc.it

Abstract. This study presents the practical validation of a non-invasive methodology for assessing the structural vulnerability of historic masonry buildings, proposed in the previous paper by the authors (see Part 1 of this volume) and based on photogrammetric data obtained using Unmanned Aerial Vehicles (UAVs). The method is tested on nine buildings in the village of Bova in Calabria, Italy. A drone survey is conducted to acquire high-resolution nadir and oblique images, which are then processed using Structure-from-Motion (SfM) techniques to generate textured 3D models, orthophotos, and dense points clouds. Visually detectable parameters, selected as significant for structural vulnerability assessment, are extracted directly from the photogrammetric outputs and used as input for a structured, Excel-based tool. Analytic Hierarchy Process (AHP) is used to weight the parameters and Excel Visual Basic for Application (Excel-VBA), enables users to select responses for each parameter via dropdown menus. Internal automation then converts these selections into numerical scores. A vulnerability index is computed for each individual building, which is classified into three vulnerability levels.

Keywords: Drone-Based 3D Modeling · Excel-VBA Decision Tool · Masonry Buildings Aggregates

1 Introduction

Historic villages in southern Italy, like Bova (Calabria), are characterized by masonry buildings aggregates that were constructed over the years without any official design documentation. These structures, which are frequently irregular in geometry and design, are especially vulnerable to structural deterioration as a result of age, seismic exposure, climate changes and the lack of systematic upkeep. The complexity of these environments is made worse by steep topography, small streets and architectural variation, all of which provide limitations for traditional structural evaluation methodologies (Giuffrida et al., 2021).

Analytical modelling and thorough on-site investigations, even necessary for a complete structural diagnosis, are often inapplicable for preliminary screening at a large

C. Nava et al. (Eds.): *Climatic and Structural Safety in Multi-Hazard Regime of Cultural and Natural Heritage*, GREEN, pp. 196–205, 2026.
https://doi.org/10.1007/978-3-032-11926-1_12

(urban) scale due to the difficulty of accessing historic locations (Sevieri et al., 2020). This situation highlights the need for rapid, non-invasive approaches that allow early-stage structural vulnerability mapping and prioritization methods (Pepe et al., 2022).

The current study uses and tests an interactive tool based on drone survey and visual detection on a digital twin (DT) model of the aggregate of a number of parameters selected as significant for structural vulnerability assessment of individual buildings. Analytic Hierarchy Process (AHP), as proposed by Harirchian et al. (2020), is used to handle the parameters within a structured Excel-VBA which incorporates automated grading and weighting algorithms based only on criteria viewable from drone photography, following the approach described by De Fino et al. (2023).

A sample of nine buildings in the historic center of Bova was examined. The methodology discussed here focuses on the interactive application of the aforementioned tool in real-world surveys. Data was collected using both nadir and oblique drone flights, which were adapted to improve visibility of needed structural characteristics, including façades and vertical discontinuities, as in Sammartano et al. (2023). The study's main goal is to evaluate the performance of the promoted tool under real field situations also demonstrating versatility and repeatability for structural vulnerability assessment in historical contexts. The findings also help to establish standard survey processes and decision-support strategies aimed at reducing seismic risk in vulnerable urban areas.

2 Survey Methodology and 3D modeling

2.1 UAV Photogrammetric Dataset

The first stage of the whole process was the acquisition of aerial data through a UAV-based photogrammetric survey of the entire historic center of Bova in Calabria. This survey employed a high-performance drone able to obtain high-resolution photos appropriate for structural analysis and morphological interpretation of the built environment. To this aim, flight paths were designed to provide extensive coverage of the urban setting while also reducing perspective distortion and obstructions. The flying approach emphasized low-altitude paths to provide meticulous examinations of façade components, roof configurations and typological anomalies—factors essential for structural vulnerability assessment.

Two complementary datasets were obtained:

- Nadir photos, oriented vertically with respect to the ground plane. These photographs are generally used for the creation of orthophotos and digital elevation models, in addition they allow the comprehensive planimetric reconstruction of the village.
- Oblique photos are taken with the camera's ~~a~~ focal plane positioned at approximately a 45° angle to the ground plane. These 45° flights were repeated, while keeping the same flight plan, along the cardinal direction and at different times of the day to avoid shadows and unfavorable light conditions. Such data resulted in the essential capture of vertical architectural elements, including façades, out-of-plane masonry walls deformations, fracture patterns and discontinuities in height or position of the buildings.

To ensure the maximum photogrammetric quality, all photos were taken with a frontal overlap of at least 80% and a lateral overlap of 70%, following architectural photogrammetry best practices.

This survey provided a large number of high-quality photos, which are required for the creation of a thorough 3D digital model as well as structural and morphological assessments of the urban structure. Figure 1 illustrates the orthophoto of the historic center of Bova, overlaid with the UAV flight grid. The drone followed two flight paths, one highlighted in blue and the other in green, to obtain a complete view of the area, as shown by the grid.

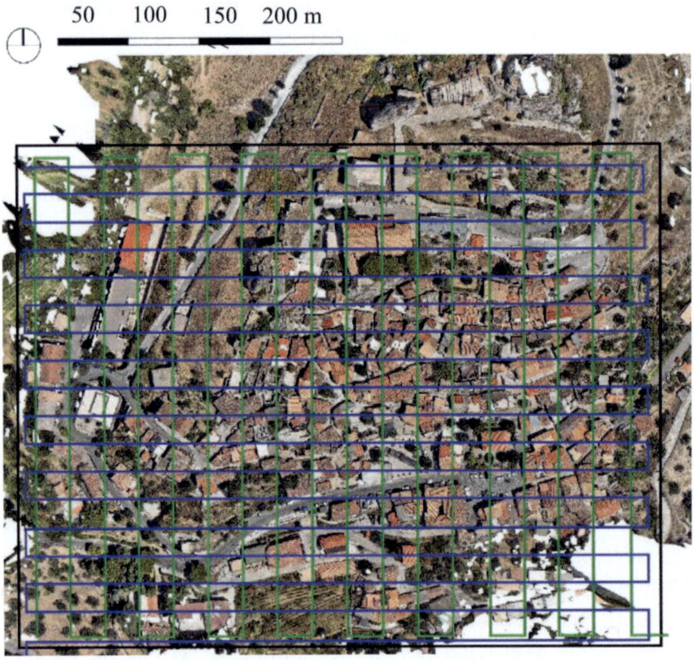

Fig. 1 . Orthophoto of the historic center of Bova with UAV flight grid overlay showing the planned paths along which both nadiral and oblique images were acquired.

2.2 3D Model Generation via Structure-from-Motion

Image data acquired from the UAV survey were processed using the Structure from-Motion (SfM) photogrammetry technique. The workflow included image alignment, generation of sparse and dense point clouds, mesh reconstruction and high-resolution texture mapping. Figure 2 shows a comparison of the dense point cloud and the final textured 3D model used.

Fig. 2. Comparison between dense point cloud generated by SfM (top image) and 3D mesh model with high-resolution texture mapping (bottom image) of Bova's historic center.

All operations were conducted in Agisoft Metashape standard version 2.2 (2025), a widely adopted platform for architectural and heritage documentation. The output dataset contains a detailed 3D model of the historic area with textures, true-scale orthophotos for flat measurements and a georeferenced points cloud for spatial referencing and additional measurements. These outputs give a complete 3D visual view of Bova's historic center which provides the geometric and visual foundation for all future steps, allowing structural and typological information.

3 Data processing and extraction

3.1 Visually detectable structural features

As said in the above quoted previous paper by the authors, the structural vulnerability assessment is grounded on the evaluation of 23 parameters, or *sub-criteria* (in the AHP logic) organized in 4 macro-categories, or *criteria.*

The sub-criteria are based on official forms provided by the Italian Civil Protection System. Precisely, reference has been made to the forms: GNDT I and II (GNDT et al., 1986), AeDES (Baggio et al., 2007) and CARTIS (Zuccaro et al., 2023) and the borrowed (here adopted) parameters were chosen for their compatibility with visual non-invasive drone photogrammetry evaluations.

Table 1 reports the criteria, the sub-criteria and the weight of each one within the process of structural vulnerability assessment.

Table 1. Criteria, sub-criteria and AHP weights used for structural vulnerability assessment of the individual building. The 23 sub-criteria are divided into four macro-categories, or criteria, and are derived from official Italian Civil Protection forms. Weights were assigned according to the AHP method to represent the relative influence of each parameter on the building vulnerability index.

Criteria	Sub-Criteria	Weight (AHP)
General Characteristics	Typological position	0,98%
	Site morphology	0,84%
	Number of total floors	1,89%
	Average building height	2,08%
	Regularity in plan	1,30%
	Regularly in elevation	1,50%
	Type of aggregate	0,77%
	Type of roof	0,64%
Structural Characteristics	Type of masonry	13,70%
	Tie rods	8,55%
	Structural bow windows	3,54%
	Asymmetrical perforations	2,20%
External Envelope Feature	Surface damage	1,04%
	Irregular vertical extension	6,40%
	Presence of dangerous elements	4,01%
	Different materials at different levels	2,40%
	Presence of consolidation	2,14%
Structural Weaknesses	Visible crack	2,35%
	Type of crack	7,04%

(*continued*)

Table 1. (*continued*)

Criteria	Sub-Criteria	Weight (AHP)
	Crack orientation	4,78%
	Out-of-plane walls deformation	16,65%
	Damage at the connection between buildings	10,58%
	Roof damage	4,60%

3.2 Extraction of the parameters for structural building vulnerability assessment

Each parameter was extracted from the 3D outputs produced by UAV-based photogrammetric reconstruction. Visual interpretation was carried out using textured meshes, orthophotos and oblique images, avoiding the necessity for access to interior areas or design documents. Each building was evaluated using a structured Excel interface that included dropdown menus for parameters selection and VBA functions for automated scoring. Parameters were evaluated on a scale of 1 to 5 according to their perceived criticality. Their contributions to the building vulnerability index were weighted using the AHP model.

It is worth noting that, while orthophotos were largely used to establish a building's planimetric shape, geometric consistency and roof conditions, oblique images and façade mesh views facilitated the identification of cracks, deformations, openings, material discontinuities, and also an estimation of average building height. Particular attention was paid to texture quality and lighting conditions, which—especially in shaded areas—way occasionally limit the accurate recognition of masonry type or minor façade damage.

The tools is based on an Excel worksheet, with one row for each single building and as many columns as the number of the pertinent sub-criteria. Each metric is associated with a dropdown menu offering a limited set of predefined qualitative responses (typically between two and five), each accompanied by a brief descriptive label. For example, for the parameter "crack orientation" (within the criterion "structural weaknesses"), the available options include:

- Not present
- Vertical
- Horizontal
- Diagonal

This logic is demonstrated in Fig. 3, which shows a sample section of the interface in which four buildings are evaluated based on crack visibility and orientation. The drop-down menu makes it simpler to provide answers, resulting in a rapid and uniform compilation that is also appropriate for inexperienced operators.

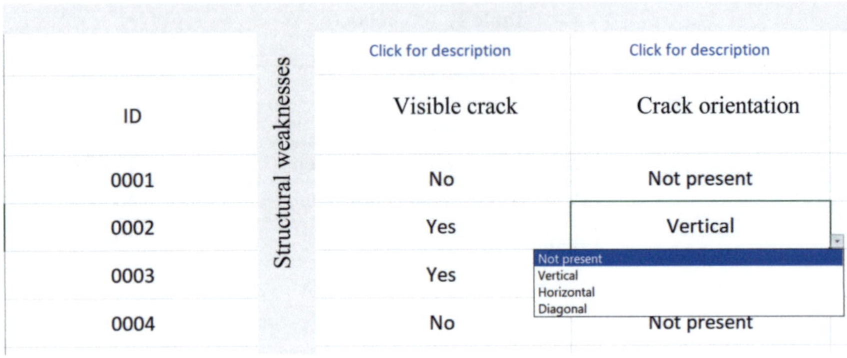

Fig. 3. Example of an Excel interface for providing parameters using a drop-down menu.

Each textual response is internally linked to a fixed numerical score, ensuring consistency across evaluations.

The Excel-based tool is fully automated through VBA functions, which convert qualitative inputs into quantitative scores and apply pre-assigned AHP weights. Once all values are entered for a given building, the tool computes the overall structural vulnerability index as a weighted sum of the individual scores. Eventually, the tool assigns the building to one of three *vulnerability classes* using fixed threshold intervals, namely:

- Low: 1.00–2.33
- Medium: 2.34–3.66
- High: 3.67–5.00

This fully visual and automated process enables rapid, consistent assessments of building structural vulnerability, while maintaining compliance with the underlying AHP model. Final scores can be exported or integrated into external platforms for mapping at urban-scale prioritizations so resulting a very useful decision support.

4 Results on the Pilot Test and Discussion

To evaluate the operational effectiveness of the proposed methodology, a pilot test was conducted on nine buildings located in the historic center of Bova (Calabria). The building structural vulnerability assessment was independently performed by three trained operators using the Excel-based evaluation tool above described. Each operator extracted visual parameters from the photogrammetric model and classified them through the guided dropdown interface, generating a final vulnerability index for each building.

Table 2 shows the structural vulnerability scores assigned by each operator at each of the nine examined buildings as well as the resulting classes.

Table 2. Structural vulnerability index and class assigned by three independent operators to the 9 benchmark buildings.

ID BUILDING	Operator 1	Operator 2	Operator 3
0001	2,30	2,83	3,17
0002	2,24	0,00	3,23
0003	2,13	2,05	2,43
0004	1,86	2,22	2,55
0005	2,50	3,20	2,97
0006	1,52	1,94	2,20
0007	3,51	2,79	2,90
0008	2,68	2,47	2,78
0009	1,99	2,15	2,04
vulnerability classes		Low: 1.00–2.33	
		Medium: 2.34–3.66	
		High: 3.67–5.00	

The results showed a consistent distribution of scores across the three evaluations, with all buildings falling within the low- to medium-vulnerability classes. Particularly, none of the structures were classified as highly vulnerable by any operator, in agreement with field observations confirming the general structural integrity of the samples. Minor variations were observed in the scores—particularly between the first and third operators—but the resulting indices remained within comparable ranges. These differences likely reflect varying interpretation thresholds, with some operators applying more conservative evaluation criteria. Nonetheless, the consistency of the final classifications confirms the method's robustness and limited sensitivity to subjective variation in judgement.

In one case (Building ID: 0002), an operator encountered difficulties in identifying specific façade features due to limited image quality and lighting conditions. This highlights a key challenge for future applications: improving image coverage, particularly in narrow alleyways and shaded areas where drone access or visibility may be limited. To address these limitations, UAV surveys might be scheduled at different times of day to mitigate lighting-related issues or supplementary ground-level image acquisition, using low-cost devices such as GoPro cameras, might be performed.

5 Concluding Remarks

This study showed the first practical use of a method for a rapid structural vulnerability assessment of historic masonry buildings based only on drone survey. The method was tested on a sample of nine buildings in the historic center of Bova (Calabria), extracting visually detectable parameters from textured 3D models, orthophotos and oblique facade views of the entire village. The tool, set up in a structured Excel format with automatic scoring and AHP weighting, showed outstanding reliability in sorting buildings into vulnerability classes using a completely non-invasive, image-based method.

The pilot test, conducted by multiple operators, yielded consistent results, confirming the method's robustness and adaptability under realistic survey conditions. No structures were classified as highly vulnerable, a result aligned with direct visual assessments and in-situ field knowledge. Minor discrepancies between operators' judgements brought attention to the need for high image quality and completeness, particularly in constrained urban environments.

Future development efforts will focus on enhancing dataset accuracy through optimized UAV flight planning, multi-time-of-day image acquisition, and the incorporation of ground-level photogrammetry using lightweight, low-cost equipment, such as GoPro cameras to be used in the areas inaccessible by drone.

Overall, the proposed methodology offers a technically sound and user-friendly solution for structural vulnerability screening in historic urban areas. Its future use in GIS environments and digital twin platforms shows great potentialities for practical application also being an effective decision-support tool for risk reduction strategies.

References

Agisoft Metashape standard edition version 2.2.1, (2025). https://www.agisoft.com/.

Baggio, C., et al.: (2007). Field manual for post-earthquake damage and safety assessment and short term countermeasures (AeDES). European Commission—Joint Research Centre—Institute for the Protection and Security of the Citizen, EUR, 22868.

De Fino, M., Galantucci, R.A., Fatiguso, F.: Condition assessment of heritage buildings via photogrammetry: a scoping review from the perspective of decision makers. Heritage. **6**(11), 7031–7066 (2023). https://doi.org/10.3390/heritage6110367

Giuffrida, D., et al.: A multi-analytical study for the enhancement and accessibility of archaeological heritage: the Churches of San Nicola and San Basilio in Motta Sant'Agata (RC, Italy). Remote Sens. **13**(18), 3738 (2021). https://doi.org/10.3390/rs13183738

GNDT, Emilia-Romagna Region, Tuscany Region: Field Manual for the completion of the building vulnerability GNDT form. *Litografia della Giunta Regionale (in Italian)* (1986)

Harirchian, E., Jadhav, K., Mohammad, K., Aghakouchaki Hosseini, S.E., Lahmer, T.: A comparative study of MCDM methods integrated with rapid visual seismic vulnerability assessment of existing RC structures. Appl. Sci. **10**(18), 6411 (2020). https://doi.org/10.3390/app10186411

Pepe, M., Alfio, V.S., Costantino, D.: UAV platforms and the SfM-MVS approach in the 3D surveys and modelling: a review in the cultural heritage field. Appl. Sci. **12**(24), 12886 (2022). https://doi.org/10.3390/app122412886

Sammartano, G., Avena, M., Fillia, E., Spanò, A.: Integrated HBIM-GIS models for multi-scale seismic vulnerability assessment of historical buildings. Remote Sens. **15**(3), 833 (2023). https://doi.org/10.3390/rs15030833

Sevieri, G., Galasso, C., D'Ayala, D., De Jesus, R., Oreta, A., Grio, M.E.D.A., Ibabao, R.: A multi-hazard risk prioritisation framework for cultural heritage assets. Nat. Hazards Earth Syst. Sci. **20**(5), 1391–1414 (2020). https://doi.org/10.5194/nhess-20-1391-2020

Zuccaro, G., Dolce, M., Perelli, F.L., De Gregorio, D., Speranza, E.: CARTIS: a method for the typological-structural characterization of Italian ordinary buildings in urban areas. Front. Built Environ. **9**, 1129176 (2023)

Assessment of Local Collapse in Masonry Walls through Visual Programming: A Full-Scale Benchmark in the Historical Center of Bova, Italy

Mariaceleste Lasorella(✉), Giulia Percolla, Aurora Angela Pisano, and Paolo Fuschi

Department of Architecture and Design – dAeD, University Mediterranean of Reggio Calabria, Reggio Calabria, Italy
{mceleste.lasorella,giulia.percolla,aurora.pisano, paolo.fuschi}@unirc.it

Abstract. This paper addresses the challenge of assessing and preserving the structural safety of traditional masonry constructions by testing a simplified numerical method for masonry walls collapse prediction, proposed by the authors in the previous paper in Part. 1 of this volume. The method, developed in a visual scripting environment, incorporates linear programming optimization within a CAD-integrated platform, Grasshopper with Python, enabling the modeling of geometry, load definition and collapse load multiplier calculation. Through a real masonry building located in Bova, the proposed tool is evaluated for its practicality and effectiveness in facilitating fast and intuitive structural assessments, particularly in post-earthquake scenarios or severe damaged urban contexts. The results demonstrate the efficiency of the proposed approach to support rapid decision-making while ensuring the reliability of safety evaluations in masonry structures.

Keywords: masonry wall collapse · limit load prediction · visual programming · full-scale benchmark

1 Introduction

The preservation of Italian cultural heritage, particularly its masonry constructions, involves different factors that affect structural integrity. Material properties, construction methods and external forces, like seismic activity and environmental conditions, all contribute to the durability of masonry buildings. Understanding the interaction of these factors is essential for assessing vulnerability and developing strategies to preserve these historic structures (Giuffrè, 1996).

In recent years, among many structural analysis approaches of the relevant literature, limit analysis has gained popularity in professional practice through its integration into visual scripting environments, enabling efficient safety assessments of masonry

C. Nava et al. (Eds.): *Climatic and Structural Safety in Multi-Hazard Regime of Cultural and Natural Heritage*, GREEN, pp. 206–218, 2026.
https://doi.org/10.1007/978-3-032-11926-1_13

structures. Early work (Block, 2009) introduced Thrust Network Analysis in CAD, later refined by Rippmann et al. (2012) for free-form vault design. In Funari et al. (2020) and Funari et al. (2021) a Grasshopper tool in Rhinoceros is developed for real-time collapse mechanism exploration. In Mousavian et al. (2021) a Grasshopper plugin for out-of-plane collapse in multi-story walls is proposed. A recent and interesting study on this topic is also proposed by Massafra et al. (2025), where a reverse engineering-based method is introduced for analyzing displacements and deformations in historical masonry buildings using terrestrial laser scanning and visual programming. These advancements highlighted the effectiveness of limit analysis in visual scripting environments for both design and heritage conservation.

This work, coupling limit analysis with CAD tools, using Grasshopper for real-time safety evaluations of masonry walls, presents a validation of a visual programming tool for limit analysis of masonry walls proposed in a previous paper by the authors in Part 1 of this volume. The promoted tool combines limit analysis with linear programming to identify collapse mechanisms and load multipliers. The parametric inputs, carried on within a visual scripting environment, allow users to quickly assess how changes in material, geometry and loading affect structural behavior, so testing numerically restoration or emergency strategies.

A full-scale benchmark is analyzed, precisely a masonry structure building located in the small village of Bova in Calabria. The surveyed structure, modeled in a CAD environment, is analyzed to assess the potential collapse mechanisms and the structural vulnerability. The obtained results seem to validate the practical applicability of the tested visual programming tool.

2 Visual Programming for Limit Analysis

2.1 Main Assumptions

By hypothesis, the crack pattern of the main walls constituting the examined building is not known a priori. Such circumstance is very common when facing real problems in which the geometry and paths of the cracks are not clear or they are not conducible to a fixed categorized collapse mechanism. This happens also after a detailed survey such as the one carried out in the present study where the benchmark building was surveyed using a drone and a laser scanner.

In such cases the promoted visual scripting results very effective as, in fact, it allows to analyze very quickly different possible collapse mechanisms.

In particular, by adopting a structural model based on rigid blocks and following Heyman's assumption, two collapse scenarios are considered for a masonry wall macro-element with fixed hinge A and roller B (Fig. 1(a,b)), reflecting crack patterns observed in situ.

As explained in the above quoted paper by the authors, in the first scenario (Fig. 1(a)), the macro-block rotates about hinge A and the collapse load multiplier, say α, can be obtained directly from the virtual-work balance equation, as demonstrated for the simple overturning mechanism of a masonry macro-element (Cifani, et al., 2006). In the second scenario (Fig. 1(b)), an internal hinge C appears at an undetermined height, dividing the macro-element into two sub-elements. Because the location of hinge C is unknown, the

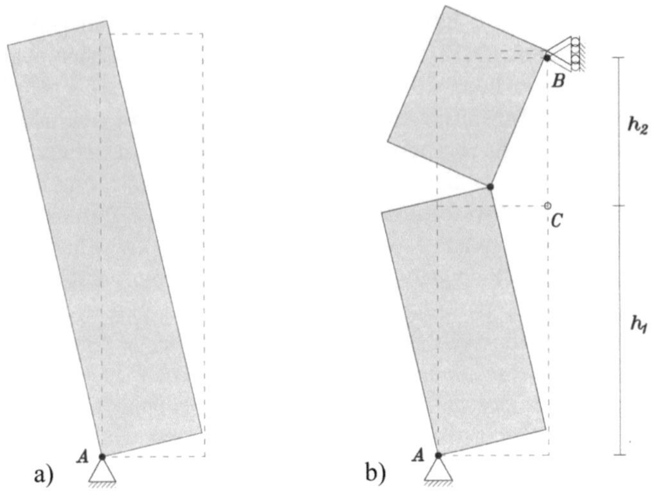

Fig. 1. Schematic representation of a masonry macro-element collapse mechanism: (a) a single macro-element rotating around the external hinge A; (b) two interconnected macro-elements rotating around the internal hinge C, whose position is unknown

virtual-work balance equation alone is not sufficient and must be supplemented by an optimization algorithm to identify the most critical collapse configuration. The resolutive optimization problem seeks the minimum collapse multiplier α under a constraint equation guaranteeing that the internal hinge lies within the block rather than at its extremity. The resulting linear-programming formulation is solved in Python using the linprog function.

2.2 Visual Programming Framework

The limit analysis optimization is implemented within Grasshopper, the visual scripting environment of Rhinoceros, enabling direct import of CAD geometries for masonry walls. This integration delivers an interactive workflow in which the kinematic limit analysis automatically identifies the most probable collapse mechanism and computes the corresponding load multiplier in real time. Figure 2 outlines the process illustrated using one possible case with the corresponding collapse mechanism identified as the vertical bending of one main wall.

The process can be divided into three main sections and begins by defining the geometry of the macro-elements by importing the CAD model into Grasshopper (left sketch in Fig. 2). The user selects the relevant wall cross-section and specifies boundary conditions, applied loads, and material properties. In the second section (central sketch of Fig. 2), the script automatically detects the wall cross-section and extracts all necessary geometric information, including centroid locations, self-weight distributions, and dimensions. Then, the Python script joins all inputs and executes the limit-analysis optimization. In the third final section (right sketch of Fig. 2), the tool outputs the collapse-load multiplier and the related collapse mechanism. It is worth noting that, users can adjust inputs

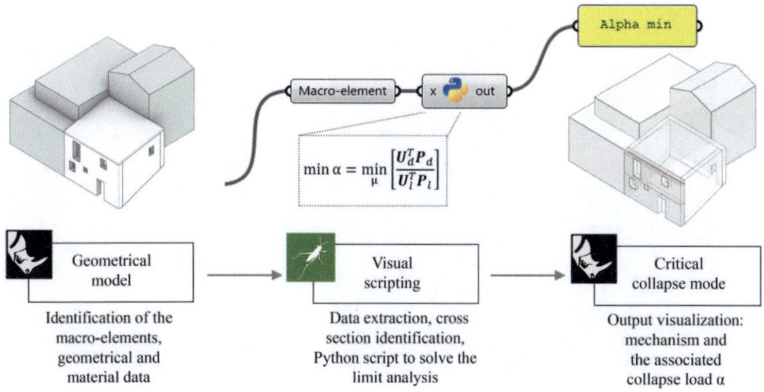

Fig. 2. Workflow of the visual scripting approach for performing limit analysis on a double-story masonry façade with a vertical bending mechanism

interactively and observe real-time updates, enabling rapid assessment of how changes in geometry, material properties, or load conditions influence collapse risk.

The mechanisms considered in the implementation of the visual scripting tool include the most common out-of-plane collapse mechanisms for masonry walls. Such mechanisms arise when external actions, most notably seismic forces, overcome the equilibrium of an entire wall or of a large wall segment. Their activation depends on factors governing out-of-plane response, including the quality of orthogonal wall connections, wall-to-wall interactions, roof configuration, opening layout, loads from upper stories, and even the position of the building within the urban fabric.

These mechanisms are here divided into *overturning mechanisms* and *bending mechanisms*. Overturning mechanisms may be *simple*, involving rotation of a single wall segment, or *composed*, in which adjacent orthogonal wall portions are also dragged into motion; each can occur at one story or span multiple stories. Bending mechanisms are categorized by the orientation of the bending plane, *vertical* or *horizontal* and likewise may affect a single level or multiple levels of the structure.

The study by Lasorella et al. (2025) schematically illustrates the eight out-of-plane collapse mechanisms, based on the classification of Milano et al. (2009): simple overturning of a single-story wall; simple overturning of a multi-story wall; composed overturning of a single-story wall; composed overturning of a multi-story wall; composed overturning of a corner wall; vertical bending of a single-story wall; vertical bending of a multi-story wall; horizontal bending of a single-story wall.

3 The Benchmark Building of Bova

The analyzed building is located in Bova a small village belonging to the so-called Grecanic area of Calabria at 820 m above sea level, (Fig. 3(a)). The benchmark building is the one evidenced in Fig. 3(b), precisely, a masonry structure located in one of the most critical areas of Bova. Following an initial inspection and degradation analysis, the building has been identified as deteriorated; it is a two-stories masonry structure that has been abandoned and in a state of evident decay.

a) b)

Fig. 3. a) municipality of Bova, with the red dot indicating the location of the case study building; b) a closer view of the case study is provided, highlighting the building's structural surrounding context.

The survey was conducted using a drone and a laser scanner, which generated a point cloud from which the 3D model was derived in a CAD environment. The macro-elements potentially involved in local collapse mechanisms were then identified looking at few visible cracks in the main walls. Regarding the material and geometric parameters; the assumed masonry specific weight is 18 kN/m^3; the wall thickness is 0.5 m. These values were estimated due to limited internal access, at only certain points, such as windows and doors, being there measurable. The building has a façade height of 5 m and a length of 7.5 m, with a lateral wall length of 6.1 m.

As shown in the following analyses, two macro-elements were identified involving the main façade and the sidewall, as they are the most exposed to possible collapse. For each macro-element, the geometric parameters and self-weights W were automatically extracted from the CAD input. The weight P_s from the roof and the second-floor slab are considered. The weights were estimated taking into account of a wooden floor with a thickness of 0.30 m, supported by the front wall. Ten collapse mechanisms have been considered. They are described hereafter with the aid of Figs. 4–13.

Figure 4 shows the *simple overturning* case involving the *front wall of the upper floor*. In this scenario, the acting loads include the self-weight of the wall, its horizontal component proportional to the collapse multiplier, and the weight of the single-pitched roof. The collapse multiplier in this case has a value of 0.16. While, Fig. 5 shows the *simple overturning* mechanism involving the *entire front façade*. In this case, the self-weight of the lower floor wall and the interfloor slab are also considered. The collapse multiplier for this scenario is 0.06.

In Fig. 6 and Fig. 7, the *composed overturning* mechanisms for the *upper floor and* the *entire façade* are analyzed. The geometry of the involved sidewall portion is assumed. The obtained collapse multipliers are 0.17 and 0.15, respectively, higher than those of the corresponding simple overturning cases.

In Fig. 8, the *overturning* mechanism *at the corner* between the walls converging from the front façade and the sidewall is analyzed, with a collapse multiplier value of 0.22. Finally, the *vertical bending* mechanism involving the *main façade* was examined

(Fig. 9). The hinge formed at the upper floor, and in this case, the collapse multiplier is 0.24.

Moreover, other analyzed mechanisms are those involving the *sidewall of the building*. As before, in Fig. 10, the *simple overturning* case is analyzed with a collapse multiplier of 0.10, and the *composed overturning* case (Fig. 11) with a multiplier of 0.12. The *corner overturning* in Fig. 12 between the sidewall and the facade wall is then examined, yielding a collapse multiplier of 0.36. The final analyzed case in Fig. 13 is the *vertical bending* that furnishes a collapse multiplier of 0.37, with the hinge forming near the top of the wall.

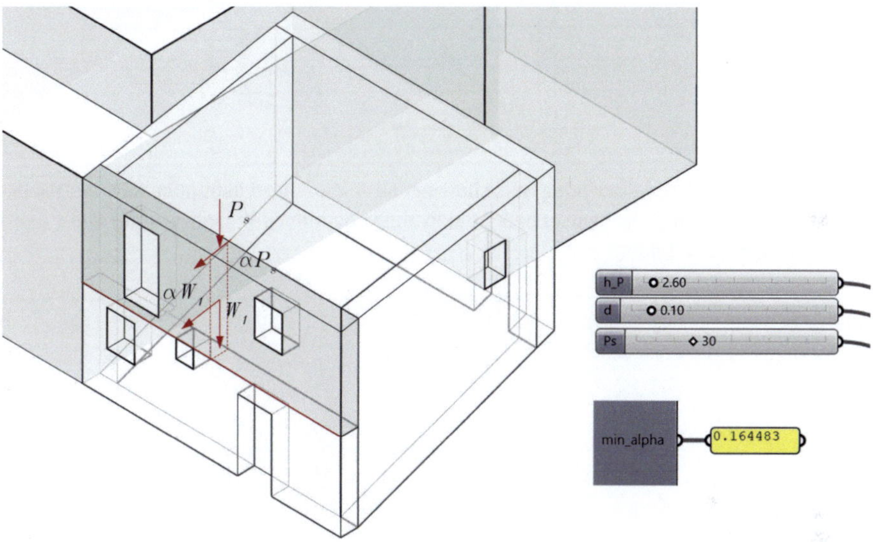

Fig. 4. Output of the visual scripting tool, showing the collapse load multiplier value for the single-story simple overturning mechanism under the loading conditions: W (self-weight) and P_S (load exerted by the roof)

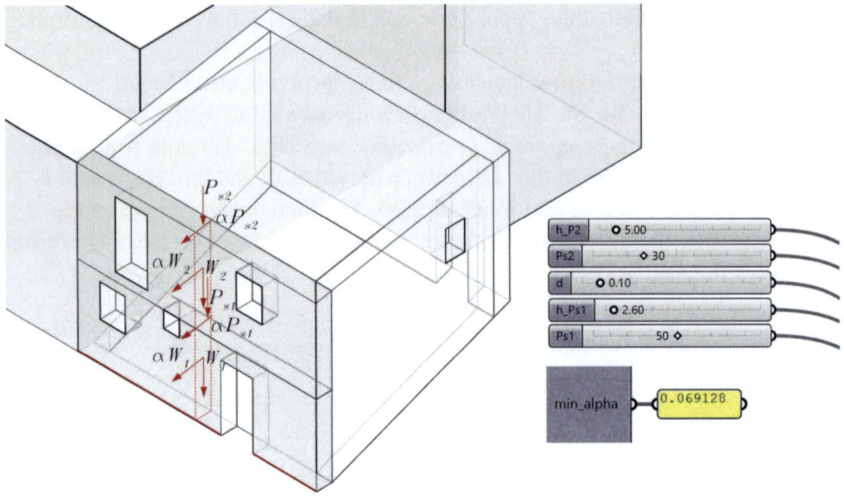

Fig. 5. Output of the visual scripting tool, showing the collapse load multiplier value for the two-story simple overturning mechanism under the loading conditions: W (self-weight) and P_s (load exerted by the roof and floor)

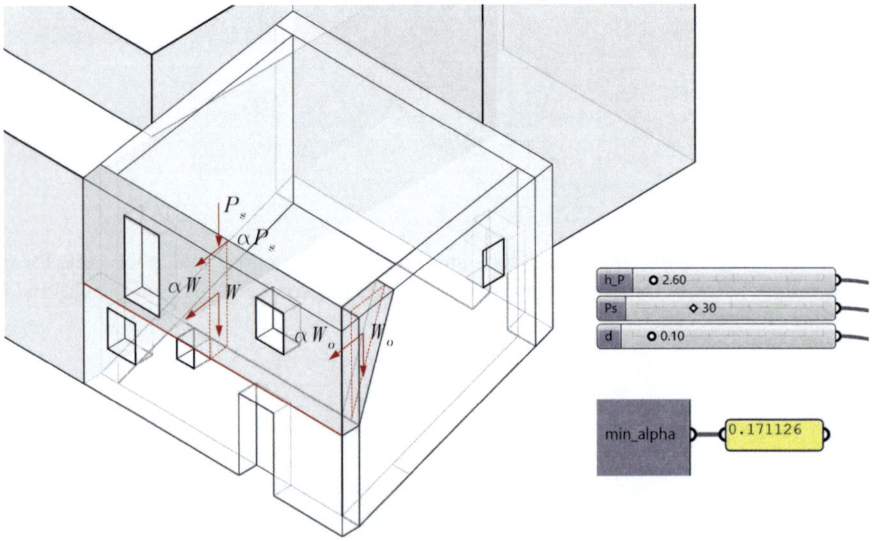

Fig. 6. . Output of the visual scripting tool, showing the collapse load multiplier value for the single-story composed overturning mechanism under the loading conditions: W (self-weight) and P_s (load exerted by the roof)

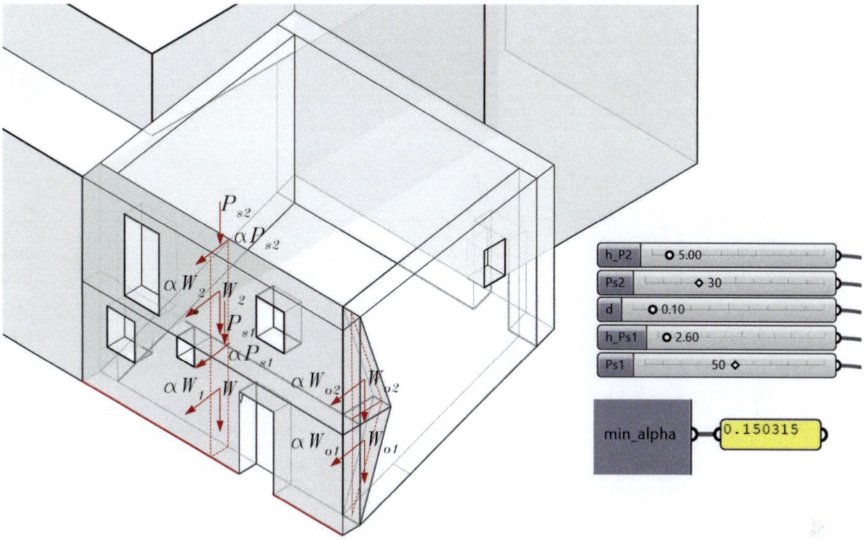

Fig. 7. Output of the visual scripting tool, showing the collapse load multiplier value for the two-story composed overturning mechanism under the loading conditions: W (self-weight) and P_s (load exerted by the roof and floor)

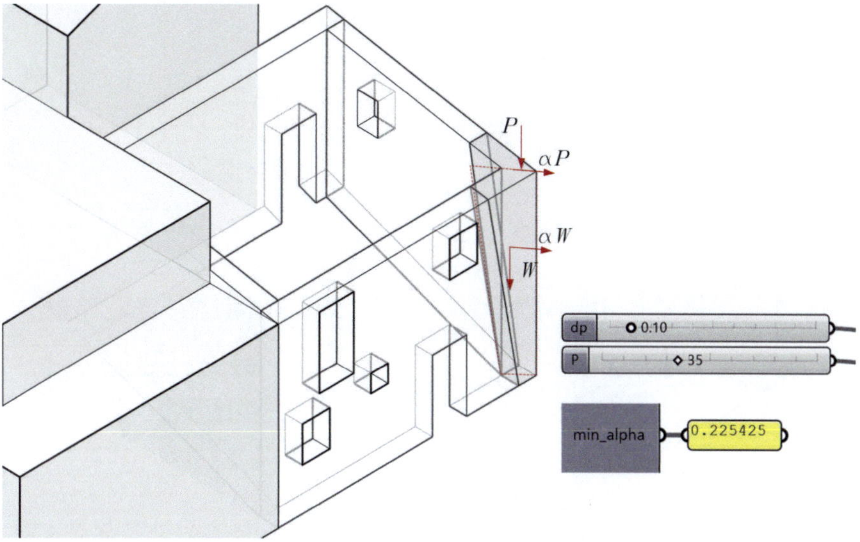

Fig. 8. Output of the visual scripting tool, showing the collapse load multiplier value for the corner overturning mechanism under the loading conditions: W (self-weight) and P_s (load exerted by the roof)

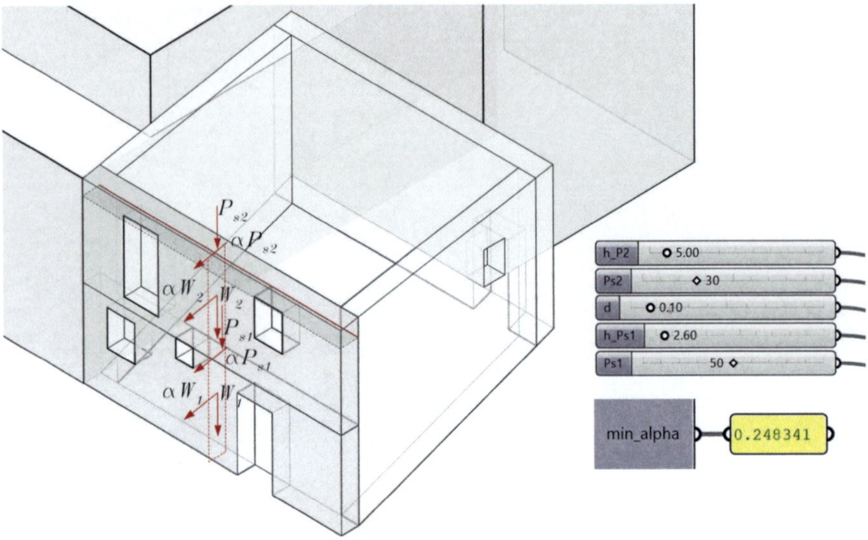

Fig. 9. Output of the visual scripting tool, showing the collapse load multiplier value for the vertical bending mechanism under the loading conditions: *W* (self-weight) and P_S (load exerted by the roof)

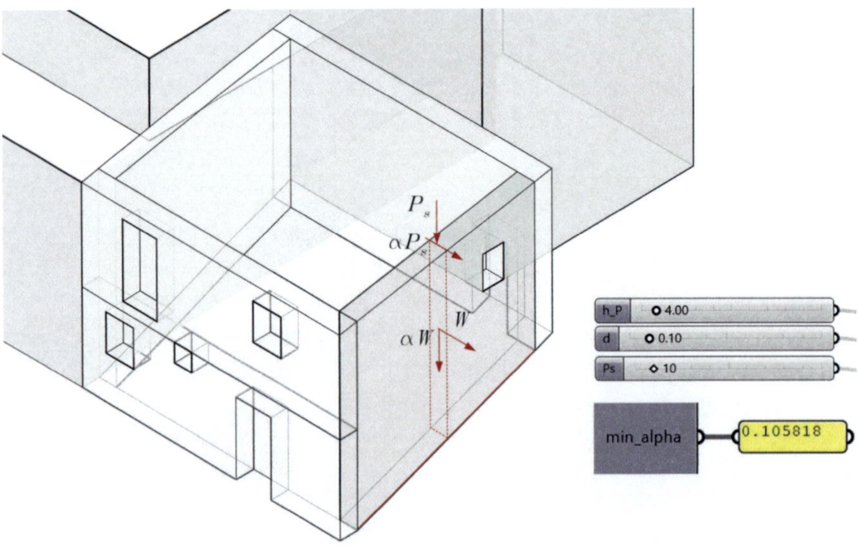

Fig. 10. Output of the visual scripting tool, showing the collapse load multiplier value for the single-story simple overturning mechanism under the loading conditions: *W* (self-weight) and P_S (load exerted by the roof)

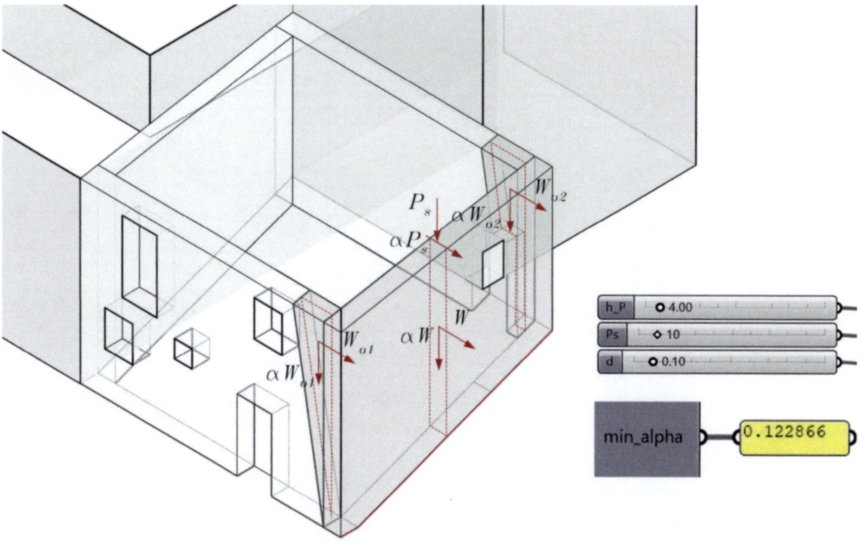

Fig. 11. Output of the visual scripting tool, showing the collapse load multiplier value for the single-story composed overturning mechanism under the loading conditions: *W* (self-weight) and P_S (load exerted by the roof)

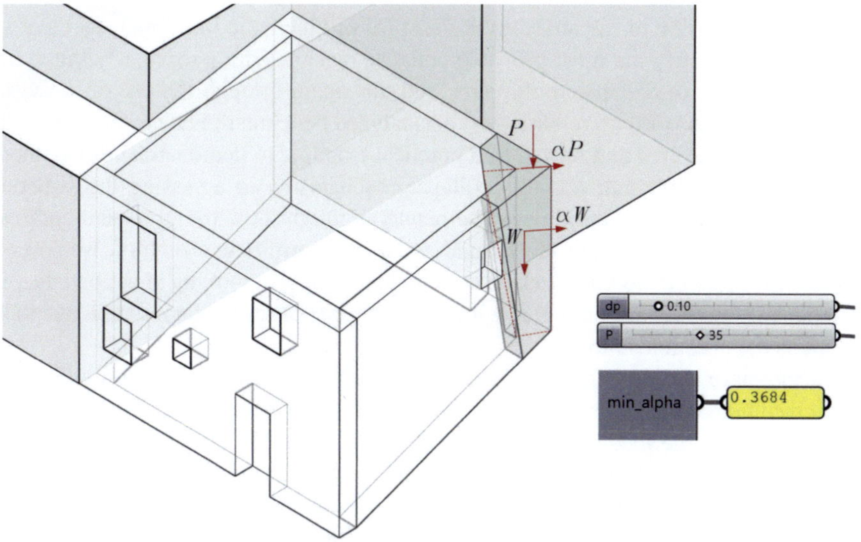

Fig. 12. Output of the visual scripting tool, showing the collapse load multiplier value for the corner overturning mechanism under the loading conditions: *W* (self-weight) and P_S (load exerted by the roof)

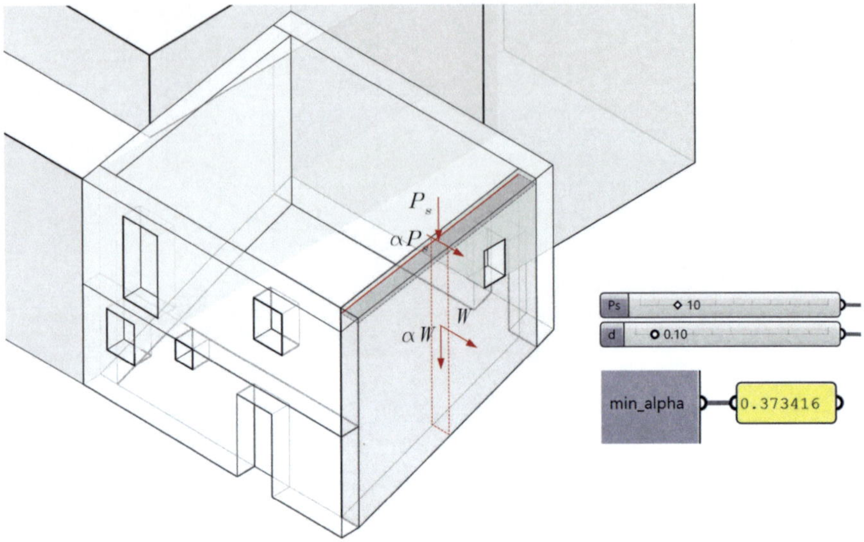

Fig. 13. Output of the visual scripting tool, showing the collapse load multiplier value for the vertical bending mechanism under the loading conditions: *W* (self-weight) and P_S (load exerted by the roof)

At the conclusion of the analyses in the spirit of kinematic limit analysis theory, it is possible to identify the most probable collapse mechanisms, particularly those with the lowest values of collapse multipliers, and this again through the use of the visual scripting tool. It is worth to remark that the analyzed benchmark is a masonry structure in a seismic-prone area and served as a practical example to demonstrate the method's effectiveness in evaluating possible collapse mechanisms so assessing the structural vulnerability of masonry buildings. The results highlight how the proposed approach can accurately predict failure modes and assist in determining structural weaknesses that require attention, not only to face seismic loads but also looking at actions due, for example, to climate change coupled with an actual state of severe degradation due to the lack of any restoration action.

Furthermore, the study confirms the effectiveness of the approach in providing reliable and computationally efficient evaluations of collapse mechanisms. Beyond the specific benchmark, the methodology shows potential for application to a wide range of masonry case studies, as it can be adapted to different geometries, bound-ary conditions and degradation scenarios. In addition, by exploiting the flexibility of the visual scripting environment, the approach opens the way to multidisciplinary applications, allowing different analyses to be performed and combined within the same framework. This general applicability reinforces its role as a versatile tool for both academic investigations and professional practice.

4 Concluding Remarks

A visual scripting tool for kinematic limit analysis of masonry walls subjected to out-of-plane loads has been tested on a real masonry building assumed as benchmark. The walls have been modeled using a rigid-block-based approach, with limit analysis applied in its standard kinematic form to predict collapse mechanisms and related load multipliers through an optimization process.

This interactive tool allows quick identification of key factors, such as geometry, material properties and applied load intensity and direction that influence the activation or prevention of collapse mechanisms in cracked walls. The tool also enables real-time evaluation of how adjustments, corresponding to structural restoration interventions, can improve safety, especially in post-seismic scenarios and/or within degraded areas.

The obtained results on a real-scale masonry building seem to validate the effectiveness of the promoted approach to be viewed as a predictive tool oriented to structural safety assessment.

An ongoing research is focusing on more complex masonry structural systems, also investigating on sliding rigid-block mechanisms to refine the modeling to better reflect real masonry behavior.

References

Baggio, C., Trovalusci, P.: Collapse behaviour of three-dimensional brick-block systems using non-linear programming. Struct. Eng. Mech. **10**(2), 181 (2000)

Block, P.P.: Thrust Network Analysis: Exploring Three-Dimensional Equilibrium. Massachusetts Institute of Technology: Doctoral Dissertation (2009)

Cifani, G., Petrucci, G., Lazzaro, D., D'Alessandro, S., Cialone, G., Martinelli, A., et al.: Repertorio dei meccanismi di danno, delle tecniche di intervento e dei relativi costi negli edifici in muratura. ITC-CNR (2006)

D'altri, A.M., Sarhosis, V., Milani, G., Rots, J., Cattari, S., Lagomarsino, S., … De Miranda, S.: Modeling strategies for the computational analysis of unreinforced masonry structures: review and classification.. Arch. Comput. Meth. Eng. (2020)

Ferris, M.C., Tin-Loi, F.: Limit analysis of frictional block assemblies as a mathematical program with complementarity constraints. Int. J. Mech. Sci. **43**(1), 209–224 (2001)

Funari, M.F., Mehrotra, A., Lourenço, P.B.: A tool for the rapid seismic assessment of historic masonry structures based on limit analysis optimisation and rocking dynamics. Appl. Sci. **11**(3), 942 (2021)

Funari, M.F., Spadea, S., Lonetti, P., Fabbrocino, F., Luciano, R.: Visual programming for structural assessment of out-of-plane mechanisms in historic masonry structures. J. Build. Eng. **31**, 101425 (2020)

Gilbert, M. (2007). Limit Analysis Applied to Masonry Arch Bridges: State-of-the-Art and Recent Developments

Giuffrè, A.: A mechanical model for statics and dynamics of historical masonry buildings. In: Protection of the Architectural Heritage against Earthquakes, pp. 71–152. Springer (1996)

Lasorella, M., Percolla, G., Pisano, A.A., Fuschi, P.: Visual scripting approach for structural safety assessment of masonry walls. Curved Layer. Struct. **12**(1), 20240025 (2025)

Massafra, A., Prati, D., Gulli, R.: Reverse engineering workflows for the structural assessment of historical buildings. Int. J. Archit. Herit. **19**(5), 702–723 (2025)

Milano, L., Mannella, A., Morisi, C., Martinelli, A.: Illustrative sheets of the main local collapse mechanisms in buildings existing masonry and related kinematic models of analysis. Annex to the repair and strengthening guidelines of structural elements, infills and partitions (2009)

MIT: Ministero delle Infrastrutture e dei Trasporti. CNTC19 – Circolare applicativa delle Norme Tecniche delle Costruzioni di cui al D.M. 17-01-2018 (NTC 2018). Gazzetta Ufficiale N. 42 del 20-02-2018. 2019 (in Italian) (2019)

Mousavian, E., Argiento, L. U., Casapulla, C.: Visual programming for macro-block analysis of multi-storey masonry buildings. In: 8th International Conference on Computational Methods in Structural Dynamics and Earthquake Engineering, COMPDYN (2021)

Nodargi, N.A., Intrigila, C., Bisegna, P.: A variational-based fixed-point algorithm for the limit analysis of dry-masonry block structures with non-associative Coulomb friction. Int. J. Mech. Sci. **161**, 105078 (2019)

Rippmann, M., Lachauer, L., Block, P.: Interactive vault design. Int. J. Space Struct. **27**(4), 219–230 (2012)

Mitigation of Seismic Liquefaction Geohazard Associated to Sea-Level Rise Through Compacted Granular Columns: A Pilot Case Study

Daniela Dominica Porcino(✉) and Giuseppe Tomasello

University Mediterranea, Reggio Calabria, Italy
{daniela.porcino,giuseppe.tomasello}@unirc.it

Abstract. Sea-level rise (*SLR*) due to climate change causes an increasing seismic liquefaction risk to coastal structures and infrastructures. This study investigates the effectiveness of compacted granular columns, specifically the rammed aggregate piers (*RAPs*), as a sustainable ground improvement technique to mitigate liquefaction risk in future *SLR* scenarios. GranularColumns v.1.2 software was specifically developed to support the design of *RAPs* under both static and seismic loading conditions. The software was used to evaluate liquefaction risk and propose mitigation solutions for a coastal site at Palizzi Marina district (Italy), where geotechnical investigations identified potentially liquefiable silty sand layers at shallow depths. Liquefaction potential was assessed using the liquefaction potential index (*LPI*), considering different earthquake scenarios, as well as three future projections of *SLR* up to the year 2100. The results show that the liquefaction geohazard is currently low, but would increase significantly due to *SLR*, with *LPI* values equal to 14.6 (high risk) under the most severe seismic and climatic scenarios. On the other hand, when *RAPs* are properly designed, a reduction of *LPI* to acceptable levels ($LPI<5$) was observed for all considered climate and seismic conditions.

Keywords: Sea-level rise · seismic liquefaction · ground improvement · rammed aggregate piers · software

1 Introduction

The general description "problematic soils" can be more precisely described as soft clay/silt soils, loose granular soils, organic and expansive soils, etc. which are likely to exacerbate some aspects of their behaviour in the future climatic changes scenarios causing damages to environment and structures. There is a general agreement in the literature that deep foundations and soil improvement can be suitable solutions to reduce the risk of damage to new buildings caused by climatic changes [1–3]. The use of vertical support columns installed in soils can be employed either as a deep foundation system or as a ground improvement technique. In fact, vertical columns system

C. Nava et al. (Eds.): *Climatic and Structural Safety in Multi-Hazard Regime of Cultural and Natural Heritage*, GREEN, pp. 219–238, 2026.
https://doi.org/10.1007/978-3-032-11926-1_14

create relatively stiff elements within soil matrix ("rigid inclusions") which allow for reduced loads on the surrounding weaker soils and improved support of the overlying structures. Vertical columns back-fill material varies greatly and includes clean natural sands and gravels dug out from quarries, cement grouted soils, crushed stones resulting from milled concrete beams, etc. To this broad range of materials corresponds a similarly wide range of construction methods, i.e. vibrofloatation (or "vibro-compaction"), vibro-replacement columns (or "vibro-stone-columns" or simply "stone columns"), rammed aggregate piers, jet grouting, compaction grouting and others of more limited use [4–7]. Among them the "compacted granular piers" or "compacted aggregate piers" (that include rammed aggregate piers and vibro-stone columns) not only provide highly stiff elements that are able to transmit structural loads to suitable soils or bedrock but also cause high lateral stress development in the soil matrix around the columns due to the lateral compression of the soils during the installation process. The surrounding soil is then improved both in strength and compressibility properties. This is an important feature which makes the real difference with other techniques. Furthermore, both stone columns and rammed aggregate piers are composed by highly draining materials, an advantageous property that allow to reduce the end of primary consolidation time (duration of delayed settlements) in cohesive soils [7–9] and the risk of cyclic liquefaction in clean and moderately silty granular soils [10–12].

Recent work have analysed the possibility of using artificial and synthetic lightweight aggregates composed respectively of expanded clay (LECA) [13] and recycled wastes (such as fly ash and reused plastic) [14] as constituent materials of compacted granular columns obtaining promising results. Even in the original version, the compacted granular piers must be considered more environmentally friendly ("low carbon") solutions compared to the concrete based deep foundations (concrete piles, jet grouting, compaction grouting); furthermore, if recycled waste materials are adopted to make the columns, the technique also meets the main goals of the circular economy, such as reducing the need for new waste disposal landfills as well as the consumption of natural resources.

Results of experimental research conducted in the last period [10, 15, 16] have shown that the lateral pre-stressing and densification of the surrounding soil caused by the construction process of rammed piers are much more pronounced than those induced by stone columns. The explanation resides in the fact that the installation process of the stone columns causes some remolding of the soil near the wall of the columns that is significantly lower in rammed aggregate piers. The consideration of the aforementioned aspects justifies the increasing recent interest of researchers for the rammed aggregate piers method [10, 16–18] in order to: i) enhance the knowledge of the basic behaviour of soil and column under static and cyclic loading; ii) define more advanced and reliable design approaches and draw up technical rules for professional engineers and construction companies.

The present paper explores the main features of a software named "Granular-Columns" developed in the context of the research project *Tech4You Spoke 4 Goal 4.7 Pilot Project 4.7.1 – Open platform "phigital space" (physical and digital) of the type "user profiling" for the advanced and dynamic codesign of interventions on the built and ex novo* for the design of compacted granular columns under both static and seismic loading conditions. Considering the link between seismic liquefaction and climate change which can increase the likelihood of liquefaction occurrences, particularly in coastal areas, an application to the pilot site of Palizzi Marina in Italy was made for mitigating seismic liquefaction risk in different climate scenarios of sea-level rise (*SLR*) and expected seismic events.

2 Installation Process of Compacted Granular Columns

The vibro-replacement stone column construction process begins with the insertion of a crane supported vibratory probe into the ground by means of the weight of the equipment, and/or water jetting and vibration (wet method) or air jetting and vibration (dry method). In some vibro-stone column processes, pre-augering of the shafts can also be performed to accelerate the construction process and increase the diameter of the stone column. Stone or aggregate backfill material is added from the surface to the bottom of the probe via the annular space created by the pressure of the jetting water between the probe and the wall of the hole (top feed system) or, alternatively, feeder tubes extending from a suspended funnel to the probe tip (bottom feed system) (Fig. 1). After a single lift of stone material has been added, the horizontal vibrator is reinserted into the stone lift multiple times. The vibration compacts the stone mass and also laterally stresses the surrounding soil. Successive lifts ranging from 0.3 to 1.2 m in thickness are placed, vibrated/compacted and continued up to the ground surface.

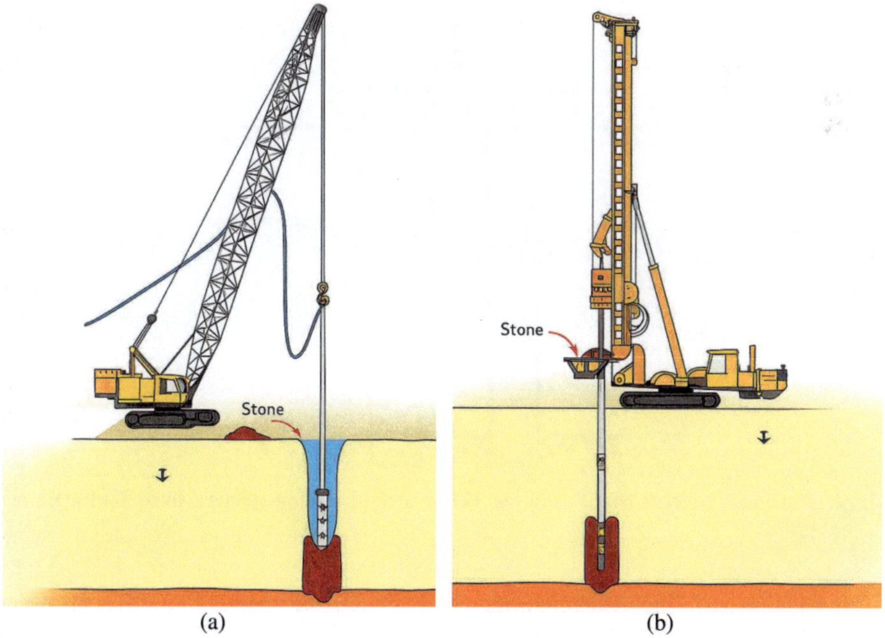

Fig. 1. Stone columns installation method: (a) wet top feed, (b) dry bottom feed (picture from Jempolslife [19]).

The lateral resistance of the surrounding soil and the vibration process combine to create a dense vertical stone column element with typical assumed friction angles on the order of 35° to 45° [20]. Under loading, stone columns with length to diameter ratios greater than 3.5 are more likely to deform in bulging near the top (Fig. 2), whereas tip deformation is more likely to occur with smaller ratios. As with other ground improvement techniques (i.e. rammed aggregate piers, jet grouting, compaction grouting, etc.) the final diameter of the columns is not known in advance but depends on soil type and accuracy of the installation procedures adopted. One of the main advantages of the aggregate piers resides in the installation process that is simpler and requires a less complex equipment compared to the vibro-replacement stone columns.

Rammed aggregate piers have been proven to be successful in increasing bearing capacities and controlling settlements in soft silt/clays, organic soils, and granular soils. The installation process begins with the auger-hole for rammed aggregate pier installation. The cavity can typically be drilled to depths up to 9 m below the ground surface at diameters ranging from approximately 600 to 900 mm. A bottom bulb is formed by placing a poorly graded crushed gravel material into the cavity and compacting with the specially designed bevelled tamper which vibrates with downward vertical force but also laterally displaces the stone. The process is continued in approximate 300 mm lifts with the continued use of open graded material, or more commonly, well graded highway aggregate base type material. Once the ground surface is reached, a bulbous shaped vertical aggregate column extends up to the desired depth. The typical equipment and schematic view of the rammed aggregate pier installation process are shown in Fig. 3.

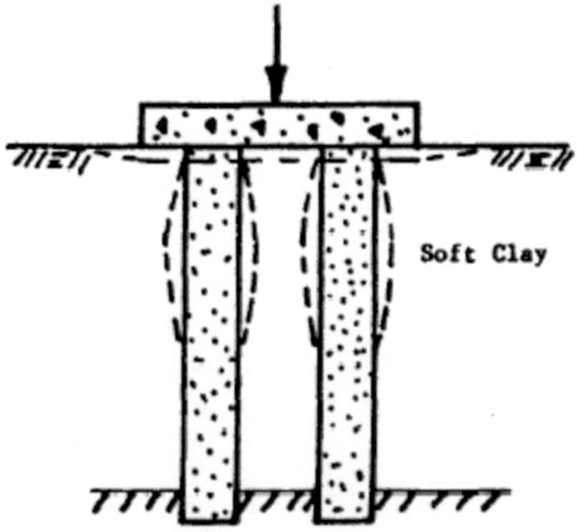

Fig. 2. Schematic view of stone columns below a rigid footing (picture from Barksdale and Bachus [20]).

In addition to the inclusion of an aggregate column, the surrounding soil is displaced laterally and compacted by the outward angle of the bevelled tamper forcing the aggregate against the soil. When open graded stone or aggregate is used, the piers act as a drainage path for rapid dissipation of excess pore water pressure. The lateral pre-stressing of the

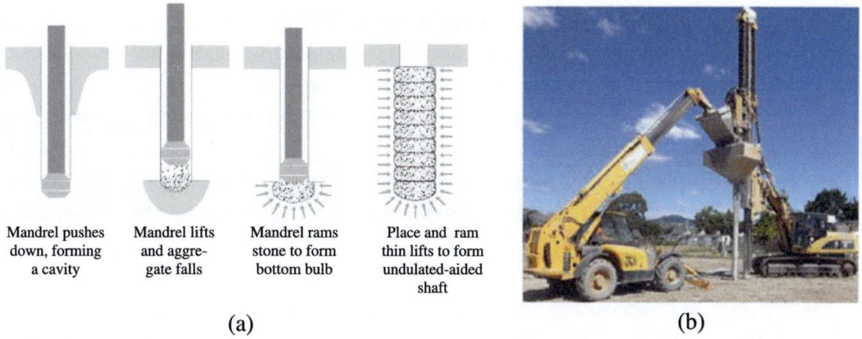

Fig. 3. Rammed aggregate pier construction: (a) installation, (b) typical equipment (pictures from Wissmann et al. [6]).

soil matrix is effective in reducing future settlements [21] and the aggregate piers have been shown to have very high angles of shearing strength, between 48° and 52° [22], due to the high level of compaction achieved. Furthermore, values of the lateral earth pressures coefficient (k_O) varying from 0.4 to 2.2 (average 1.2) for stone columns and from 0.4 to 4 (average 2.1) for rammed piers [22].

3 Design of Compacted Granular Columns: The Software "GranularColumns v.1.2"

GranularColumns v.1.2 is an easy-to-use compacted granular columns design software implemented in an Excel spreadsheet developed and validated in the context of the research project *Tech4You Spoke 4 Goal 4.7 Pilot Project 4.7.1 – Open platform "phigital space" (physical and digital) of the type "user profiling" for the advanced and dynamic codesign of interventions on the built and ex novo*. The software is an effective tool for the design of compacted granular columns (stone columns or rammed piers technique) under both static and seismic loading conditions, i.e. for liquefaction risk mitigation.

The main characteristics of the software are the following:

- Capability to account for possible effects of column installation on shear strength and compressibility properties of the soil.
- Performing settlement calculations of load areas overlying: a) an unimproved soft soil using the basic theory of elasticity and b) the same soil improved by installation of compacted granular columns, according to the methodologies proposed by Priebe [23] and Van Impe and De Beer [24] for stone columns and Lawton et al. [25], Lawton and Fox [26], and Wissmann et al. [27] for rammed aggregate piers.
- Performing time rate calculations of settlements on unimproved and improved soils taking into account the combined effect of vertical and radial flow according to Carillo's [28] approach through the vertical consolidation theory of Terzaghi [29] and the radial consolidation theory of Barron [30].
- Performing short term and long term bearing capacity calculations of single compacted granular columns considering bulging and shear failure modes according to the methodologies proposed by Hughes and Withers [31], Vesic [32] and Brauns [33].

- Capability to analyse footings or loaded areas having triangular, rectangular and hexagonal shape in plan. Load stress induced by the footing on the improved soil can be either uniform or trapezoidal (i.e. an embankment).
- Soft soil can be either homogeneous or stratified and all columns are assumed to rest on a firm end bearing stratum.
- Performing uplift resistance calculations of single rammed aggregate piers, relevant especially in swelling soils, according to the methodology proposed by Wissmann et. al. [34] and Caskey [35].
- Performing liquefaction potential calculations on improved soils with compacted granular columns, according to the methodologies proposed by Seed and Booker [36], Baez and Martin [37], Priebe [38], Goughnour and Pestana [39], Rayamajhi et al. [40], Ben Salem et al. [12] and Amoroso et al. [16].
- Generating the main plots of results and an exhaustive report.

Furthermore, a guideline to help the practitioners for selecting input parameters has been foreseen.

Under shallow foundations the stone columns are generally installed in group formed by a certain number arranged in such a way that they influence each other in the ground when loaded; this is described as "group effect". For design purposes, a simplifying approach has been introduced that is based on the unit cell concept and it is largely adopted (Fig. 4). The unit cell represents an elementary volume of the improved ground whose behaviour can be assumed as representative of the whole columns grid, i.e. all columns and all portions of soil in the grid behave like the column and the soil of the single unit cell. This statement does not apply to small numbers of columns or for columns placed at the edge of a large group since they are not laterally confined by the presence of a sufficient number of columns. The theoretical methods implemented in the software for settlement calculations under static loads and liquefaction analyses are all based on the concept of "unit cell" too.

The failure mechanisms of a single isolated compacted granular column considered in the calculation of bearing capacity in the software are shown in Fig. 5. Since most constructed granular columns have length to diameter ratios greater than 3 a bulging failure usually develops (Fig. 5a) whether the tip of the column is resting on a firm bearing stratum (end bearing) or floating in soft soil. A short granular column bearing on a firm support will undergo a local shearing failure mechanism at the surface as shown in Fig. 5b.

Finally, a floating granular column less than about 2 or 3 diameters in length may fail in punching in the weak underlying layer before a bulging failure can develop (Fig. 5c).

The software was validated by comparing the results with the corresponding ones obtained in previous valued studies in which the same theoretical methods were applied. Table 1 reports some data regarding the validation activity of the developed software. Data refer to n. 4 case studies reported in the literature where settlement data were available from calculations using Priebe [23] and Van Impe and De Beer [24] methods as well as, in some cases, in situ direct measurements.

Seismic liquefaction is a common phenomenon associated with earthquakes, which causes severe building and infrastructure damages due to loss of serviceability or even collapse. Mainly three conditions are required for liquefaction to occur: i) loose, granular

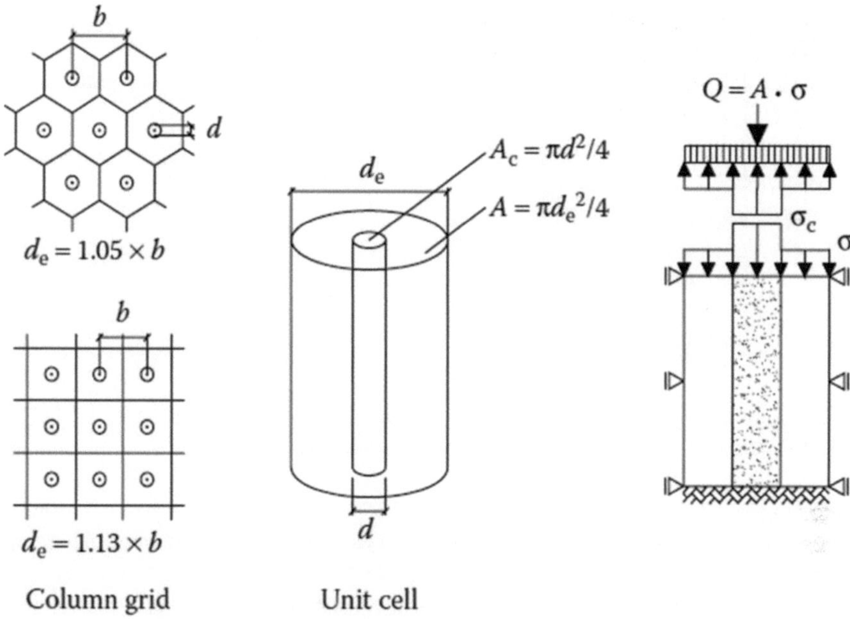

Fig. 4. Unit cell idealisation (picture from Kirsch and Kirsch [7]).

Table 1. Comparison between the results obtained by literature studies and by GranularCol-umns v.1.2 software.

Case study	Description	Reference	Theoretical approach	Results in terms of β obtained from	
				Literature	GranularColumns v.1.2
1	Tank foundation at Canvey Island	[23]	[23]	2.55	2.58
2	Oil storage tank at Zarzis	[42]	[23]	3.77	3.54
3	Foundation of a road embankment	[7]	[23]	2.47	2.47
4	Foundation load over a large area	[24]	[24]	2.19	2.20

Note: β=settlement improvement factor (i.e. ratio between settlement of untreated and treated soil).

sediment, such as relatively young, predominantly sandy, non-cohesive materials that do not contain (or contain relatively poor) plastic fines content. Typically, reclaimed land and artificial fill, as well as beach and stream deposits that are young enough (late Holocene), satisfy this condition; ii) strong shaking: the earthquake shaking must reach

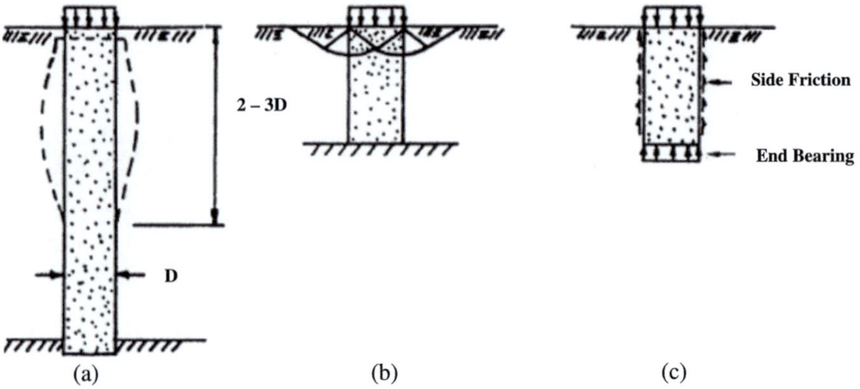

Fig. 5. Failure mechanisms of a single stone column in a homogeneous soft layer: (a) bulging, (b) local shearing, (c) punching (modified from Barksdale and Bachus [20]).

sufficient intensity and duration so that liquefaction may occur; iii) degree of saturation of the sediment due to groundwater water: the water fills the spaces between sand and silt grains and consequently increases the liquefaction susceptibility. Considering this last factor, with rising sea levels, groundwater level may increase in some coastal areas, causing unsaturated sand and silt layers to become saturated. In this case, during seismic events, the excess pore water pressure increases leading to a corresponding decrease in the effective confining stresses and consequently in the shear strength and stiffness of soil with the development of high shear strains in the ground (Fig. 6). For this reason, areas with higher groundwater levels are more easily liquefied than those with lower levels.

Compacted granular columns mitigate the risk of seismic liquefaction through three main fundamental mechanisms, namely:

a) providing drainage which dissipates the excess pore water pressure induced by earthquake almost as fast as it is generated (i.e. "drainage effect") [36];
b) introducing stiff elements which potentially can carry higher shear stress levels, causing a reduction in the cyclic shear stress levels in the surrounding soil (i.e. "shear reinforcement or stiffening effect") [37–40];
c) increasing the packing density of the soil around the column during the installation process of stone columns (i.e. "densification effect") [12, 42].

Fig. 6. Effects of liquefaction phenomena after the 1989 Loma Prieta earthquake in coastal areas (picture from Poitras et al. [42]).

Such mechanisms with the relative design methods were implemented in the developed software with the availability for the users to design compacted granular columns against liquefaction according to one or more approaches proposed in the literature. In the following, the software has been adopted for an application to a pilot case at Palizzi Marina site located in Calabria region, Italy.

4 Application to the pilot case study at Palizzi Marina site, Italy

In a previous paper [44] aimed at providing a methodological approach to the problem, the authors conducted a sensitivity analysis on the most significant parameters influencing the effect of *SLR* on liquefaction potential at Palizzi Marina site evidencing that liquefaction potential tends to be more sensitive to the future increases of *SLR* with decreasing fines content and density state of soils and increasing earthquake peak ground acceleration (*PGA*). The study was based on information gained from previous geotechnical site investigations carried out in the district area for buildings and infrastructures construction purposes. The present contribution represents an advancement of the previous work since it is based on data gained from new specifically addressed in situ and laboratory tests carried out in a more recent geotechnical survey at Palizzi Marina site (Fig. 7).

Fig. 7. Aerial view of the Palizzi Marina pilot area showing the location of surveys conducted in the coastal zone (picture modified from Google Earth).

4.1 Description of Palizzi Marina Site

The Palizzi Marina district constitutes the coastal area of the municipality of Palizzi, located on the south-eastern coast of the Calabria region (Italy). The analysis of technical reports provided by the municipal Authority revealed the presence of several geological formations characterising the subsoil of area. These include Quaternary deposits such as ancient and recent alluvial sediments (*as*, *ar*); Tertiary sedimentary formations, including Trubi (*Tr*), sand and conglomerate blocks (*Scg*), marly clays (*Am*), and varicolor clays (*Ap*); and Paleozoic rocks, represented by biotic and phylladic shales (*Sb*). Among these geological formations, particular emphasis should be placed on the *as* and *ar* units, which, in coastal areas, are composed of sands and gravels with fractions of silt that may be susceptible to seismic liquefaction when located below groundwater table. As a matter of fact, liquefaction phenomena at Palizzi Marina were observed during the 1978 earthquake at a site called "Cascina" where a copious ejection of sandy-fines mud rising from the subsoil formed a typical small "volcano" [45].

The list of the main historical earthquakes occurred over the past few centuries at Palizzi Marina, reported in Table 2, shows the importance of taking into consideration seismic risk due to *SLR* at the site (Table 2).

Table 2. Main seismic events that occurred at Palizzi Marina.

Day	Month	Year	Latitude	Longitude	I_{loc}	I_{max}	M	Epicentral area
5	2	1783	38.30	15.97	7.5	11	6.9	Southern Calabria
16	11	1894	38.28	15.87	7	9	5.9	Bagnara Calabria
8	9	1905	38.67	16.07	7	10.5	6.8	Gulf of S. Eufemia
12	10	1907	38.13	16.00	8.5	9	6	Ferruzzano
28	12	1908	38.15	15.68	7	11	7.1	Southern Calabria-Messina
16	1	1975	38.12	15.65	4.5	7.5	5.4	Strait of Messina
15	4	1978	38.12	15.02	5	8	5.8	Gulf of Patti
11	3	1978	38.02	15.98	8	8	5.5	Ferruzzano
13	12	1990	37.27	14.98	3.5	7.5	5.4	South-eastern Sicily

Note: I_{loc} = local maximum intensity of the earthquake, I_{max} = maximum intensity of the earthquake, M = magnitude.

4.2 Geotechnical Investigation at the Site

At the central southern part of Palizzi Marina district, where the construction of a naturalistic and cultural park is planned, a series of in situ investigations, including boreholes drilling (*B*), piezometers, standard penetration tests (*SPT*), multichannel analysis of surface waves (*MASW*), and undisturbed soil sampling, were carried out on behalf of the municipality of Palizzi, as shown in Fig. 7. The same figure also highlights the location of one site where a *MASW* test was conducted by the authors (*MASW1*). Collected samples from the boreholes were tested in laboratory to determine their physical properties, including unit weight (γ), grain-size distribution, water content (w), liquid limit (*LL*), and plasticity index (*PI*) (Fig. 8). Based on these tests, two main geotechnical units were identified: Unit A, consisting of silty gravelly sand (fines content $f_c \approx 9\%$) extending to a depth ranging from 2.7 to 7.1 m, and Unit B, encountered at greater depths, characterised by sandy silty clay with a high fines content ($f_c = 78–96\%$). The soil profile shown in Fig. 8a is based on the interpretation of the boreholes *B1* and *B2*, which represent the most severe stratigraphic conditions against liquefaction for the investigated area. Notably, groundwater level measurements in borehole *B2* revealed the presence of a water table at approximately 1.3 m below the ground surface.

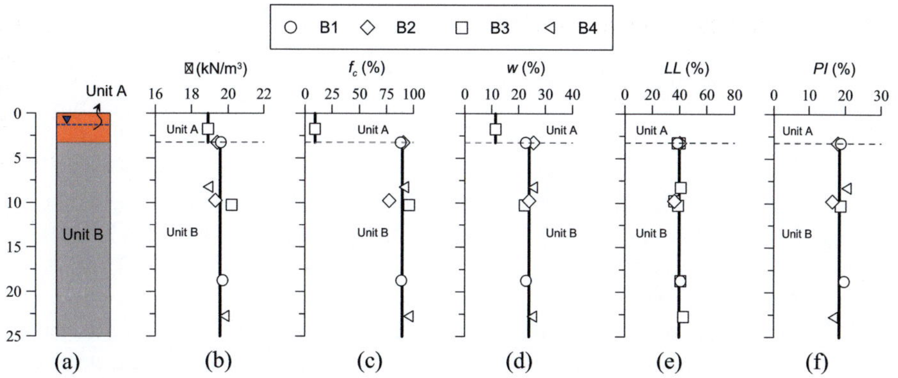

Fig. 8. Stratigraphic profile and physical properties of soils at Palizzi Marina coastal area: (a) soil column and variation with depth of (b) unit weight, (c) fines content, (d) natural water content, (e) liquid limit, (f) plasticity index.

The analysis of the physical properties displayed in Fig. 8 indicates no liquefaction risk for Unit B, according to the criteria provided by Seed et al. [46]. In contrast, Unit A, that is characterised by a high sand content (~87%), appears to be potentially susceptible to liquefaction requiring further more precise evaluations. Liquefaction potential of unit A was assessed using both *SPT* and *MASW* tests (Fig. 9). The results of these tests showed that Unit A has low blow counts and relatively low shear wave velocities, while Unit B, classified as non-liquefiable, exhibits higher blow counts and shear wave velocities both increasing with depth.

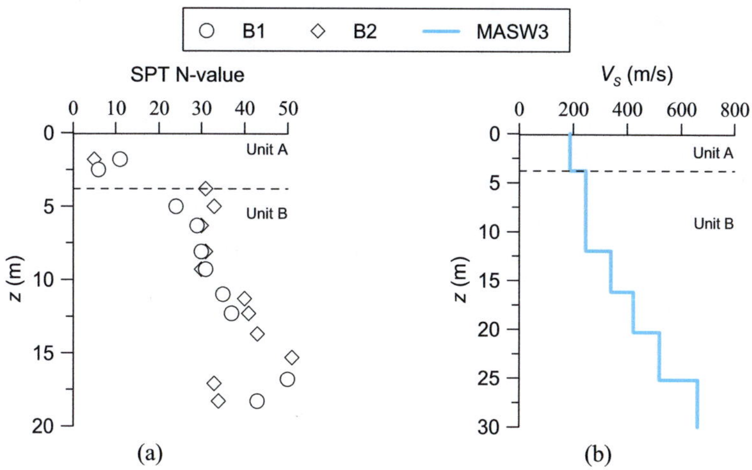

Fig. 9. Variation with depth of (a) SPT N-value and (b) shear wave velocity at Palizzi Marina site.

4.3 Evaluation of Liquefaction Risk Due to Sea Level Rise

The assessment of liquefaction hazard required the evaluation of *PGA* and earthquake magnitude (*M*) representative of the seismicity of the site. Three seismic loading scenarios (corresponding to different earthquake return periods T_R) were considered in the analyses. According to the Italian Building Code [47], these T_R refer to life safety limit state (*SLV*), which is the reference condition for liquefaction risk assessment, and are deemed representative for the seismic design of the selected building types listed in the following: 1) residential buildings ($T_R = 475$ years), 2) schools ($T_R = 712$ years), and 3) hospitals ($T_R = 1898$ years).

Table 3 presents the seismic hazard parameters for the three considered scenarios. The results were obtained using the "Spettri NTC ver.1.0.3" software and the INGV online platform. The values of *M* represent modal values for the pilot case area, taking into account the specific T_R and the geographic location of the boreholes *B1* and *B2*. Additionally, liquefaction analyses were carried out considering different climate change scenarios. In particular, Fig. 10 summarises the projected *SLR* at the Palizzi Marina site under various greenhouse gas emission scenarios (*RCP4.5* and *RCP8.5*) and associated outcome probabilities (medium-luck and bad-luck cases) [44, 48]. Among these, the precautionary *RCP8.5* scenario, corresponding to the bad-luck outcome, was adopted for the subsequent analyses.

Table 3. Seismic hazard parameters considered in the analysis of liquefaction risk.

Scenario	T_R (year)	a_g (g)	M	S_S	a_{max} (g)
1	475	0.171	6.00	1.452	0.248
2	712	0.202	6.50	1.407	0.284
3	1898	0.294	6.50	1.267	0.373

Note: T_R=return period, a_g=expected peak ground acceleration, M=magnitude, S_S=stratigraphic seismic amplification factor, a_{max}=peak ground acceleration at surface.

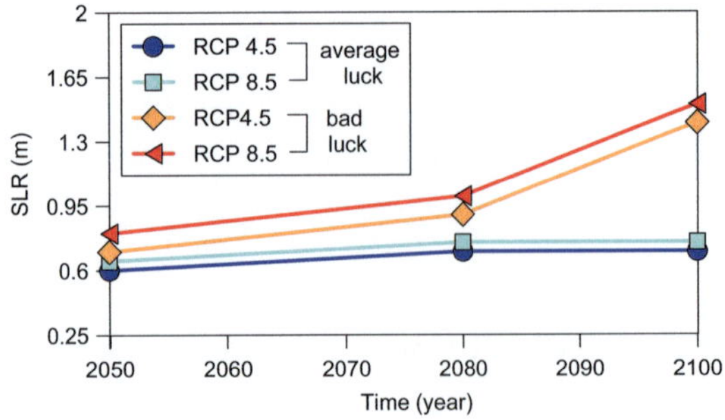

Fig. 10. Projected *SLR* for Palizzi Marina under different CO_2 emission scenarios and outcome probabilities [44, 48].

The results of the liquefaction analyses are presented in Fig. 11. The seismic liquefaction potential was evaluated using the "GranularColumns" software, considering the liquefaction potential index (*LPI*) proposed by Iwasaki et al. [49] as a control parameter:

$$LPI = \int_0^{20} F \bullet W(z)dz \tag{1}$$

in which F=1-FS for $FS \leq 1$ and F=0 for $FS > 1$ and $w(z)$=10-0.5·z (z in meter).

The *FS*, namely the factor of safety against liquefaction, required for *LPI* calculation, was determined using the simplified procedure developed by Seed and Idriss [50], following the methodology based on *SPT* data of Idriss and Boulanger [51] to evaluate liquefaction resistance. *LPI* index is a widely used tool for estimating the likelihood of surface manifestations of liquefaction at a given site [52, 53]. In particular, *LPI* equal to 0 indicates very low risk, values between 0 and 5 correspond to low risk, whereas *LPI* between 5 and 15 indicate high risk. A very high risk of liquefaction-related damage would occur when *LPI* are greater than 15.

Figure 11a shows that, under current sea level conditions, the liquefaction risk is low for all the considered T_R of seismic action. This range includes most of the T_R typically used for designing common structures. However, an increase in liquefaction vulnerability, due to the saturation of shallow layers induced by *SLR*, was obtained in all seismic scenarios (Fig. 11b). Specifically, Fig. 11b shows that, under the worst climate scenario (*RCP8.5* line with bad-luck probability in Fig. 10), the liquefaction risk is expected to become high within the next 25 years in the area of study for seismic events with a T_R exceeding 712 years. In contrast, for events with shorter return periods (T_R=475 year), a high risk of seismic liquefaction is expected to be reached only within the next 75 years.

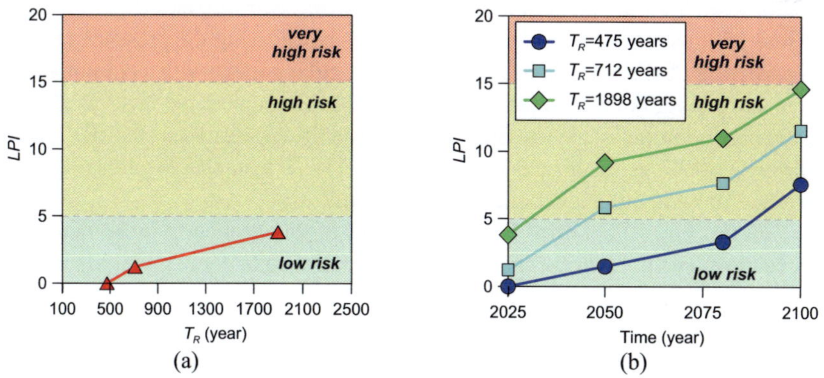

Fig. 11. Liquefaction potential index at Palizzi Marina site under (a) current and (b) future *SLR* scenarios (2050, 2080, 2100)

4.4 Mitigation of Liquefaction Risk Due to Sea Level Rise by Rammed Aggregate Piers

A ground improvement technique based on *RAPs* was proposed to mitigate liquefaction geohazard, taking into account future climate scenarios. As discussed in Sect. 3, although *RAPs* mitigate liquefaction risk through three main potential mechanisms, only two, i.e. drainage and stiffening, can be directly considered in the preliminary design phase. This is because the effects of densification require in situ tests before and after *RAP* installation to be properly assessed. Among the two usable mechanisms, only the stiffening effect currently has a well-established design approach for *RAPs*. This method involves the introduction of a cyclic stress reduction factor, K_G, which entails a reduction of the cyclic stress ratio induced by the earthquake (*CSR*) to be used in the evaluation of *FS* against liquefaction. The underlying principle is that, in analogy with the static loading condition, the higher stiffness of *RAPs* compared to the surrounding soil, leads to a concentration of seismic shear stresses τ_{cycl} within the *RAPs* while reducing correspondingly τ_{cycl} transmitted to the soil matrix. This effect can be interpreted as a reduction of the *CSR* in the treated soil relative to the untreated condition, thereby mitigating the risk of liquefaction and earthquake-induced settlements. K_G factor was calculated using the formulation proposed by Baez and Martin [37], which requires the knowledge of the area replacement ratio (A_r) and the shear modulus ratio (G_r) of the *RAPs*, where:

- $A_r = f$ (diameter, spacing, layout of *RAPs*) is the ratio between the area of *RAP* column and the total area of the unit cell surrounding it;
- G_r is the ratio between the shear modulus of *RAP* material (G_{sc}) and that of the surrounding soil (G_s).

For G_{sc} evaluation, a significant amount of data from site-specific compression modulus load tests is available [27], whereas G_s can be estimated from N_{SPT} values using empirical correlations, such as that one proposed by Burland and Burbridge [54].

Figure 12 illustrates the changes in *LPI* values considering the presence of *RAPs*.

In particular, *RAPs* were designed using a triangular pattern with a column diameter of 0.6 m and a length of 3 m to reach the underlying non-liquefiable layer. Based on the geotechnical conditions at the site (Fig. 9) and the analysed seismic scenarios (Table 3), a spacing equal to 3.2 m, 2.2 m, and 1.6 m for T_R of 475, 715, and 1898 years respectively, are required to maintain a low liquefaction risk until 2100.

To sum up, Fig. 12 confirms the effectiveness of this ground improvement method to reduce the expected increase in liquefaction risk due to *SLR* at Palizzi Marina.

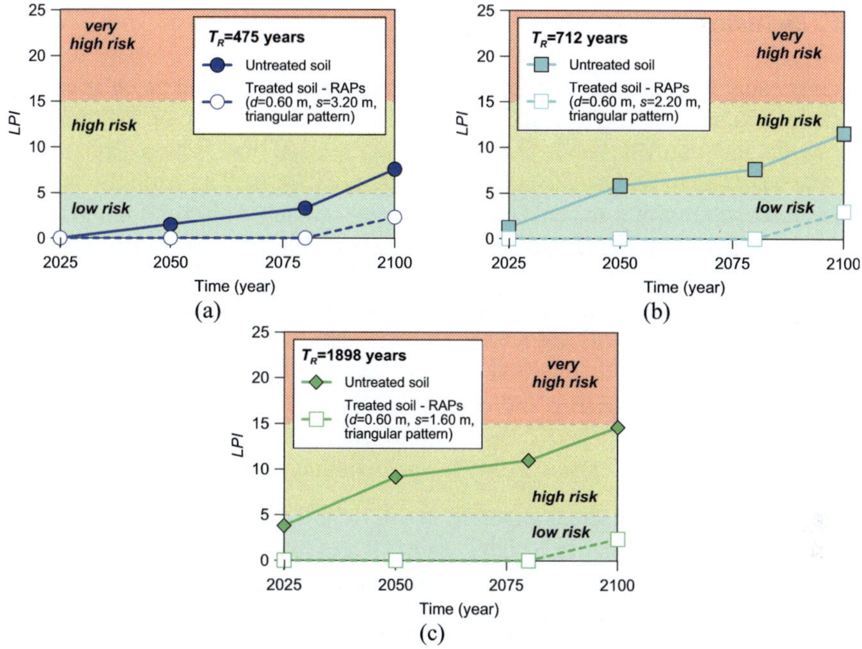

Fig. 12. *LPI* values for both untreated and treated soil by *RAPs*: (a) $T_R = 475$ years, (b) $T_R = 712$ years, (c) $T_R = 1898$ years.

5 Recommendations for Regulations

To effectively incorporate the impact of *SLR* due to climate change into liquefaction susceptibility analyses of coastal areas, regulations should be appropriately revised [42, 55]. The following recommendations are suggested:

- Update building codes to promote integration of comprehensive site-specific geotechnical investigations with groundwater level variations induced by *SLR*.
- Restrict land use prohibiting the realisation of critical infrastructure in zones with high liquefaction potential in future climate change scenarios.
- Develop guidelines to be applied to existing structures located in vulnerable areas to ensure that they meet safety conditions in a changing climate.
- Encourage interdisciplinary collaboration among geotechnical engineers, climate scientists, and urban planners to inform regulatory updates with the latest scientific and technical insights.
- Introduce continuous monitoring and early warning systems through protocols in order to ensure timely risk detection and consequent mitigation.
- Integrate education and public awareness initiatives to promote community engagement and preparedness.

Adaptive regulations are essential to ensure long-term resilience of structure and infrastructure in the face of possible future climate scenarios.

6 Conclusions

A user-friendly software was developed for the design of compacted granular columns, i.e. stone columns and *RAPs*, under both static and seismic loading conditions. Considering the link between sea-level rise (*SLR*) and seismic liquefaction geohazard, the software was used to analyse the potential impact of future *SLR* scenarios on the liquefaction susceptibility of a selected coastal pilot site, namely Palizzi Marina (Calabria, Italy). The analysis was based on detailed geotechnical investigations consisting in boreholes, piezometers, *SPTs*, *MASWs*, and laboratory tests on disturbed and undisturbed soil samples, provided by local municipality. The results identified a potentially liquefiable upper layer consisting of silty sands with approximately 10% fines. Liquefaction hazard was quantified using the liquefaction potential index (*LPI*), derived from *SPT*-based data, for three seismic scenarios with return periods (T_R) of 475, 712, and 1898 years, respectively. The rise of sea levels was quantified for three future projections deadlines, namely the years 2050, 2080 and 2100. The main findings are as follows:

- Sea level at the Palizzi site is expected to rise from 0.8 m in the year 2050 to 1.5 m in the year 2100, considering a high greenhouse gas emission scenario with bad-luck probability (RCP8.5 line with bad-luck probability in Fig. 10).
- *LPI* values ranged from 0 (very low risk) to 4 (low risk) in the current sea-level condition. The analysis indicated that projections of *SLR* up to year 2100 under the considered climate scenarios indicate a high risk of liquefaction with *LPI* values ranging from 7.6 to 14.6 for different return periods of seismic event, requiring mitigation measures in the area of study.
- Several ground improvement techniques can mitigate the liquefaction risk exacerbated by *SLR*, including vibro-compaction, stone columns, jet grouting, and *RAPs*. *RAPs* were adopted in this study due to their simpler installation, which produced columns with high internal friction angles and stiffness that markedly densify the surrounding soil. In addition, the lower environmental impact of the *RAPs* installation process and the possibility to use recycled materials align with circular economy principles, promoting sustainability in geotechnical design.
- *RAPs* were designed, considering the stiffening effect induced by the columns [37], using a triangular layout with a diameter of 0.6 m and a length of 3 m to reach the underlying non-liquefiable layer. To maintain a low liquefaction risk at the Palizzi Marina site under the most severe greenhouse gas emission and bad luck scenario, column spacings from 1.6 to 3.2 m were found to be suitable.
- To ensure long-term resilience of coastal infrastructure in the face of evolving climate scenarios, it is essential to integrate *SLR* considerations into liquefaction hazard assessments. This includes updating building codes, encouraging the use of ground improvement techniques like *RAPs*, and limiting critical development in high-risk areas. A coordinated, interdisciplinary strategy linking engineering, regulatory and community efforts, can represent an effective approach to mitigate liquefaction geohazard.

Acknowledgements. This research was funded by the National Recovery and Resilience Plan (PNRR), Mission 4, Component 2, Investment 1.5, funded by the European Union—NextGenerationEU for the realization of Innovation Ecosystem Tech4You "Technologies for climate change adaptation and quality of life improvement", Pilot Project 4.7.1—Open platform "phigital space" (physical and digital) of the type "user profiling" for the advanced and dynamic co-design of interventions on the built and ex novo (Project code: ECS_00000009, CUP: C33C22000290006).

The authors sincerely acknowledge technical offices of the municipality of Palizzi (Reggio Calabria, Italy) for their support.

References

1. Dixit, M.S.: Damage mechanism in problematic soils. Int. J. Civil Eng. Technol. **7**(5), 232–241 (2016)
2. Gariano, S.L., Guzzetti, F.: Landslides in a changing climate. Earth-Sci. Rev. **162**, 227–252 (2016). https://doi.org/10.1016/j.earscirev.2016.08.011
3. Monk, C.B., van Ballegooy, S., Hughes, M., Villenueve, M.: Liquefaction vulnerability increase at North New Brighton due to subsidence, sea level rise and reduction in thickness on the non-liquefying layer. Bull. New Zealand Soc. Earthq. Eng. **49**(4), 334–340 (2016)
4. Mitchell, J.: Soil improvement – State-of-the-art report. In: Proceedings of 10th International Conference on Soil Mechanics and Foundation Engineering, Stockholm, 15–19 June 1981
5. Han, J.: Principles and practices of ground improvement. Wiley, Hoboken NJ (2015)
6. Wissmann, K.J., van Bellegooy, S., Metcalfe, B.C., Dismuke, J., Anderson, C.: Rammed aggregate pier ground improvement as a liquefaction mitigation method in sandy and silty soils. In: Proceedings 6th International Conference on Earthquake Geotechnical Engineering, Christchurch, 1–4 Nov 2015
7. Kirsch, K., Kirsch, F.: Ground Improvement by Deep Vibratory Methods. CRC Press, Boca Raton (2017)
8. Han, J., Ye, S.-L.: Simplified method for consolidation rate of stone column reinforced foundations. J. Geotech. Geoenviron. Eng. **127**(7), 597–603 (2001)
9. Bouassida, M., Ellouze, S., Chenche, K., Bouali, M.F.: Optimized design of foundations on soft soil reinforced by floating granular columns. Int. J. Geomech. **24**(5), 04024082 (2024)
10. Rollins, K.M., Amoroso, S., Anderson, P., Tonni, P., Wissmann, K.: Liquefaction mitigation of silty sands using rammed aggregate piers based on blast-induced liquefaction testing. J. Geotech. Geoenviron. Eng. **147**(9), 04021085 (2021)
11. Zalachoris, G., Zekkos, D., Yerro, A., Bowers, G.A., Wissmann, K.J.: 3D numerical assessment of rammed aggregate pier performance under dynamic loading in liquefiable soils. J. Geotech. Geoenviron. Eng. **149**(3), 04023001 (2023)
12. BenSalem, Z., Frikha, W., Bouassida, M.: Liquefaction potential of reinforced soil by stone columns. Geomech. Geoeng. **16**(6), 915–932 (2024)
13. Zukri, A., Nazir, R., Shien, N.K.: Settlement prediction of a group of lightweight aggregate (LECA) columns using finite element modelling. Int. J. Eng. Technol. **7**, 59–65 (2018)
14. Porcino, D.D., Tomasello, G., Giunta, M.S.: Performance of synthetic aggregates for road embankment construction on improved soft soil. In: Ferrari, A., Rosone, M., Ziccarelli, M., Gottardi, G. (eds.) Geotechnical Engineering in the Digital and Technological Innovation Era. CNRIG 2023 Springer Series in Geomechanics and Geoengineering. Springer, Cham (2023)
15. Saftner, D.A., Zheng, J., Green, R.A., Hryciw, R., Wissmann, K.: Rammed aggregate pier installation effect on soil properties. Proc. Inst. Civil Eng. Ground Improv. **171**(2), 63–73 (2018)

16. Amoroso, S., Rollins, K.M., Minarelli, L., Monaco, P., Wissmann, K.J.: Improved liquefaction resistance with rammed aggregate piers resulting from increased earth pressure coefficient and density. J. Geotech. Geoenviron. Eng. **150**(6), 04024040 (2024)
17. Thum, T.S., Yerro, A., Saade, A., Ye, E., Wissmann, K.J., Green, R.A.: Numerical modelling of rammed aggregate pier (RAP) in liquefiable soil. Soil Dyn. Earthq. Eng. **153**, 107088 (2022)
18. Wang, X., Guo, W., Ren, Y., Li, S.: PFC-FLAC3D coupled modeling on rammed stone columns. Int. J. Geomech. **25**(7), 04025135 (2025)
19. Jempolslife: Method of vibro replacement stone column. https://jempolslife.blogspot.com/2012/02/method-statement-of-vibro-replacement.html (2012). Accessed 21 May 2025
20. Barksdale, R.D., Bachus, R.C.: Design and Construction of Stone Columns (Volume II, Appendices). USA, Final Report, no. FHWA7RD-83/027. Federal Highway Administration, Washington, DC (1983)
21. Handy, R.L.: Does lateral stress really influence settlements? J. Geotech. Geoenviron. Eng. **127**(7), 623–626 (2001)
22. Pitt, J.M., White, D.J., Gaul, A., Hoevelkamp, K.: Highway Applications for Rammed Aggregate Piers in Iowa Soils. Report No. TR-443, p. 154. Iawa Highway Research Board, Ames, Iowa (2003)
23. Priebe, H.J.: The design of vibro replacement. Ground Eng. **28**(10), 31–37 (1995)
24. Van Impe, W., De Beer, E.: Improvement of settlement behaviour of soft layers by means of stone columns. In: Proceedings of the 8th European Conference on Soil Mechanics and Foundation Engineering Organised by the Finnish Geotecnical Society, Helsinki (Finland), 23–26 May 1983
25. Lawton, E.C., Fox, N.S., Handy, R.L.: Control of settlement and uplift of structures using short aggregate piers. In: Situ Deep Soil Improvement, Atlanta (Georgia), 9–13 Oct 1994
26. Lawton, E.C., Fox, N.S.: Settlement of structures supported on marginal or inadequate soils stiffened with short aggregate piers. In: Vertical and Horizontal Deformations of Foundations and Embankments, Texas A&M University, 16–18 June 1994
27. Wissmann, K.J., FitzPatrick, B.T., White, D.J., Lien, B.H.: Improving global stability and controlling settlement with geopier soil reinforcing elements. In: Proceedings of the 4th International Conference on Ground Improvement, 26–28 March 2002
28. Carillo, N.: Simple two and three dimensional cases in the theory of consolidation of soils. J. Math. Phys. **21**(1), 1–5 (1942)
29. Terzaghi, K.: Die berechnung der durchlassigkeitzifer des tones aus dem verlauf der hydrodynamischen spannungserscheinungen, Mathematish-naturwissenschaftliche, pp. 125–138. Klasse. Akademie der Wissenschaften, Vienna (1923)
30. Barron, R.A.: Consolidation of fine-grained soils by drain wells. Trans. ASCE. **113**, 718–742 (1948)
31. Hughes, J.M.O., Withers, N.J.: Reinforcing of soft cohesive soils with stone columns. Ground Eng. **4**(3), 42–49 (1974)
32. Vesic, A.S.: Expansion of cavities in infinite soil mass. J. Soil Mech. Found. Div. **98**(3), 265–290 (1972)
33. Brauns, J.: Initial bearing capacity of stone columns and sand piles. In: Proceedings of the International Symposium on Soil Reinforcing and Stabilizing Techniques in Engineering Practice, Sidney (Australia), 16–18 Oct 1978
34. Wissmann, K.J., Williamson, T., Jean, C., Ringholz, R.: Use of Geopier soil reinforcing elements to support a large aboveground storage tank facility in Texas. In: Proceedings of the Independent Liquid Terminals Association 2001 Annual Operating Conference, Houston (Texas), 11–14 June 2001
35. Caskey, J.M.: Uplift capacity of rammed aggregate pier soil reinforcing elements. MSc thesis, University of Memphis (2001)

36. Seed, H.B., Booker, J.R.: Stabilization of potentially liquefiable sand deposits using gravel drains. J. Geotech. Geoenviron. Eng. **103**, 757–768 (1977)
37. Baez, J.I., Martin, G.R.: Advances in the design of vibro systems for improvement of liquefaction resistance. In: Proceedings of the Symposium on Ground Improvement, 28 May 1993
38. Priebe, H.J.: Vibro replacement to prevent earthquake induced liquefaction. In: Proceedings of the Geotechnique-Colloquium, Darmstadt (Germany), 19 Mar 1998
39. Goughnour, R.R., Pestana, J.M.: Mechanical behavior of stone columns under seismic loading. In: Proceedings of the 2nd International Conference on Ground Improvement Techniques, Singapore, 8–9 Oct 1998
40. Rayamajhi, D., Nguyen, T.V., Ashford, S.A., Boulanger, R.W., Lu, J., Elgamal, A., Shao, L.: Numerical study of shear stress distribution for discrete columns in liquefable soils. J. Geotech. Geoenviron. Eng. **140**, 04013034 (2014)
41. Ellouze, S., Bouassida, M., Hazzar, L., Mroueh, H.: On settlement of stone column foundation by Priebe's methos. Proc. Instit. Civil Eng.-Ground Improve. **163**(2), 101–107 (2010)
42. Poitras, T., Grant, A., Wein, A., Knudnes, K., Befus, K., Erdman, M., Petersen, K.: Liquefaction and sea-level rise. Sensitivity to liquefaction hazards from sea-level rise in the San Francisco Bay area, California. Available online: https://geonarrative.usgs.gov/liquefactionandsealevelrise/ (2022). Accessed 7 February 2022
43. Vautherin, E., Barry-Macaulay, D., Smith, M.: Performance of Rammed Aggregate Piers as a soil densification method in sandy and silty soils: experience from the Christchurch rebuild. In: Proceedings of the 3rd International Conference on Performed-based Design in Earthquake Geotechnical Engineering, Vancouver (Canada), 16–19 July 2017
44. Tomasello, G., Porcino, D.D., Lucanto, D., Mangano, G.: The impact of sea-level rise on seismic liquefaction susceptibility of coastal areas and risk mitigation. In: Sánchez, A., Tunio, M.N., Longato, D., Mohammad, P. (eds) Mediterranean landscapes Infrastructure: A Data-Driven Analysis of Housing and Design. CROSSMED 2024. Springer Series in Sustainable Civil Infrastructure. Springer Nature Switzerland AG (in press)
45. Guerricchio, A.: Aspetti geologici sull'erosione dei litorali e la loro influenza nel campo applicativo. Geologia Applicata e Idrogeologia. **23**, 29–78 (1988)
46. Seed, R.B., et al.: Recent Advances in Soil Liquefaction Engineering: A Unified and Consistent Framework. EERC-2003-06, Earthquake Engineering Research Center, Univ. of California, Berkeley, CA (2003)
47. NTC: Aggiornamento delle "Norme Tecniche per le Costruzioni". Official Journal of the Italian Republic no 42 (in Italian) (2018)
48. Lucanto, D., Nava, C., Mangano, G.: Digital prototyping and regenerative design toward carbon-neutrality and a climate resilient built environment: a multi-scale assessment of environment multi-risks. Buildings. **14**, 3934 (2024)
49. Iwasaki, T., Tatsuoka, F., Tokida, K., Yasuda, S.: A practical method for assessing soil liquefaction potential based on case studies at various sites in Japan. In: Proceedings of the Second International Conference on Microzonation for Safer Construction Research and Application, 26 Nov–1 Dec 1978
50. Seed, H.B., Idriss, I.M.: Simplified procedure for evaluating soil liquefaction potential. J. Soil Mech. Found. Div. **97**(9), 1249–1273 (1971)
51. Idriss, I.M., Boulanger, R.W.: Soil Liquefaction During Earthquakes. Monograph MNO-12, pp. 1–235. Earthquake Engineering Research Institute (2008)
52. Iwasaki, T., Arakawa, T., Tokita, K.-I.: Simplified procedures for assessing soil liquefaction during earthquakes. Soil Dyn. Earthq. Eng. **3**(1), 49–58 (1984)
53. Toprak, S., Holzer, T.L.: Liquefaction potential index: field assessment. J. Geotech. Geoenviron. Eng. **129**(4), 315–322 (2003)

54. Burland, J.B., Burbridge M.C.: Settlement of foundations on sand and gravel. In: Proceedings of the Institution of Civil Engineers, 6 Dec 1985
55. Quilter, P.W., van Ballegooy, S., Russ, M.: The effect of sea level rise on liquefaction vulnerability. In: Proceedings of the 6th International Conference on Earthquake Geotechnical Engineering, Christchurch (New Zealand), 1–4 Nov 2015

Optimization of Sensor Network for Mode Shape Identification of a Reinforced Concrete Building

Daniele Giunta and Eugenio Chioccarelli(✉)

Department of Civil Engineering, Energy, Environment and Materials, Università Mediterranea, Reggio Calabria, Italy
{daniele.giunta,eugenio.chioccarelli}@unirc.it

Abstract. The design of a vibration-based monitoring system for civil structures requires not only the deployment of accelerometers but also the careful selection of their number and locations to ensure reliable identification of the structural dynamic properties. To this end, an optimization of sensor network (OSN) framework was formulated in previous works, with the objective of accurately identifying modal frequencies. In this paper, a modified framework is proposed and assessed. The objective is the accurate reconstruction of mode shapes, measured by the modal assurance criterion, which replaces modal frequencies as the metric for evaluating the effectiveness of the sensor network design. The proposed framework involves three established sensor placement strategies, namely, the modal kinetic energy (MKE) method, the effective independence (EFI) method, and the information entropy (IE) method. They are compared by applying each method separately within the OSN framework. A retrofitted, four-story reinforced concrete building exhibiting spatial irregularity is considered as case study structure. Comparative results demonstrate that both EFI and IE strategies can achieve high accuracy in mode shape reconstruction using a sparse sensor network, while the MKE approach requires a larger number of sensors to reach a similar result. All analyses are based on numerical simulations and offer a useful basis for the design of actual dynamic tests and experimental validations.

Keywords: Optimal Sensor Placement · Operational Modal Analysis · Modal Properties · Modal Assurance Criterion

1 Introduction

Operational modal analysis (OMA) allows to extract the dynamic properties of structures under ambient excitation. The accuracy and efficiency of OMA depend on the sensor network configuration, which should balance information quality with installation and operational costs. Optimal sensor placement (OSP) methods help addressing this challenge by ranking candidate sensor locations according to their expected contribution to modal identification. Widely adopted OSP methods include the modal kinetic energy (MKE) method (Heo et al. 1997), the effective independence (EFI) method (Kammer 1991), and the information entropy (IE) method (Papadimitriou 2004). These methods

C. Nava et al. (Eds.): *Climatic and Structural Safety in Multi-Hazard Regime of Cultural and Natural Heritage*, GREEN, pp. 239–255, 2026.
https://doi.org/10.1007/978-3-032-11926-1_15

rely on modal displacements obtained from numerical models, such as finite element (FE) models. Beyond these classical ranking strategies, recent OSP studies have broadened toward different optimization algorithm such as bio-inspired algorithms, physics-inspired algorithms and the geography-based techniques, as well as hybrid and machine learning-based approaches (Behera et al. 2015); these algorithms enable multi-objective optimization but often require problem-specific tuning and significant computational effort (Ostachowicz et al. 2019). Thus, in this study, interpretable traditional methods (i.e., MKE, EFI and IE methods) are retained to enable a controlled comparison within the OSN framework.

Traditional OSP strategies provide a ranking of sensor locations without determining the minimum number of sensors required to accomplish OMA objectives. To overcome this limitation, previous works from the same authors (Chioccarelli et al. 2024; Giunta et al. 2025) introduced an optimization of sensor network (OSN) algorithm which integrates traditional OSP rankings with a performance metric to define the minimum number of sensors required. The choice of the OSP method and the performance metric depends on the objectives of the modal identification. For instance, in earlier applications focused on frequency identification, the OSN procedure employed the cumulative frequency error (CFE) as the performance metric and was validated on a bridge structure. In the present study, the optimization objective shifts to the capability of a sensor configuration to reconstruct mode shapes. Hence, a widely-used mode shape comparison metric, known as modal assurance criterion (MAC) (Allemang and Brown 1982), is embedded in the OSN algorithm. Furthermore, the OSN procedure is herein applied to a numerical model of a three-dimensional building with spatial irregularity, thus extending previous applications from bridge structures.

As pertaining to the OSP context, previous studies have applied the MAC-based approach to simple numerical models, such as plates (Chang and Pakzad 2014) and planar frames (Leyder et al. 2018). The present work adopts the MAC metric within the proposed OSN algorithm to a more complex and practically relevant context. The results are intended to inform and support the design of experimental tests on the actual structure.

With respect to the previous papers of these authors on the topic, the main contributions of this study are: (i) the use of the MAC as the performance metric within the OSN framework, thus enabling the optimization of both sensor number and placement based on mode shape reconstruction; (ii) a comparative analysis, in which the OSN procedure is applied separately using three widely used OSP methods, i.e., the EFI, the MKE and the IE methods, with their performance compared; (iii) the application of the OSN algorithm to support the design of an actual dynamic test on an existing building.

In the following, a theoretical formulation of the OSN algorithm is first provided together with an overview of the applied OSP method and the formulation of the MAC-based metrics. Then, the case study building and its finite element model are described, followed by the analysis of the OSN results. A section of conclusions summarizes the main findings of this study.

2 Optimization of Sensor Network Algorithm

The OSN framework is based on a backward sequential sensor placement (BSSP) algorithm. The optimization procedure consists in the steps that are summarized here to provide a comprehensive view of the entire procedure. However, steps 3 and 4 require additional details that are discussed in Sect. 2.1 and Sect. 2.2, respectively.

1. Develop a FE model of the structure to obtain modal properties. The primary inputs are: (i) a set of N target modes for the dynamic identification; (ii) a set of S candidate sensor locations, with $S \geq N$ as identifying a set of N modes requires at least N sensors (Meo and Zumpano 2005). All the sensors are considered as uniaxial accelerometers, meaning that each i-th sensor defines both its position and its measurement direction.
2. Define an index, say k, identifying the number of the iteration and set k equal to one.
3. For the current sensor configuration (i.e., the configuration with $S - k + 1$ sensors), rank the sensor locations according to their expected contribution to the identification of the target modes, as determined by a selected ranking criterion, e.g., MKE, EFI, or IE (see Sect. 2.1).
4. For the current sensor configuration, compute the value of the adopted performance metric; i.e., the MAC (see Sect. 2.2).
5. Check two conditions: (i) the number of remaining sensors is greater than N, this happens if $k < S - N + 1$; (ii) the performance metric has not yet reached its optimal value or a predefined threshold.
6. If both conditions are satisfied, remove the sensor with the least contribution to the modal identification based on the selected ranking criterion, set $k = k + 1$, and repeat from step 3. Otherwise, stop the BSSP algorithm and identify the sensor configuration as the optimal.

This iterative procedure for the OSN algorithm is sketched in Fig. 1.

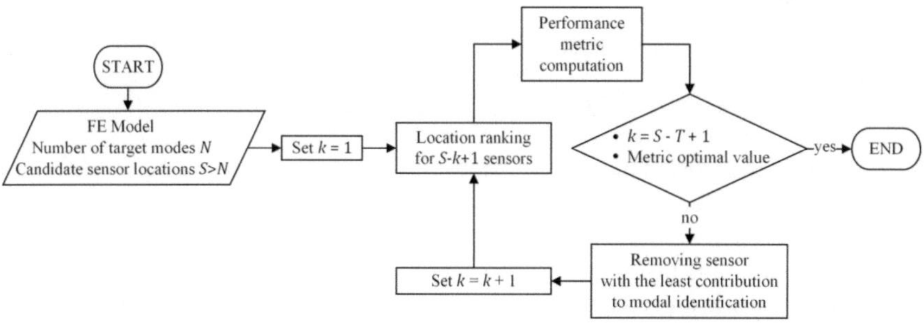

Fig. 1. Flowchart describing the OSN algorithm.

Before proceeding any further, it should be noted that the absence of a frequency-based metric in the optimization algorithm may be considered a limitation of the framework. However, this work intentionally focuses solely on mode shape reconstruction, in order to highlight the peculiarities of the results arising from this choice. Future developments, however, will require the development of an algorithm capable of incorporating both frequency-based and mode shape-based metrics, which must be evaluated through the analysis of numerical results and actual data acquired from in situ experiments.

2.1 Sensor Location Ranking

In this study, the adopted ranks of the sensors position are MKE, EFI, and IE. The MKE method prioritizes sensors at locations that maximize the kinetic energy contribution of the target modes (Heo et al. 1997; Papadopoulos and Garcia 1998). The rank for each sensor is computed as the sum of its kinetic energy contribution over all target modes. The kinetic energy matrix [MKE] is defined according to Eq. (1):

$$[MKE] = [\Phi] \odot ([M] \cdot [\Phi]), \tag{1}$$

where $[\Phi]$ is the mode shape matrix of dimensions $[S \times N]$, representing the displacement of the i-th sensor ($i = 1, \ldots S$) when the j-th mode ($j = 1, \ldots N$) is activated; $[M]$, with dimensions $[S \times S]$, is a diagonal matrix which collects the values of the masses associated with each monitored node; $\odot$ denotes the element-wise product, and $\cdot$ is the matrix product. Each element of [MKE], MKE_{ij}, quantifies the contribution of the i-th sensor to the kinetic energy of the j-th mode. To assess the effectiveness of each sensor at maximizing the kinetic energy of all the target modes, MKE_i is defined in Eq. (2):

$$MKE_i = \sum_{j=1}^{N} MKE_{ij}. \tag{2}$$

The EFI method quantifies the contribution of each sensor to the linear independence of the mode shapes of the target modes, based on the Fisher Information Matrix [FIM], with dimensions $[N \times N]$, defined as per Eq. (3):

$$[FIM] = [\Phi]^T \cdot [\Phi]. \tag{3}$$

The effective independence value, EFI_i, for the i-th sensor is defined in Eq. (4):

$$EFI_i = \{\varphi\}_i \cdot [FIM]^{-1} \cdot \{\varphi\}_i^T, \tag{4}$$

where $\{\varphi\}_i$ is the i-th row of the $[\Phi]$, containing the modal displacements of the i-th sensor for each of the N modes, and $[FIM]^{-1}$ is the inverse of the Fisher information matrix. The EFI_i index ranges from 0 to 1, with higher values indicating a greater contribution of the i-th sensor to the linear independence of the target mode shapes (Kammer 1991).

The IE index measures the uncertainty associated with the system parameters $\{\theta\} \in \mathbb{R}^{N_\theta}$ in a Bayesian framework, where N_θ is the number of parameters to be estimated (Papadimitriou 2004). In this application, the parameter vector $\{\theta\}$ consists of the N target mode shapes, thus, $N_\theta = N$. The objective of the IE method is to minimize the Shannon entropy IE of $\{\theta\}$, with lower entropy corresponding to reduced uncertainty in the mode shape estimation of the target modes (Leyder et al. 2018). IE for a given sensor configuration is defined according to Eq. (5):

$$IE = \frac{1}{2} N \ln(2\pi) - \frac{1}{2} \ln(\det([FIM])), \tag{5}$$

in which $[FIM]$ is the Fisher information matrix computed as described in Eq. (3). Both the EFI method and the IE method rely on the $[FIM]$ as an indicator of the information content in the data.

2.2 Performance Evaluation and Optimal Sensor Configuration

While the sensor ranking methods serve to iteratively reduce the candidate set based on sensor locations, the final evaluation and selection of the optimal sensor network is governed by a performance metric, chosen in accordance with the identification objectives. In previous applications, the optimization relied on the CFE, a metric that measures the relative deviation between modal frequencies identified via simulation and those predicted by the FE model. Although this frequency-based approach is effective for optimizing frequency estimation, it may neglect crucial aspects in OMA, such as the accurate reconstruction of mode shapes (Chioccarelli et al. 2024). To overcome this issue, the present study adopts the MAC (Allemang and Brown 1982) as the performance metric within the OSN framework. MAC provides a quantitative measure of the correspondence between two sets of mode shapes. More specifically, given two sets of R mode shapes, let $\{\varphi_p\}$ and $\{\varphi_q\}$, with $p = 1, \ldots R$ e $q = 1, \ldots R$, be column vectors describing the mode shapes of the two sets. The element of the MAC matrix at row p and column q, MAC_{pq}, is computed as defined in Eq. (6):

$$MAC_{pq} = \frac{\left|\{\varphi_p\}^T \cdot \{\varphi_q\}\right|^2}{\left[\{\varphi_p\}^T \cdot \{\varphi_p\}\right] \cdot \left[\{\varphi_q\}^T \cdot \{\varphi_q\}\right]}. \tag{6}$$

The value of MAC_{pq} ranges from 1, in the case of complete correspondence, to 0 when there is no correspondence between mode shapes p and q (which are, in fact, orthogonal).

In the context of this study, the MAC is computed to compare the target mode shapes (i.e., those obtained from the FE model) and the mode shapes (potentially) identified using the reduced number of sensors. In other words, Eq. (6) is applied in a way that $\{\varphi_p\}$ is the vector collecting the displacements due to mode p, provided by the FE model, for all the sensor locations identified before applying the BSSP algorithm; $\{\varphi_q\}$ is the mode shape q that can be reconstructed by the sensor network at each iteration of the BSSP algorithm. Thus, since modal displacements are available only at sensor locations, an interpolation is required to estimate the modal displacements at the locations where sensors were removed (Meo and Zumpano 2005). This step ensures that both the reconstructed mode shape via interpolation and the FE mode shape are defined on the same set of degrees of freedom, allowing for comparison of the MAC values among different sensor configurations.

For the case study building, the performed interpolation is based on a set of structural assumptions. First, (i) nodes at the edges of the columns are constrained against vertical displacement. (ii) For each story, in-plane displacements are reconstructed by imposing a rigid diaphragm constraint, i.e., assuming rigid body motion of the floor. The displacements at measured locations are used to determine the rigid body displacement parameters (i.e., in-plane translation and rotation) which are then propagated to unmeasured locations. If fewer than three sensors are available on a floor, only in-plane translations along the measured directions can be reconstructed; with three or more appropriately placed sensors, both translation and rotation can be determined for the rigid diaphragm. The rigid diaphragm assumption is commonly accepted for RC buildings with continuous slabs and stabilizes the mode shape reconstruction under sparse sensor layouts; however, its suitability should be verified for irregular structures or when significant in-plane floor deformability is expected. For the investigated case study (presented below), it was verified that replacing the constraint with a physical model of the floor slabs did not significantly affect the results, and therefore the assumption was retained. (iii) When the sensor configuration does not allow direct reconstruction for certain unmeasured positions (e.g., if no sensors are placed at a given story), spline interpolation is applied along the columns to estimate remaining modal displacements (Chang and Pakzad 2014; Leyder et al. 2018). With this approach, mode shapes can be estimated despite a limited number of sensors.

The MAC matrix between the target mode shapes from the FE model and the reconstructed mode shapes is used to assess how effectively a given sensor configuration can reproduce the structural modal behavior. The diagonal elements of the MAC matrix (i.e., the elements MAC_{pq} with $p = q$, which compare each target mode shape with its reconstructed counterpart) are used as accuracy indicators for the mode shape reconstruction. A threshold value for the diagonal element of the MAC matrix-e.g., 0.80 as suggested in (Brigante et al. 2017; Domaneschi et al. 2021)-can be adopted as a criterion for acceptable mode shape reconstruction within the OSN application. The overall quality of a sensor configuration is summarized by the root mean square of the diagonal elements of the MAC matrix, Ω, computed across all target modes, N in number, as per Eq. (7):

$$\Omega = \sqrt[2]{\frac{1}{N}\sum_{j=1}^{N} MAC_{jj}^2}. \tag{7}$$

The optimal sensor configuration is then selected as the last achieving the desired performance (i.e., Ω exceeding the specified threshold) before removing the subsequent sensors in the BSSP algorithm.

3 Case Study Structure and Finite Element Model Description

The case study is a four-story, isolated reinforced concrete (RC) frame building constructed in the early 1980s and located in southern Italy. Perspective views of the structure are represented in Fig. 2. The building features story elevations of 3.10 m, 6.20 m and 9.30 m, for the first three stories, while the fourth one, characterized by a double-pitched RC slab supported on perimeter beams, reaches a maximum elevation of 13.90 m.

Each floor presents a 0.21 m-thick RC slab incorporating hollow clay blocks. The external infills and all interior partitions consist of hollow clay masonry. The ground floor covers an area of approximately 520 m^2, while the upper floors are set back irregularly, resulting in varying plan geometry and stiffness along the building height. Plan views for each story are provided in Fig. 3. The typical bay spans range from 3.00 to 6.00 m. The structure also includes a 0.30 m-thick RC C-shaped wall extending for the full height of the building and positioned slightly off-center with respect to the plan centroid.

Fig. 2. Perspective views of the RC building.

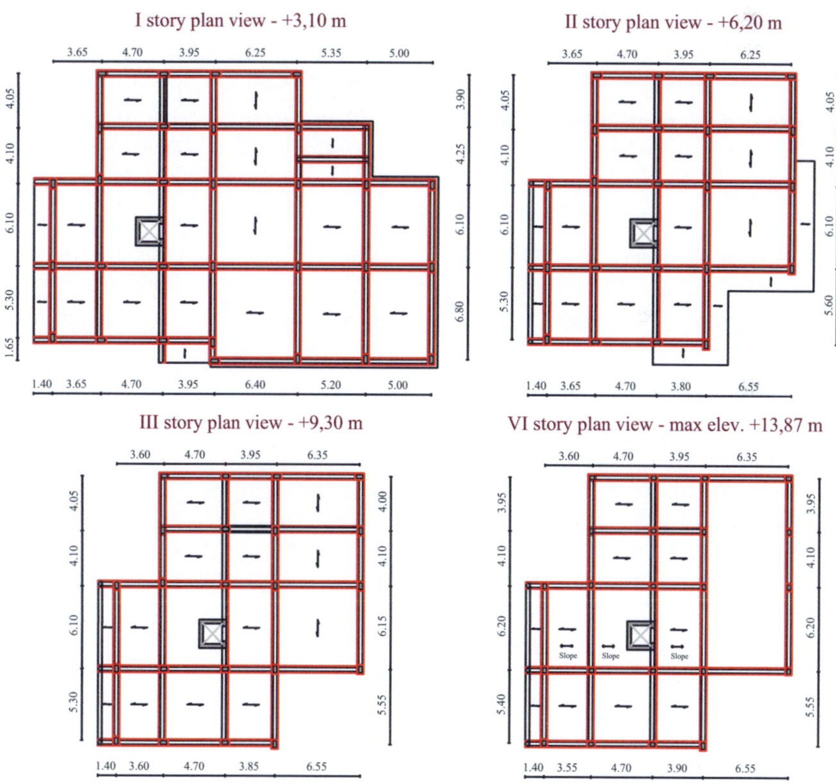

Fig. 3. Plan views of the stories of the case study building (measures in meters).

A change in use of the structure led to the implementation of retrofit interventions. The retrofit strategy, limited to the gravity-load and lateral-force resisting systems, involved RC jacketing of columns and cross-sectional enlargement of the main beams by casting a new outer shell using higher-strength concrete. The jacketing of columns enhances axial and shear capacity and improves plastic hinge rotation capacity through confinement, while the enlargement of beams increases both flexural and shear strength. Table 1 summarizes the cross-sectional dimensions of the main structural elements before and after retrofitting. Further details on the FE modeling and mechanical characterization of the structural elements are provided in the following section.

Table 1. Cross-section dimensions before and after jacketing.

Story	Element	Section dimensions before jacketing		Section dimensions after jacketing		Jacketing type
		base (m)	height (m)	base (m)	height (m)	
1st	Rectangular column	0.30	0.60	0.50	0.80	+0.10 m (all faces)
	Square column	0.30	0.30	0.50	0.50	+0.10 m (all faces)
	Rectangular beam	0.30	0.60	0.50	0.60	+0.20 m web +0.20 m base
2nd	Rectangular column	0.30	0.60	0.50	0.80	+0.10 m (all faces)
	Rectangular beam	0.30	0.60	0.50	0.60	+0.20 m web +0.20 m base
3rd	Rectangular column	0.30	0.50	0.50	0.70	+0.10 m (all faces)
	Rectangular beam	0.30	0.50	0.50	0.50	+0.20 m web +0.20 m base
4th	Rectangular column	0.30	0.40	0.50	0.60	+0.10 m (all faces)
	Rectangular beam (type 1)	0.30	0.40	0.50	0.40	No jacketing
	Rectangular beam (type 2)	0.30	0.50	0.50	0.50	+0.20 m web +0.20 m base

3.1 Finite Element Model

A linear elastic three-dimensional FE model of the building was developed using SAP2000 (Computers & Structures 2024). The model, shown in Fig. 4, was set up based on the available design drawings and the results of an extensive in-situ investigation campaign, which included direct surveys, core sampling, and destructive tests.

Table 2 summarizes the material properties adopted in the modeling. For all concrete elements, the density and Poisson's ratio were assigned according to Eurocode 1 (European Committee for Standardization 2002) and Eurocode 2 (European Committee for Standardization 2005), respectively; the Young's modulus was determined as a function of the compressive strength of each element, following the prescriptions of Eurocode 2. The mechanical properties of the materials used for retrofitting were taken from the design values specified for the intervention.

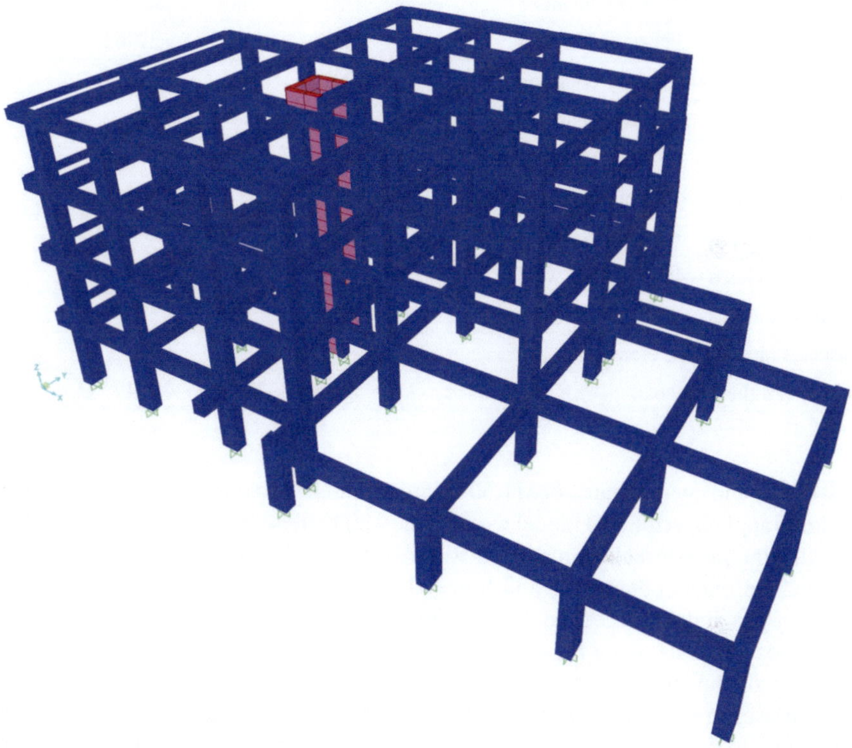

Fig. 4. Three-dimensional view of the FE model of the structure (slab omitted for the sake of clarity).

In the adopted model, columns and beams are represented by frame elements, whereas both the slabs and the RC wall are modeled with shell elements. Cross-sectional geometries and material properties correspond to the jacketed, i.e., retrofitted, state of the structural members. The geometry of the fourth-floor roof was simplified by modeling the columns with an effective height equal to the average elevation of the double-pitched slab. The bases of all columns and the RC wall are fully fixed; diaphragm constraints are applied to represent the rigid in-plane behavior of the floor slabs. The self-weight of all structural members is automatically computed by the software. Additional permanent loads, comprising the weight of non-structural elements supported by the slab

Table 2. Material properties used in the FE model.

Property	Unit	Value
Young's modulus – 1st story columns	MPa	27085
Young's modulus – 1st story beams	MPa	28608
Young's modulus – 1st story RC wall	MPa	29962
Young's modulus – 2nd story columns	MPa	29962
Young's modulus – 2nd story beams	MPa	28608
Young's modulus – 2nd story RC wall	MPa	29962
Young's modulus – 3rd story columns	MPa	28608
Young's modulus – 3rd story beams	MPa	29962
Young's modulus – 3rd story RC wall	MPa	27085
Young's modulus – 4th story columns	MPa	29962
Young's modulus – 4th story beams	MPa	28608
Young's modulus – 4th story RC wall	MPa	28608
Young's modulus of intervention concrete	MPa	33346
Poisson's ratio (all the elements)	–	0.2
Density (all the elements)	kg/m^3	2549

(1.50 kN/m) and internal partitions (1.60 kN/m) are modeled as uniformly distributed vertical loads applied on the slab model whereas the weight of the external infills (7.32 kN/m) is applied on the external beams. Infill stiffness was neglected in the presented model. The final FE model consists of 333 frame elements, 205 shell elements, and 40 fully fixed supports at the basis of the columns and RC wall.

A modal analysis of the FE model was performed. Table 3 lists, for each of the first six modes of the structure, the modal frequencies, f, the corresponding participating modal masses in x-, y- and z-direction, M_x, M_y, M_z, and a brief description of the mode shapes. Due to the presence of an off-center RC wall and the irregularity in elevation, the translational modes along both main directions display significant torsional components, as demonstrated by the significant values of the rotational participating mass.

The developed FE model provides a comprehensive basis for the OSN algorithm application, enabling the assessment of the elastic dynamic response of the retrofitted structure and the definition of candidate sensor locations. As the procedure relies on this numerical model, uncertainties in material parameters and modeling assumptions may influence the results. However, the comparison among location ranking methods is performed under a common FE model. The effect of model uncertainties related to, e.g., material properties, mass distribution, boundary conditions, can be verified after an in-situ identification test and incorporated into subsequent updates of the OSN procedure.

Table 3. Modal frequencies, f, modal masses M_x, M_y, M_z, and description of the mode shapes obtained from the FE model.

N.	f (Hz)	M_x (%)	M_y (%)	R_z (%)	Mode shape description
1	3.87	6.57E+01	2.41E+00	3.90E+00	1st translation – x
2	4.12	4.37E+00	5.40E+01	1.84E+01	1st translation – y
3	4.26	4.19E-01	1.49E+01	3.89E+01	1st global torsion
4	11.06	2.91E-01	6.72E+00	2.20E+01	2nd global torsion
5	12.00	1.93E+01	2.40E+00	3.64E-01	2nd translation – x
6	12.45	2.14E+00	1.18E+01	3.76E+00	2nd translation – y

4 Optimized Sensor Network

To implement the OSN algorithm, $N = 6$ target modes were selected, i.e., those shown in Table 3. These modes capture the global dynamic behavior of the structure under seismic loading, including the primary translational and torsional responses.

The candidate sensor locations were defined using the FE model, resulting in S=396 sensors, corresponding to 132 nodes spatially distributed at beam-column intersections. The axes of the columns and beams composing the building model are sketched in Fig. 5. The red dots are the selected nodes for the computation of the OSN procedure. At each node, it is assumed that three uniaxial sensors are considered to capture the three orthogonal components of motion.

For each ranking method (i.e., MKE, EFI, and IE), the iterative BSSP procedure presented in Sect. 2 was performed by sequentially removing the least informative sensor positions according to the respective ranking indices. This yields, for each method, to several alternative networks characterized by a decreasing number of sensors ranging from 396 to 6 (for MKE, Eq. (2) was implemented assuming the same mass value applied to each monitored node). For each configuration, the performance in terms of mode shape reconstruction was quantified by computing the MAC matrix as illustrated in Sect. 2.2.

Figure 6 shows the trend of each diagonal term of the MAC matrix, i.e., MAC_{ii} with $i = 1, \ldots 6$, as the number of sensors decreases (for the sake of simplicity, configurations up to 54 sensors are represented). For all modes, both the EFI and IE methods achieve MAC values close to 1.00 with the minimum number of sensors, showing a strong correspondence between the reconstructed and target mode shapes. This means that adding additional sensors does not significantly increase the MAC values, as the mode shape reconstruction is already accurate. On the other hand, for the MKE method, the first three modes are well reproduced as nine sensors are included in the sensor configuration; however, accurate reconstruction of higher-order mode shapes requires approximately 50 sensors.

To summarize these outcomes, the value of Ω, see Eq. (7), varying the number of sensors is presented in Fig. 7. These results confirm that the EFI and IE methods are the most robust for mode shape reconstruction, achieving high performance even with a sparse sensor network. The reduced effectiveness of the MKE method can be attributed

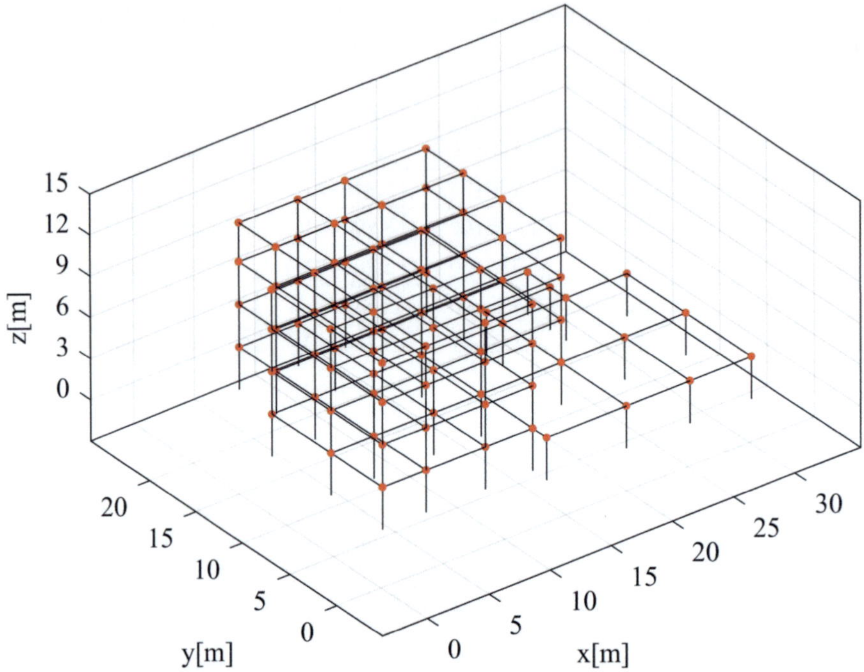

Fig. 5. Candidate sensor locations defined on the FE model.

to its tendency to deploy sensors in positions of high kinetic energy. This aspect may lead to neglect important locations for capturing torsional or higher-order mode shapes.

Hence, the optimal configurations (i.e., those providing $\Omega > 0.80$) should be selected at 6 sensors ($\Omega = 0.94$) for the EFI and IE methods. For the MKE method, a configuration with 53 sensors is required to achieve the same level of accuracy. Although this is the optimal result for MKE, displaying such a configuration would not be representative of a typical real-world monitoring scenario and would make interpretation difficult. Thus, Fig. 8 illustrates the spatial distribution of sensor networks selected by the three methods when 6 sensors are selected. Specifically, Fig. 8a–c present three-dimensional views of the sensor locations for each method, while Fig. 8d–i show the corresponding plan views at two elevations (z = 6.20 m and z = 12.15 m). The provided representation allows for a direct visual comparison of sensor placement strategies when considering the same number of sensors. The MKE method (Fig. 8a, d and g) allocates sensors only on the uppermost floor and along the x-axis, allowing accurate reconstruction of the first mode shape only. Thus, since the 6-sensor configuration provided by the MKE method is not suitable for the intended mode-shape reconstruction, it is shown in Fig. 8 only to enable direct comparison with the other methods. Both EFI (Fig. 8b, e and h) and IE (Fig. 8c, f and i) present the same sensor locations as they both rely on the $[FIM]$ matrix for selecting optimal sensor locations (see Eqs. (4) and (5)). The optimal sensor configuration for such methods distributes sensors across both the second and top floors and along both principal directions, enabling reconstruction of all six target mode shapes.

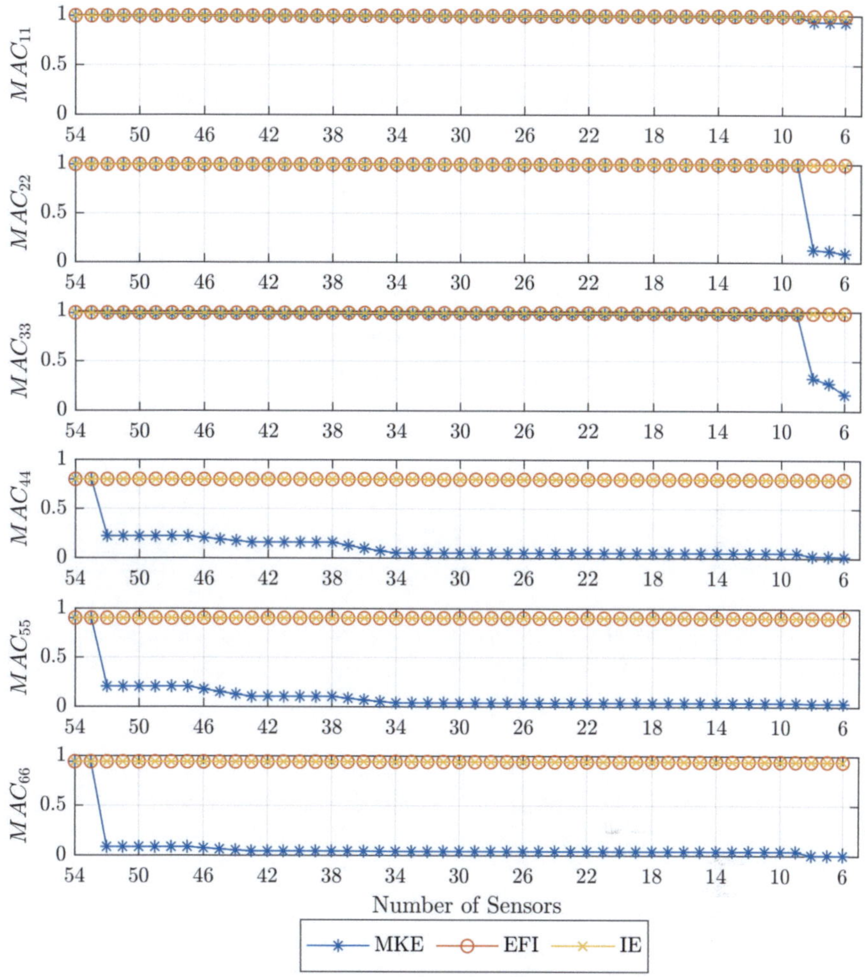

Fig. 6. MAC values for the six target modes varying the number of sensors.

A limitation of the present OSN application, not mentioned before, is that certain practical aspects, such as the actual distance between sensor positions, installation feasibility, accessibility, and wiring constraints (which may be relevant for large or irregular structures), are not explicitly considered in the optimization. The current approach assumes that any combination of sensor locations is equally viable, whereas, in real-world applications, factors such as cable length (in wired systems), power supply, and data acquisition logistics may impose additional constraints or influence the optimal network design. Furthermore, the accuracy of mode shape interpolation may be affected by local irregularities or modeling simplifications, especially when only a sparse network is used.

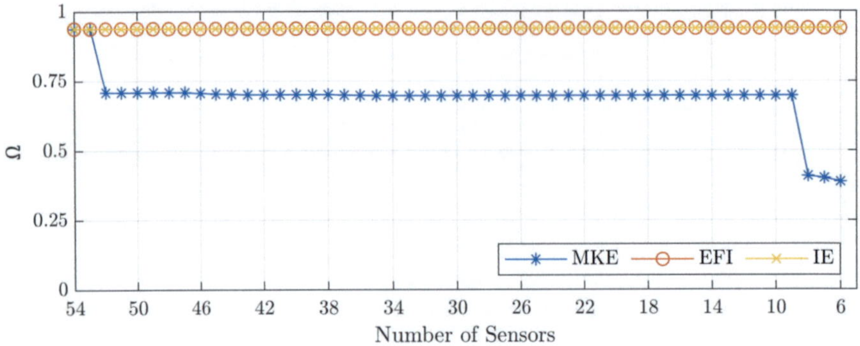

Fig. 7. Root mean square of the diagonal elements of the MAC matrix, Ω varying the number of sensors.

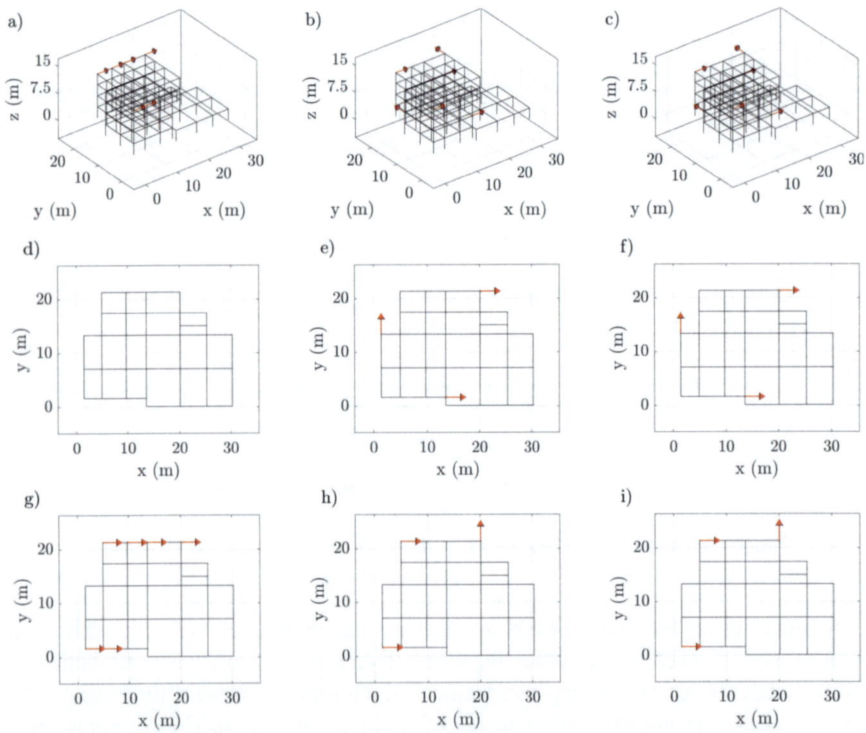

Fig. 8. 6-sensor configurations obtained by using the three sensor ranking strategies: (a–c) three-dimensional views of the configuration derived from the MKE, the EFI and the IE methods, respectively; (d–f) corresponding plan views at elevation z = 6.20 m; (g–i) plan views at elevation z = 12.15 m. Red arrows indicate the locations and measurement directions of the selected sensor positions.

5 Conclusions

This study presented an application of the OSN algorithm for the dynamic identification of a reinforced concrete building, comparing three widely adopted OSP methods: the MKE method, the EFI method, and the IE method. In contrast to frequency-based metrics used in previous works, the present study integrated the MAC into the OSN framework, allowing both the number and the sensor locations to be optimized with respect to mode shape reconstruction. In this approach, the reconstruction of mode shapes for reduced sensor networks was performed by interpolating modal displacements at unmeasured locations based on the numerical model and structural assumptions. The main findings of this work can be summarized as follows.

Both the EFI and IE methods proved highly effective, enabling accurate reconstruction of the six target mode shapes with only 6 sensors. The sensor configurations identified by these methods were broadly distributed across both horizontal directions and different elevations, reflecting a balanced coverage of the structural response.

The MKE method tended to cluster sensors in areas with high kinetic energy, primarily along the top story and in one direction, thus requiring a larger number of sensors, i.e., 53 sensors, to achieve an accuracy in mode shape reconstruction comparable with that of EFI and IE.

The analysis of this case study highlights that, among those considered, methods based on maximizing independence or information content (EFI, IE) are preferable to the one considering kinetic energy (MKE). This is also related to the objective of the optimization algorithm that is to reliably reconstruct mode shapes with a minimal number of sensors.

Moreover, this preliminary study suggests that the OSN algorithm proposed in previous works to drive experimental dynamic identification campaigns, could also be adapted, in a modified manner, to consider alternative performance metrics. Furthermore, the OSN can be extended to a multi-objective formulation that combines MAC with, e.g., a frequency-error metric such as the CFE, for cases where both frequency and mode shapes are the monitoring objectives.

It should be noted that, as in most numerical optimization studies, the present application does not consider certain practical constraints, such as sensor accessibility or wiring requirements, which may affect the feasibility of the proposed configurations in a real-world setting. In subsequent developments, such practical constraints may be embedded as explicit design variables within the OSN procedure. Furthermore, the present results are based on a numerical model and rely on assumptions adopted in the interpolation of reconstructed mode shapes. The accuracy of the reconstruction may be affected by these modeling hypotheses, for instance, when using sparse sensor networks. However, the results provide a basis for designing the experimental dynamic identification test to be performed on the real structure. For an actual field test, a 6-sensor layout identified by the EFI and IE strategies, with sensors distributed across the second and top floors and along both principal directions, is recommended to capture both translational and torsional modes of interest.

Acknowledgments. The presented work has been supported by the project NGEU PNRR– MUR– –Call for tender for the presentation of proposals for the Strengthening of research structures and creation of R&D "innovation ecosystems", set up of "territorial leaders in R&D"—to be funded under the National Recovery and Resilience Plan (NRRP), Mission 4, Component 2 Investment 1.4, funded from the European Union—NextGenerationEU. PROPOSAL: Tech4You—Technologies for climate change adaptation and quality of life improvement. NRRP THEMATIC AREA: CLIMATE, ENERGY, SUSTAINABLE MOBILITY TERRITORY OF REFERENCE: Calabrian and Lucanian territorial areas. Spoke 4—Technologies for resilient and accessible cultural and natural heritage: Unibas Affiliates: Unical, Unirc, CNR—Goal 4.7: Protection and enhancement of natural and cultural heritage and identity of the inner territories. Pilot Project 4.7.1—Open platform "phigital space" (physical and digital) of the type "user profiling" for the advanced and dynamic codesign of interventions on the built and ex novo.

Competing Interests The authors have no conflicts of interest to declare that are relevant to the content of this chapter.

References

Allemang, R.J., Brown, D.L.: A correlation coefficient for modal vector analysis. In: Proceedings of the 1st International Modal Analysis Conference. Orlando, FL, USA (1982)

Behera, S., Sahoo, S., Pati, B.B.: A review on optimization algorithms and application to wind energy integration to grid. Renew. Sustain. Energy Rev. **48**, 214–227 (2015). https://doi.org/10.1016/j.rser.2015.03.066

Brigante, D., Rainieri, C., Fabbrocino, G.: The role of the Modal Assurance Criterion in the interpretation and validation of models for seismic analysis of architectural complexes. Proc. Eng. Elsevier Ltd. (2017). https://doi.org/10.1016/j.proeng.2017.09.484

Chang, M., Pakzad, S.N.: Optimal sensor placement for modal identification of bridge systems considering number of sensing nodes. J. Bridge Eng. **19** (2014). https://doi.org/10.1061/(ASCE)BE.1943-5592.0000594

Chioccarelli, E., Giunta, D., De Angelis, A., Bilotta, A.: Discussing optimal sensor placement for ambient vibration test of an existing bridge. Eng. Struct. **310** (2024). https://doi.org/10.1016/j.engstruct.2024.118000

Computers & Structures Inc: SAP2000 v26, Integrated Solution for Structural Analysis and Design. Walnut Creek, CA, USA (2024)

Domaneschi, M., Zamani Noori, A., Pietropinto, M.V., Cimellaro, G.P.: Seismic vulnerability assessment of existing school buildings. Comput. Struct. **248** (2021). https://doi.org/10.1016/j.compstruc.2021.106522

European Committee for Standardization: Eurocode 1: Actions on structures – Part 1-1: General actions – Densities, self-weight, imposed loads for buildings (EN 1991-1-1:2002) (2002)

European Committee for Standardization: Eurocode 2: Design of Concrete Structures – Part 1-1: General Rules and Rules for Buildings (EN 1992-1-1:2004) (2005)

Giunta, D., Chioccarelli, E., Chatzi, E.: Optimal sensor placement for vibration monitoring of bridge structures. In: Proceedings of the 6th International Conference on Uncertainty Quantification in Computational Science and Engineering. Rhodes, Greece (2025). https://doi.org/10.7712/120225.12354.21227

Heo, G., Wang, M.L., Satpathi, D.: Optimal transducer placement for health monitoring of long span bridge. Soil Dyn. Earthq. Eng. **16**, 495–502 (1997). https://doi.org/10.1016/S0267-7261(97)00010-9

Kammer, D.C.: Sensor placement for on-orbit modal identification and correlation of large space structures. J. Guid. Control Dyn. **14**, 251–259 (1991). https://doi.org/10.2514/3.20635

Leyder, C., Dertimanis, V., Frangi, A., Chatzi, E., Lombaert, G.: Optimal sensor placement methods and metrics–comparison and implementation on a timber frame structure. Struct. Infrastruct. Eng. **14**, 997–1010 (2018). https://doi.org/10.1080/15732479.2018.1438483

Meo, M., Zumpano, G.: On the optimal sensor placement techniques for a bridge structure. Eng. Struct. **27**, 1488–1497 (2005). https://doi.org/10.1016/j.engstruct.2005.03.015

Ostachowicz, W., Soman, R., Malinowski, P.: Optimization of sensor placement for structural health monitoring: a review. Struct. Health. Monit. **18**, 963–988 (2019). https://doi.org/10.1177/1475921719825601

Papadimitriou, C.: Optimal sensor placement methodology for parametric identification of structural systems. J. Sound Vib. **278**, 923–947 (2004). https://doi.org/10.1016/j.jsv.2003.10.063

Papadopoulos, M., Garcia, E.: Sensor placement methodologies for dynamic testing. AIAA J. **36**, 256–263 (1998). https://doi.org/10.2514/2.7509

Inerter-Based Base Isolation System for Masonry Structures: Numerical Models for Experimental Testing

Giuseppe Failla(✉), Andrea Francesco Russillo, and Gioacchino Alotta

Department of Civil, Energy, Environmental and Materials Engineering (DICEAM), University of Reggio Calabria, Reggio Calabria, Italy
{giuseppe.failla,andreaf.russillo,gioacchino.alotta}@unirc.it

Abstract. This paper presents the results obtained from numerical simulations of shaking-table experimental tests on a benchmark masonry structure, base-isolated by a hybrid base isolation system. The latter combines standard base isolators with a tuned inerter damper, consisting of a grounded rack-and-pinion inerter connected in series to a set of linearly-elastic springs and a viscous dashpot in parallel. Finite element analyses implemented in SAP2000 demonstrate the effectiveness of the proposed hybrid base isolation system.

1 Introduction

Base isolation (Clemente and Martelli 2019; De Luca and Guidi 2019) is certainly very effective and particularly suitable for protecting historical masonry structures (Clemente et al. 2012, 2016; Osgooei et al. 2015; Habieb et al. 2019a, b, c; Basar et al. 2021; Losanno et al. 2021, 2022; Zhang et al. 2021; Faiella et al. 2023), as it is deployed at the foundation level without invasive or unaesthetic modifications to the structure above. Several shaking-table experimental tests were conducted on base-isolated masonry structures (Basar et al. 2021), to assess the performances of different types of isolators. Moreover, experimental tests under real earthquake records were also simulated numerically (Losanno et al. 2021, 2022; Zoheb et al. 2023, 2025).

Although base isolation is certainly a well-established technique for seismic protection, with several applications worldwide in areas at high seismic risk, research studies are ongoing to overcome some inherent shortcomings as, for instance, the high displacement demand at the isolation floor, which requires resorting to large-size isolators capable of withstanding the design displacements and, in some cases, allowing an adequate separation distance between buildings to prevent mutual collisions or structural pounding (De Domenico and Ricciardi 2018). To reduce the displacement demand on the base isolation system, some studies on steel and reinforced concrete structures (Tsai 1995; Taniguchi et al. 2008; Xiang and Nishitani 2014; Mele et al. 2021) proposed to couple the base isolation system with a tuned mass device targeting the first natural frequency of the isolated structure; in this context, a typical example of tuned mass device is a tuned mass damper (TMD), i.e., a secondary mass connected to the structure by

C. Nava et al. (Eds.): *Climatic and Structural Safety in Multi-Hazard Regime of Cultural and Natural Heritage*, GREEN, pp. 256–274, 2026.
https://doi.org/10.1007/978-3-032-11926-1_16

a linearly-elastic spring and a viscous dashpot acting in parallel at the isolation floor of the structure, where the maximum relative displacements occur. The effectiveness of this strategy, however, is strictly related to the amount of the TMD mass that can be used; in general, very large masses are needed to attain the sought performances, requiring a very large space and inevitable costs (De Domenico and Ricciardi 2018). Another important limitation is that the TMD stroke may be unacceptably large, requiring a large space/clearance in the building to accommodate the displacement demand of this secondary mass/subsystem, sometimes even exceeding the maximum admissible displacement of the base isolation system.

Since numerical evidence demonstrates that the larger the TMD mass is, the larger the benefits are in reducing the displacements of the base isolation system, a very promising alternative to the classical TMD are tuned mass devices based on the so-called inerter (Smith 2002). The inerter is a two-terminal mechanical device whose inertia provides a force proportional to the relative translational acceleration at the terminals (Smith 2002). Its distinctive feature is the capability of providing a very large inertia with a relatively small mass; this is possible by exploiting specific mechanisms as, for example, rack-and-pinion or ball-screw mechanisms, which transform the translational kinetic energy associated with the relative motion of the device terminals into rotational kinetic energy of a lightweight flywheel (Papageorgiou et al. 2009; Marian and Giaralis 2017; Ge and Wang 2018). A very impressive number of recent studies demonstrate the great potential of the inerter in a wide range of problems (Marian and Giaralis 2014; Hu and Chen 2015; Alotta and Failla 2021; Russillo et al. 2022; Alotta et al. 2023; Sciutteri et al. 2024). However, extensive effort is needed for experimental validation of inerters in structural engineering applications.

To contribute to the development of innovative and effective strategies for the protection of monumental heritage assets, this paper presents the results obtained from numerical simulations of shaking-table experimental tests on a benchmark masonry structure, base-isolated by a hybrid base isolation system combining classical base isolators with a tuned-inerter-damper (TID). The TID consists of a rack-and-pinion inerter in series with a set of linearly-elastic springs and a viscous dashpot in parallel. The TID is installed at the level of the isolators, between the isolation floor of the masonry structure and the ground, to exploit the relative motion between the two for activating the inerter and leveraging its large inertia effects. Numerical simulations are carried out on 3D finite element models of the masonry structure, adopting a simplified yet accurate representation of the TID and isolators. Results show that the hybrid base isolation system is effective in reducing the displacement demand on the isolators and absolute acceleration on the masonry structure, with advantages over the classical base isolation system without TID. It is noteworthy that the inerter is designed to be re-tunable, as re-tuning is important to make the proposed control system adaptable to potential variations in the structural properties of the masonry structure to be protected. Indeed, non-negligible variations in mechanical parameters such as stiffness and strength may be envisaged in masonry structures as a result of climate-change related phenomena: e.g., chemical degradation due to the increase of CO2 in the atmosphere and/or high thermal stresses during increasingly frequent heat waves, as recommended by the “National Climate

Change Adaptation Plan" released by the Italian Ministry of Environment and Energy Security in December 2022 (Italian Ministry of Environment and Energy Security 2022).

The paper is structured as follows. The system under study is described in Sect. 2. Numerical results are reported in Sect. 3.

2 System Under Study

The system under study is a benchmark two-floor masonry structure, base-isolated by the proposed hybrid base isolation system combining classical base isolators with a TID. The system is designed for 2:3 scale experimental tests on shaking table under real earthquake records.

The masonry structure is depicted in Figs. 1, 2 and 3, with main geometry and material parameters given in Table 1. The plan has a rectangular shape. The foundation beam, the first-floor bond beam and roof bond beam are made by reinforced concrete (rc). The foundation slab is made by rc, the first-floor and roof slab by wood. Six wood beams support the first-floor slab and the roof slab, respectively.

The TID consists of a rack-and-pinion inerter in series with a set of linearly-elastic springs and a viscous dashpot in parallel. The TID, installed at the level of the isolators between the isolation floor of the masonry structure and the ground, leverages the relative motion between the two to activate the inerter and exploit its large inertia effects. Therefore, the rack-and-pinion inerter is grounded, while the springs and the dashpot are connected to the isolation floor (see Fig. 4). All components of the TID are made of steel. The TID was realized by Metalmeccanica Falcone, Reggio Calabria, Italy; a detailed list of its components is reported in the Appendix, for completeness.

Figure 5 illustrates the main components of the proposed rack-and-pinion inerter: a flywheel, a gearbox and a rack, all supported by a grounded support structure. The rack slides on two parallel rails and is connected in series to the set of linearly-elastic springs and the viscous dashpot in parallel by an end stiffened plate. As a result of the relative motion between structure and ground, the translational motion of the rack on the rails is converted into rotational motion of the flywheel via the gearbox. By this mechanism, the rack-and-pinion inerter can provide large inertia effects with a relatively small mass. The inertia effects are measured by the so-called inertance (kg), which depends on the number of gears, the gearing ratio and the mass moment of inertia of the flywheel attached to the output gearbox shaft. Therefore, changing mass/diameter of the flywheel allows a straightforward re-tuning of the device. It is also noteworthy that the physical mass of the inerter contributing to the motion of the device relative to the ground is rather small, as it only includes the masses of the rack and the end stiffened plate above the rack. On the other hand, the flywheel is mounted to the lower part of the device, which can be considered to be almost non-deformable and rigidly fixed to the ground. Therefore, the device meets the fundamental requisites of an inerter, as its mass contributing to motion relative to the ground is small and independent of the nominal inertance (Pietrosanti et al. 2021). For the purposes of this study, the inerter is designed to provide a nominal inertance equal to 12 t, see Table 2.

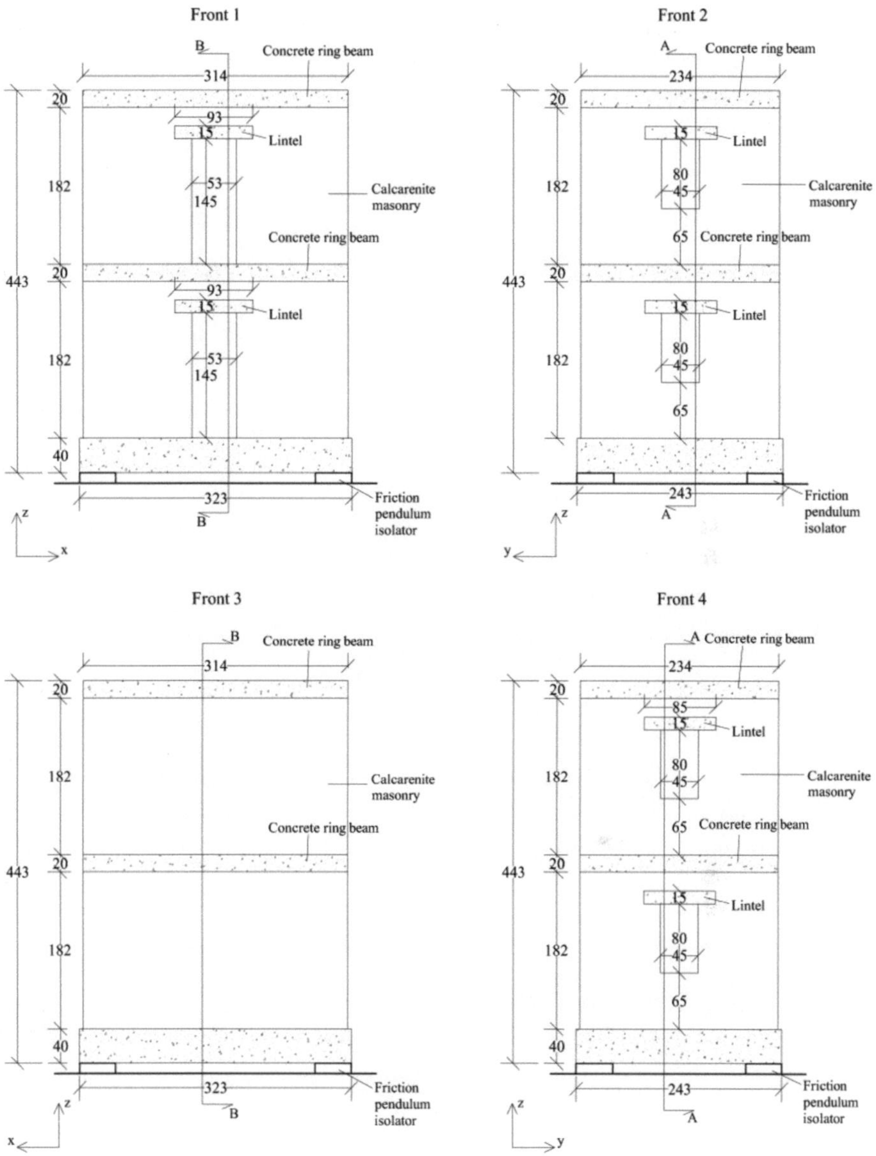

Fig. 1. Side views of benchmark two-floor masonry structure.

The springs are linearly-elastic compression springs, the dashpot is a fluid viscous damper, they are all connected to a support attached to the structure. Being compression springs, two sets of springs are used, placed on two opposite sides of the support. They are designed to provide a stiffness equal to 20300 N/m, see Table 2. It is assumed that the fluid viscous damper behaves linearly with viscous damping coefficient equal to 11950 Ns/m, again see Table 2.

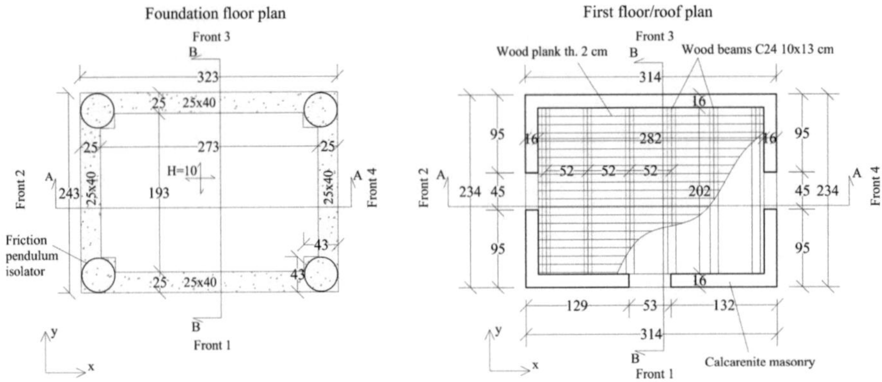

Fig. 2. Foundation, first-floor and roof plans of benchmark two-floor masonry structure.

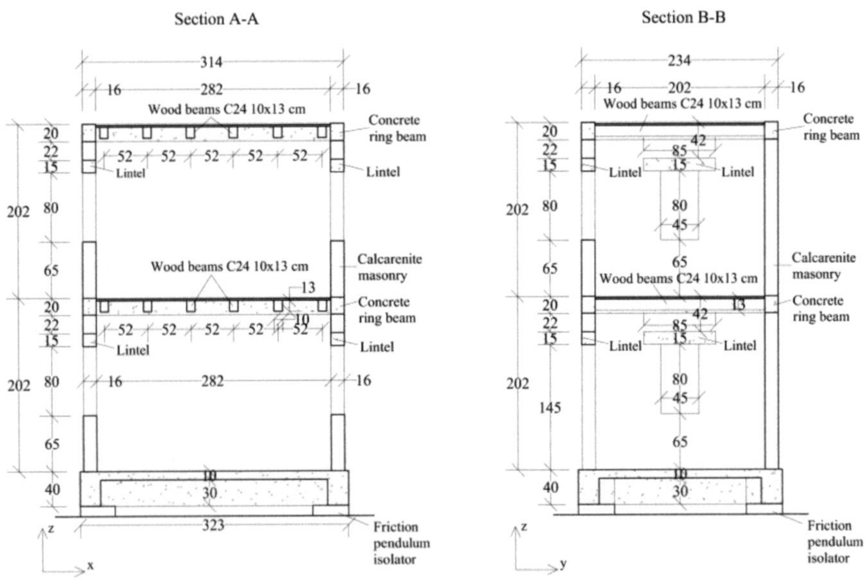

Fig. 3. Vertical plane cross sections of benchmark two-floor masonry structure.

Finally, the isolators considered in this study are friction pendulum isolators, with main parameters in Table 2.

Table 1. Geometry and material parameters of benchmark two-floor masonry structure

First floor	
Wall sides length	2.34 × 3.23 m
Wall height	2.02 m
Wall thickness	16 cm
Ring beam cross section	16 × 20 cm
Slab thickness	2 cm
Beams cross section	10 × 13 cm
Second floor	
Wall sides length	2.34 × 3.23 m
Wall height	2.02 m
Wall thickness	16 cm
Roof ring beam cross section	16 × 20 cm
Roof slab thickness	2 cm
Roof beams cross section	10 × 13 cm
Foundation beam cross section	25 × 40 cm
Foundation slab thickness	10 cm
Masonry Young modulus	1410 MPa
Masonry mass density	1800 kg/m^3
Reinforced concrete Young modulus	30 GPa
Reinforced concrete mass density	2500 kg/m^3
Wood Young modulus	11 GPa
Wood mass density	420 kg/m^3

At this stage, some remarks are in order on criteria adopted to select the parameters of the TID-base isolation system. The isolators are taken as available on the market, while the inertance of the inerter is calculated from the gearbox characteristics and mass moment of inertia of the flywheel attached to the output gearbox shaft. The choice of the isolator parameters and the design of gearbox/flywheel are driven based on engineering judgement. On the other hand, the parameters of the set of linearly-elastic springs and viscous dashpot are obtained by a trial-and-error tuning procedure, on adopting a reference two-degree-of-freedom model of the system where: (a) the base-isolated masonry structure is represented as a rigid mass oscillating in the horizontal plane; (b) the friction pendulum isolators provide linearly-elastic restoring forces and linear viscous damping forces with stiffness $k = 4 \times 29000$ N/m and damping coefficient $c = 4 \times 2664$ Ns/m, estimated based on their geometrical and mechanical characteristics; (c) the inerter is idealized as a linear massless mechanical element; (d) the TID is modelled as a linear system with one degree of freedom associated with the displacement of the

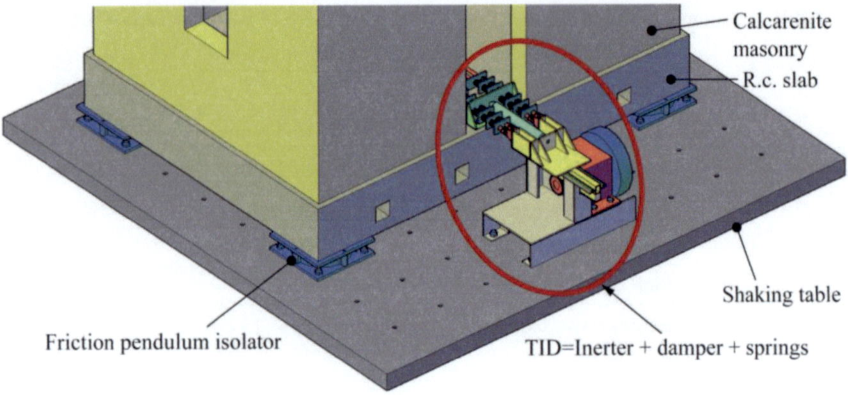

Fig. 4. Proposed TID-base isolation system for masonry structure.

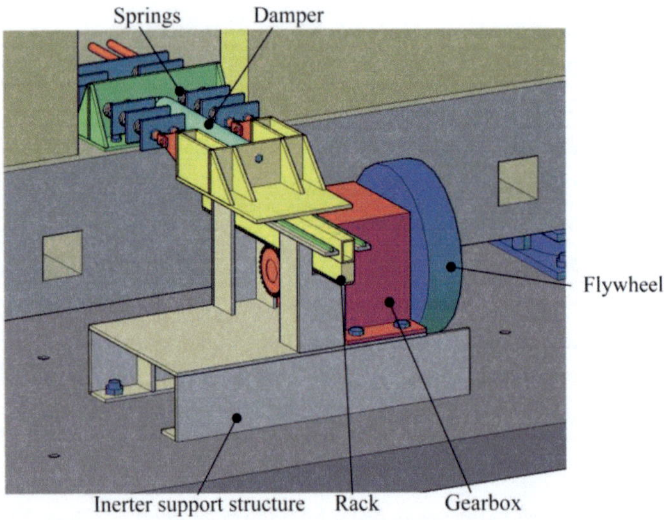

Fig. 5. Main components of TID.

connection point between the inerter and the springs/dashpot. The first step of the trial-and-error procedure hinges upon the well-known fixed-point theory (Krenk 2019; Krenk and Høgsberg 2016), which delivers first tentative values for the optimal stiffness of the set of linearly-elastic springs and the optimal viscous damping coefficient of the dashpot. Starting from its first tentative value, the optimal stiffness is sought as the stiffness that provides two peaks of the same height in the frequency response function of the system under study, in the vicinity of the natural frequency ω_0 of the base-isolated masonry structure without TID. Obviously, the two peaks are lower than the resonance peak at ω_0 of the based-isolated masonry structure without TID, ensuring vibration reduction with respect to the classical base isolation system without TID.

Table 2. Parameters of TID and base isolators.

Rack-and-pinion inerter	
Inertance	12 t
Compression springs	
Stiffness	20300 N/m
Dashpot	
Viscous damping coefficient	11950 Ns/m
Friction pendulum isolators	
Maximum payload	80 kN
Nominal friction coefficient	2%
Equivalent curvature radius	2500 mm
Maximum horizontal displacement	±300 mm

3 Numerical Results

A finite-element model of the system under study is implemented in SAP2000 (see Fig. 6).

The masonry panels, rc foundation slab, wood first-floor and roof slabs are modelled using shell elements, rc bond/foundation beams are modelled using beam elements; wood first-floor/roof beams are not modelled but related masses are distributed over the slabs. Friction pendulum isolators are modelled using the built-in option in SAP2000. The TID is modelled as in previous studies of the authors (Russillo et al. 2022; Alotta et al. 2023). The parallel spring-dashpot part of the TID is modelled as a horizontal two-joint link, which is a standard element in the library of SAP2000. This link has one joint connected to the structure and the other joint to the inerter element. The latter is modelled using a virtual vertical massless rigid beam element, with one end hinge-connected to the two-joint link and the other end hinge-connected to ground. The rigid beam element is equipped with a lumped mass with rotational inertia, depending on which the sought inertance can readily be calculated. Indeed, as the rigid beam element rotates due to horizontal relative motion between its two ends, the rotational inertia of the lumped mass induces transverse forces at the beam ends for equilibrium; specifically, the force at the end connected to the two-joint link represents the inertia force developed by the inerter.

Fig. 6. Finite element model of benchmark two-floor masonry structure with TID-base isolation system.

Numerical analyses are implemented under two real earthquake records with one horizontal component acting in the y-direction of the reference system (see Figs. 1, 2, 3 and Fig. 6). The two earthquake records are taken from PEER Ground Motion Database (2013) and are properly scaled to simulate 2:3 scale experimental testing. A 5% damping factor is attributed to masonry and rc, while no damping is attributed to wood for conservativeness.

Figures 7, 8, and 9 show the absolute y-acceleration of the center of the roof top side (top acceleration), the absolute y-acceleration and the y-displacement relative to ground of the center of the foundation slab bottom side (base acceleration, base displacement), considering two configurations for the base-isolated masonry structure: (a) the classical base-isolation system without TID and (b) the proposed TID-base isolation system. It is evident that the TID has positive effects, as reduces the accelerations on the structure as well as the base displacement, which is a measure of the displacement demand on the isolators. Specifically, the reductions obtained are reported in Table 3. Based on these results, it can be concluded that the proposed TID-base isolation system is very promising for seismic protection of masonry structures, offering advantages over the classical base isolation system.

Table 3. Accelerations and displacements of benchmark two-floor masonry structure under real earthquake records

	RSN 292	RSN 753
Top acceleration without TID (m/s^2)	0.91	0.87
Top acceleration with TID (m/s^2)	0.65	0.68
Top acceleration reduction (%)	29	22
Base acceleration without TID (m/s^2)	0.90	0.78
Base acceleration with TID (m/s^2)	0.62	0.53
Base acceleration reduction (%)	31	32
Base displacement without TID (m)	0.23	0.18
Base displacement with TID (m)	0.13	0.08
Base displacement reduction (%)	43	55

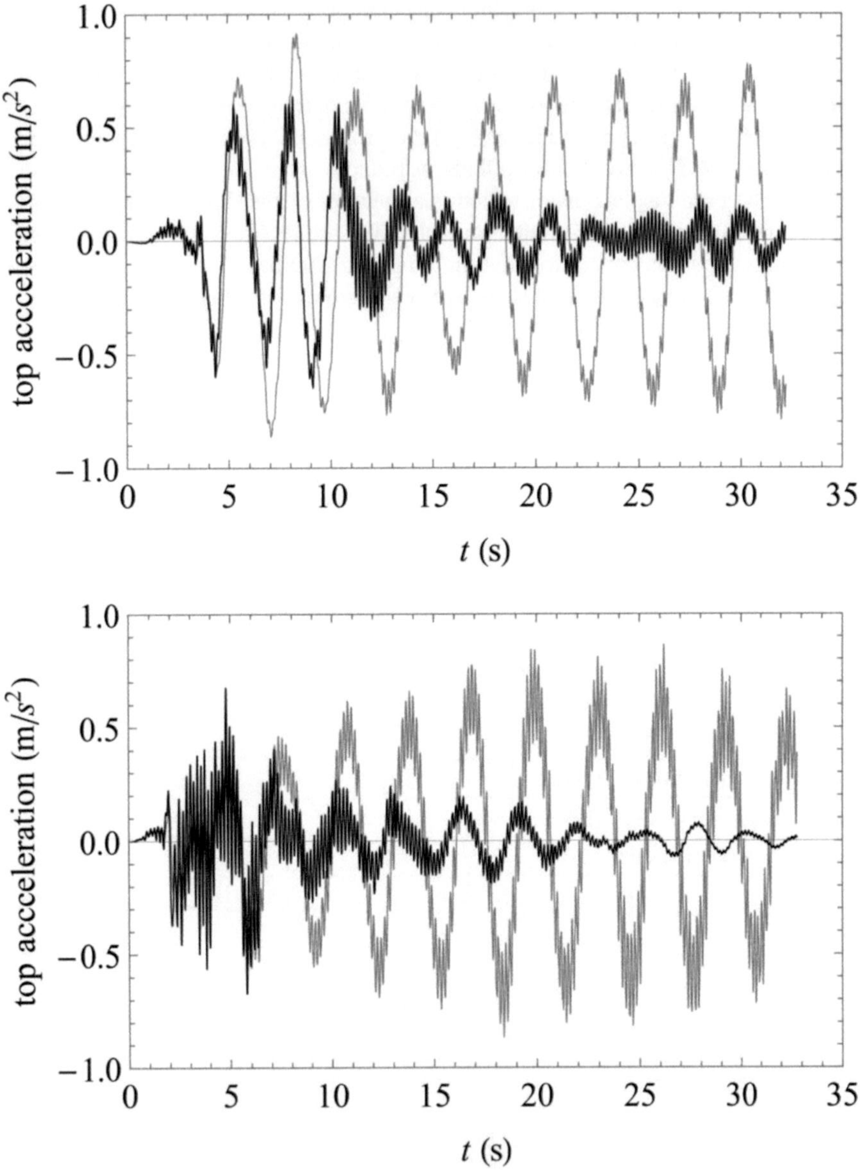

Fig. 7. Absolute top acceleration of benchmark two-floor masonry structure under real earthquake records: Irpinia, 1980 RSN 292 (top); Loma Prieta 1989 RSN 753 (bottom). Base isolation system (gray line), TID-base isolation system (black line).

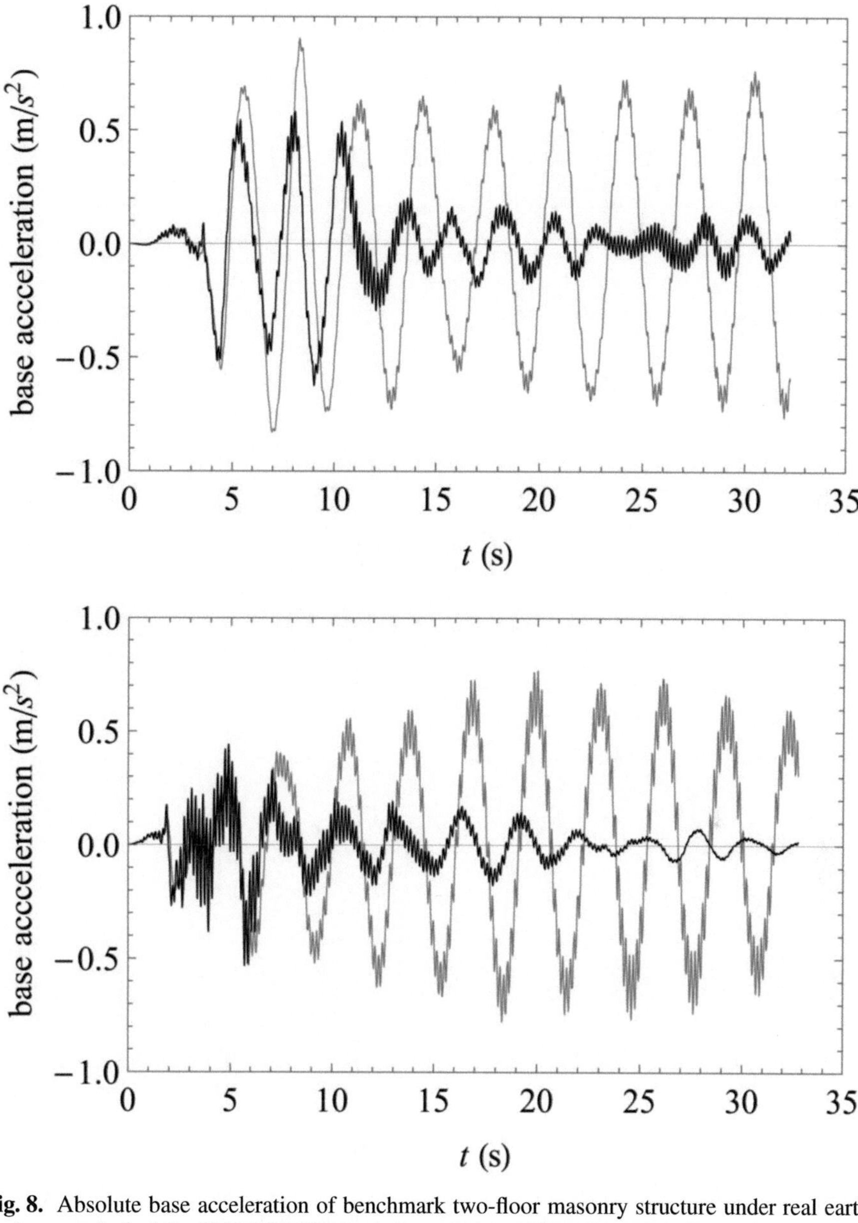

Fig. 8. Absolute base acceleration of benchmark two-floor masonry structure under real earthquake records: Irpinia, 1980 RSN 292 (top); Loma Prieta 1989 RSN 753 (bottom). Base isolation system (gray line), TID-base isolation system (black line).

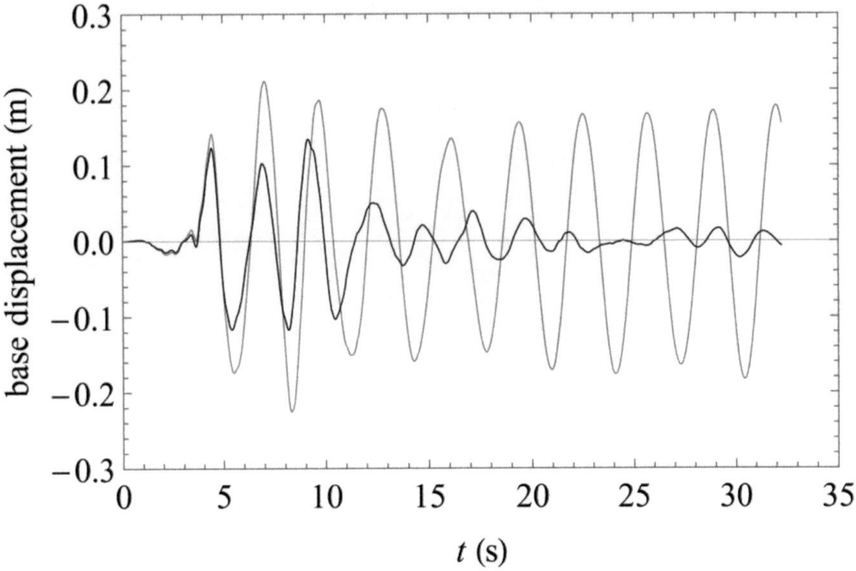

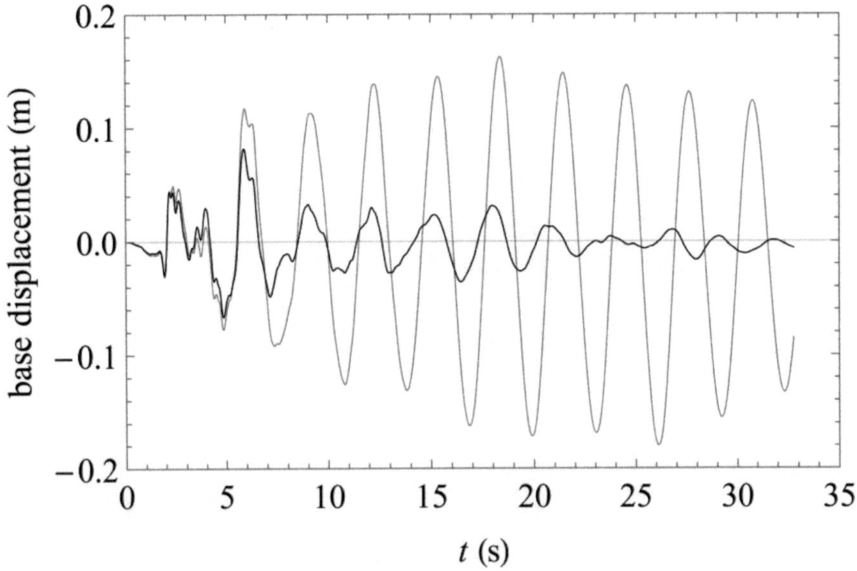

Fig. 9. Base displacement of benchmark two-floor masonry structure under real earthquake records: Irpinia, 1980 RSN 292 (top); Loma Prieta 1989 RSN 753 (bottom). Base isolation system (gray line), TID-base isolation system (black line).

4 Concluding Remarks

This paper has discussed the results obtained from numerical simulations of shaking-table experimental tests on a benchmark masonry structure, base-isolated by a hybrid base isolation system. The latter combines standard base isolators with a TID, consisting of a grounded rack-and-pinion inerter connected in series to a set of linearly-elastic springs and a viscous dashpot in parallel. Finite element analyses implemented in SAP2000 demonstrate the effectiveness of the proposed hybrid base isolation system in reducing structural accelerations and displacement demand on the isolators.

5 Appendix

Here follows a detailed list of all components of the TID along with some images of the TID, as realized by Metalmeccanica Falcone, Reggio Calabria, Italy.

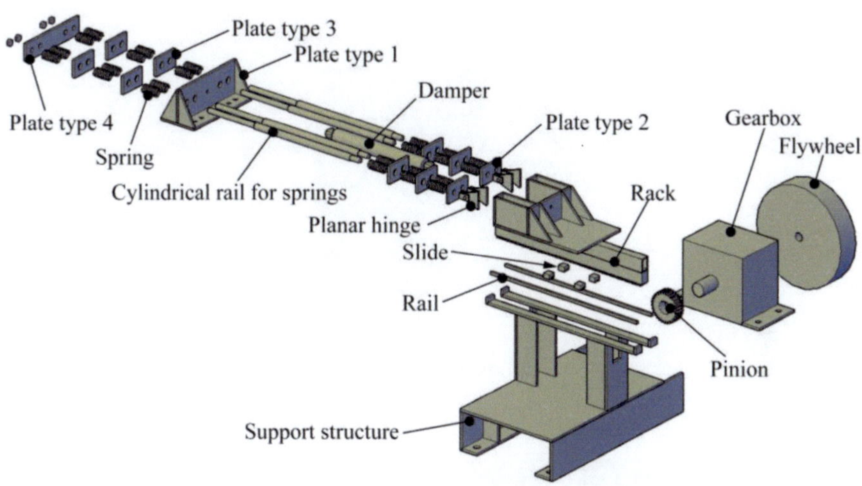

Fig. 10. Overview of all components of TID.

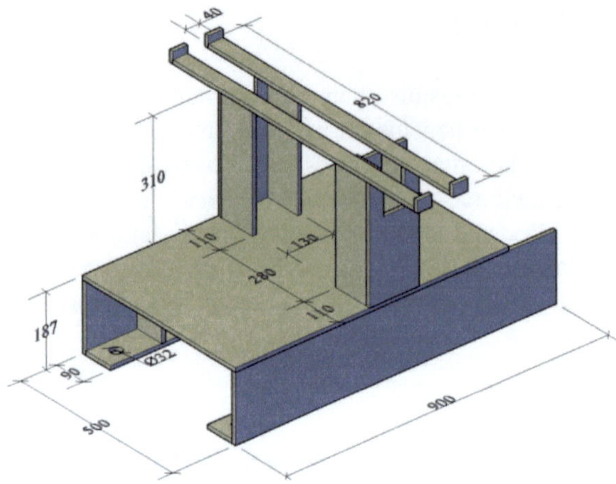

Fig. 11. Support structure.

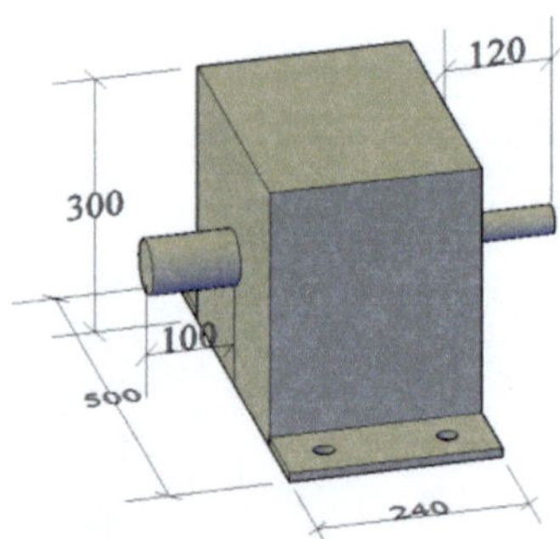

Fig. 12. Gearbox (schematic representation).

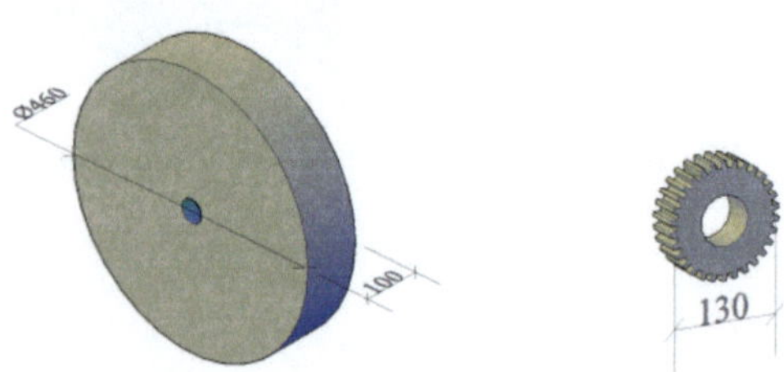

Fig. 13. Flywheel (left) and pinion (right).

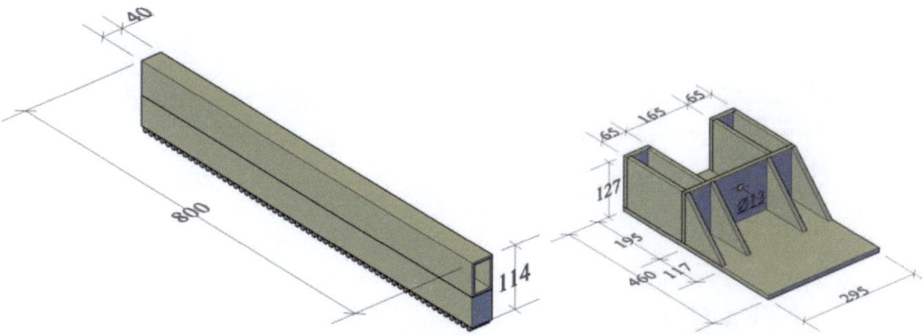

Fig. 14. Rack (left) and end stiffened plate connecting rack to springs/dashpot (right).

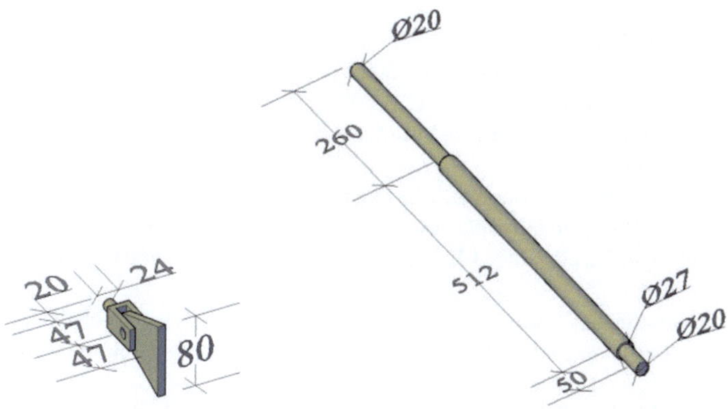

Fig. 15. Plane hinge (left) and cylindrical rail for springs (right).

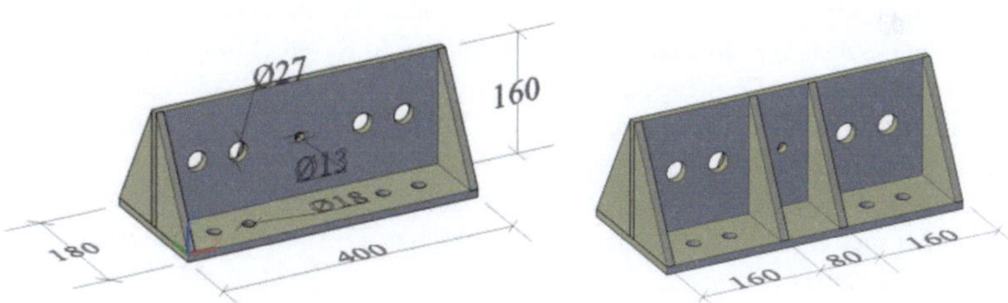

Fig. 16. Plate type 1 (opposite views).

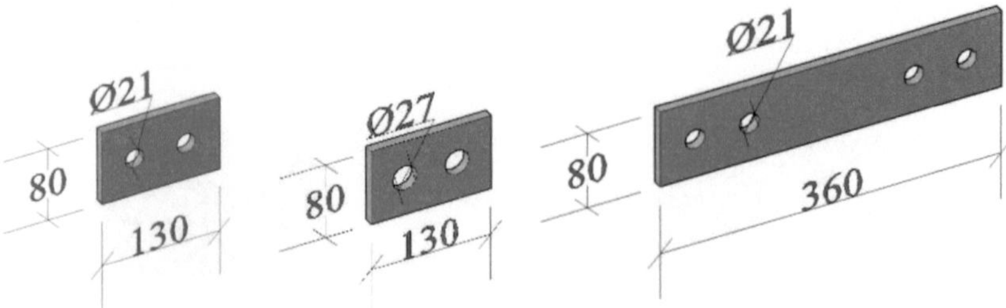

Fig. 17. Plates type 2 (left), type 3 (center), type 4 (right).

References

Alotta, G., Failla, G.: Improved inerter-based vibration absorbers. Int. J. Mech. Sci. **192**, 106087 (2021). https://doi.org/10.1016/j.ijmecsci.2020.106087

Alotta, G., Biondo, C., Giaralis, A., Failla, G.: Seismic protection of land-based wind turbine towers using the tuned inerter damper. Structures. **51**, 640–656 (2023). https://doi.org/10.1016/j.istruc.2023.03.004

Basar, T., Deb, S.K., Das, P.J., Sarmah, M.: Seismic response control of low-rise unreinforced masonry building test model using low-cost and sustainable un-bonded scrap tyre isolator (U-STI). Soil Dyn. Earthq. Eng. **142**, 106561 (2021). https://doi.org/10.1016/j.soildyn.2020.106561

Clemente, P., Martelli, A.: Seismically isolated buildings in Italy: state-of-the-art review and applications. Soil Dyn. Earthq. Eng. **119**, 471–487 (2019). https://doi.org/10.1016/j.soildyn.2017.12.029

Clemente, P., Bontempi, F., Boccamazzo, A.: Base isolated masonry buildings. In: Proceedings of the 15th World Conference on Earthquake Engineering (15WCEE), Portugal, Lisbon, 24–28 Sept 2012

Clemente, P., Bontempi, F., Boccamazzo, A.: Seismic isolation in masonry buildings: technological and economic issues. In: Modena, C., F, d.P., Valluzzi, M.R. (eds.) Brick and Block Masonry: Trends, Innovation and Challenges. 16th International Brick and Block Masonry Conference (IB2MAC), Padua, Italy, June 2016, pp. 2207–2215. Taylor & Francis Group, London, UK (2016)

De Domenico, D., Ricciardi, G.: An enhanced base isolation system equipped with optimal tuned mass damper inerter (TMDI). Earthq. Eng. Struct. Dyn. **47**(5), 1169–1192 (2018). https://doi.org/10.1002/eqe.3011

De Luca, A., Guidi, L.G.: State of art in the worldwide evolution of base isolation design. Soil Dyn. Earthq. Eng. **125**, 105722 (2019). https://doi.org/10.1016/j.soildyn.2019.105722

Faiella, D., Argenziano, M., Esposito, F., Brandonisio, G., Fraldi, M., Mele, E.: Effectiveness of isolated vertical extension of masonry buildings as nonconventional TMD. Soil Dyn. Earthq. Eng. **165**, 107675 (2023). https://doi.org/10.1016/j.soildyn.2022.107675

Ge, Z., Wang, W.: Modeling, testing, and characteristic analysis of a planetary flywheel inerter. Shock. Vib. **2018**, 2631539 (2018). https://doi.org/10.1155/2018/2631539

Habieb, A.B., Valente, M., Milani, G.: Effectiveness of different base isolation systems for seismic protection: Numerical insights into an existing masonry bell tower. Soil Dyn. Earthq. Eng. **125**, 105752 (2019a). https://doi.org/10.1016/j.soildyn.2019.105752

Habieb, A.B., Valente, M., Milani, G.: Hybrid seismic base isolation of a historical masonry church using unbonded fiber reinforced elastomeric isolators and shape memory alloy wires. Eng. Struct. **196**, 109281 (2019b). https://doi.org/10.1016/j.engstruct.2019.109281

Habieb, A.B., Valente, M., Milani, G.: Base seismic isolation of a historical masonry church using fiber reinforced elastomeric isolators. Soil Dyn. Earthq. Eng. **120**, 127–145 (2019c). https://doi.org/10.1016/j.soildyn.2019.01.022

Hu, Y., Chen, M.Z.: Performance evaluation for inerter-based dynamic vibration absorbers. Int. J. Mech. Sci. **99**, 297–307 (2015). https://doi.org/10.1016/j.ijmecsci.2015.06.003

Italian Ministry of Environment and Energy Security: Piano Nazionale di Adattamento ai Cambiamenti Climatici. https://www.mase.gov.it/sites/default/files/archivio/allegati/clima/PNACC_versione_dicembre2022.pdf (2022). Accessed 14 June 2025

Krenk, S.: Resonant inerter based vibration absorbers on flexible structures. J. Franklin Inst. **356**(14), 7704–7730 (2019). https://doi.org/10.1016/j.jfranklin.2018.11.038

Krenk, S., Høgsberg, J.: Tuned resonant mass or inerter-based absorbers: Unified calibration with quasi-dynamic flexibility and inertia correction. Proc. R. Soc. A. **472**(2185), 20150718 (2016). https://doi.org/10.1098/rspa.2015.0718

Losanno, D., Ravichandran, N., Parisi, F., Calabrese, A., Serino, G.: Seismic performance of a Low-Cost base isolation system for unreinforced brick Masonry buildings in developing countries. Soil Dyn. Earthq. Eng. **141**, 106501 (2021). https://doi.org/10.1016/j.soildyn.2020.106501

Losanno, D., Ravichandran, N., Parisi, F.: Seismic fragility of base-isolated single-storey unreinforced masonry buildings equipped with classical and recycled rubber bearings in Himalayan regions. J. Build. Eng. **45**, 103648 (2022). https://doi.org/10.1016/j.jobe.2021.103648

Marian, L., Giaralis, A.: Optimal design of a novel tuned mass-damper-inerter (TMDI) passive vibration control configuration for stochastically support-excited structural systems. Probab. Eng. Mech. **38**, 156–164 (2014). https://doi.org/10.1016/j.probengmech.2014.03.007

Marian, L., Giaralis, A.: The tuned mass-damper-inerter for harmonic vibrations suppression, attached mass reduction, and energy harvesting. Smart Struct. Syst. **19**(6), 665–678 (2017) https://doi.org/10.12989/sss.2017.19.6.665

Mele, E., Faiella, D., Argenziano, M.: Innovative mass-damping based approaches for seismic design of tall buildings. Eccomas Proceedia COMPDYN: 1778–1792 (2021). https://doi.org/10.7712/120121.8598.18959

Nauman Masoom, M., Qul Ain karim, Q., Badar, I., Khushnood, R.A., Najam, F.A., Naseer, A.: Development of a new base isolation system using the concept of metamaterials. Eng. Struct. **286**, 116151 (2023). https://doi.org/10.1016/j.engstruct.2023.116151

Osgooei, P.M., Van Engelen, N.C., Konstantinidis, D., Tait, M.J.: Experimental and finite element study on the lateral response of modified rectangular fiber-reinforced elastomeric isolators (MR-FREIs). Eng. Struct. **85**, 293–303 (2015). https://doi.org/10.1016/j.engstruct.2014.11.037

Papageorgiou, C., Houghton, N.E., Smith, M.C.: Experimental testing and analysis of inerter devices. J. Dyn. Sys. Meas. Control. **131**(1), 011001 (2009). https://doi.org/10.1115/1.3023120

Pietrosanti, D., De Angelis, M., Giaralis, A.: Experimental seismic performance assessment and numerical modelling of nonlinear inerter vibration absorber (IVA)-equipped base isolated structures tested on shaking table. Earthq. Eng. Struct. Dyn. **50**(10), 2732–2753 (2021). https://doi.org/10.1002/eqe.3469

Russillo, A.F., Failla, G., Alotta, G.: Ultra-wide low-frequency band gap in locally-resonant plates with tunable inerter-based resonators. Appl. Math. Model. **106**, 682–695 (2022). https://doi.org/10.1016/j.apm.2022.02.015

Sciutteri, S., Russillo, A.F., Santoro, R., Ricciardi, G., Failla, G.: An inerter-based concept of locally resonant fluid-conveying pipe. Eur. J. Mech. A/Solids. **106**, 105316 (2024). https://doi.org/10.1016/j.euromechsol.2024.105316

Smith, M.C.: Synthesis of mechanical networks: the inerter. IEEE Trans. Autom. Control. **47**(10), 1648–1662 (2002). https://doi.org/10.1109/TAC.2002.803532

Taniguchi, T., Der Kiureghian, A., Melkumyan, M.: Effect of tuned mass damper on displacement demand of base-isolated structures. Eng Struct. **30**, 3478–3488 (2008). https://doi.org/10.1016/j.engstruct.2008.05.027

Tsai, H.C.: The effect of tuned-mass dampers on the seismic response of base-isolated structures. Int. J. Solids Struct. **32**, 1195–1210 (1995). https://doi.org/10.1016/0020-7683(94)00150-U

Xiang, P., Nishitani, A.: Optimum design for more effective tuned mass damper system and its application to base-isolated buildings. Struct. Control Health Monit. **21**, 98–114 (2014). https://doi.org/10.1002/stc.1556

Zhang, C., Ali, A., Sun, L.: Investigation on low-cost friction-based isolation systems for masonry building structures: experimental and numerical studies. Eng. Struct. **243**, 112645 (2021). https://doi.org/10.1016/j.engstruct.2021.112645

Zoheb, N.M., Mohan, S.C., Jyosyula, S.K.R.: Development of low-cost base isolation technique using multi-criteria optimization and its application to masonry building. Soil Dyn. Earthq. Eng. **172**, 108024 (2023). https://doi.org/10.1016/j.soildyn.2023.108024

Zoheb, N.M., Mohan, S.C., Jyosyula, S.K.R.: Shake-table testing of unreinforced fly ash brick masonry model with unreinforced elastomeric isolator. Sci. Rep. **15**, 19650 (2025). https://doi.org/10.1038/s41598-025-04549-5

Locally Resonant Metamaterial Foundation for Masonry Structures: Numerical Models for Experimental Testing

Giuseppe Failla(✉), Luigi Cartone, Gioacchino Alotta, and Andrea Francesco Russillo

Department of Civil, Energy, Environmental and Materials Engineering (DICEAM), University of Reggio Calabria, Reggio Calabria, Italy
{giuseppe.failla,luigi.cartone,gioacchino.alotta, andreaf.russillo}@unirc.it

Abstract. This study presents the results obtained from numerical simulations of shaking-table experimental tests on a benchmark masonry structure, base-isolated by a novel concept of locally resonant metamaterial foundation. The foundation is an assembly of steel cells of circular shape, vertically stacked and mutually connected by circular steel laminated elastomeric bearings. The steel cell consists of a hollow steel cylinder, connected by a rubber ring to an annular steel ring, which acts as resonant unit oscillating in the horizontal plane. Detailed 3D finite element models of foundation and masonry structure are implemented in ABAQUS. Numerical results confirm the effectiveness of the proposed foundation.

1 Introduction

Locally resonant metamaterial (LRM) structural components are an emerging concept in engineering, with a remarkable number of potential applications (Chen et al. 2011; Xiao et al. 2012, 2013; Zhou et al. 2012; Baravelli and Ruzzene 2013; Peng and Pai 2014; Qian and Shi 2016; Al Ba'ba'a et al. 2017; Miranda et al. 2019; Wang et al. 2019; Failla et al. 2020; Russillo and Failla 2020, 2022, 2026; Russillo et al. 2021, 2022; Burlon and Failla 2022; Burlon et al. 2023; Santoro et al. 2023; Failla et al. 2024a, b, c; Russillo et al. 2024a, b; Sciutteri et al. 2024). The term metamaterial indicates, in a broad sense, an artificially engineered material with exotic properties not found in nature, obtained by appropriate architecture either at macro-scale or at micro-/nano-scales. The distinctive feature of LRM structural components stems from the capability of infinite media with periodically arranged resonant units to forbid the propagation of elastic waves within frequency ranges called bandgaps. The bandgaps arise from coupling between elastic waves propagating through the infinite medium and vibrations of the resonant units in the vicinities of the respective resonant frequencies; as a result of coupling, the opening frequencies of the bandgaps occur approximately at the tuning frequencies of the resonant units. Correspondingly, the transmittance of (finite) LRM structural components features strong attenuation zones matching the bandgaps predicted by wave dispersion analysis, where the propagation of elastic waves is prevented with obvious benefits in terms of vibration mitigation.

C. Nava et al. (Eds.): *Climatic and Structural Safety in Multi-Hazard Regime of Cultural and Natural Heritage*, GREEN, pp. 275–294, 2026.
https://doi.org/10.1007/978-3-032-11926-1_17

In the context of ongoing research on LRM structural components, LRM foundations are gaining increasing attention. Indeed, LRM foundations can act as mechanical filters between soil and structure, inhibiting earthquake-induced structural vibrations at frequencies falling within the theoretical bandgaps predicted by wave dispersion analysis. Several types of LRM foundations were proposed differing by the design of the unit cell (Jia and Shi 2010; Cheng and Shi 2013; Shi and Huang 2013; Yan et al. 2014; Casablanca et al. 2018; Cheng and Shi 2018; Colombi et al. 2020; Naumann Masoom et al. 2023; Li et al. 2023, 2024; Zheng et al. 2024). Some studies carried out numerical as well as experimental tests. For example, Naumann Masoom et al. (2023) conducted scale experimental tests on a LRM foundation made of high-density concrete panels embedding steel-silicone resonators, placed on top of each other and mutually connected by frictionless layers; for shear waves caused by horizontal ground motion, the experimental tests revealed an attenuation zone ranging from 2.54 Hz to 8.08 Hz, while numerical simulations envisaged a bandgap from 0.3 Hz to 7.35 Hz for the corresponding full-scale LRM foundation. Yan et al. (2014) performed experimental tests on a LRM foundation with unit cell resonating in two directions, made by a steel circular cylinder with rubber coating and embedded in concrete; the foundation was installed at the base of a small-scale frame structure and subjected to ground motions in one horizontal direction and in the vertical direction as well; Li et al. (2023, 2024) carried out scale experimental tests on a LRM foundation with unit cell made of a rubber layer and a concrete layer with resonators acting in three directions, each made up of silicone/silica gel and metallic mass; and tested it numerically and experimentally at small scale, under vertical ground motion and horizontal ground motion in one direction; as in ref. (Naumann Masoom et al. 2023), the LRM foundation was tested alone, without any base isolated structure.

The studies described above demonstrate a strong interest in LRM foundations for base isolation. However, advancing research in this field requires tackling relevant challenges. When dealing with LRM foundations and LRM structural components in general, a main and well-known issue is to obtain bandgaps of considerable size at low frequency because, for a given resonator-to-cell mass ratio, the size generally reduces as the tuning frequency decreases. Moreover, not only ensuring an adequate size at low frequency but also enhancing the attenuation within the bandgaps is a fundamental research target in this field. On the other hand, there are further aspects to be investigated for potential advancements in the current design of LRM foundations. For example, an interesting aspect is to ensure considerable attenuation not only in the low frequency range typical of seismic waves but also in higher frequency ranges, which are relevant to mitigate the effects of other ground-borne vibration sources as railway and road traffic.

This paper focuses on seismic protection of masonry structures, presenting the results obtained from numerical simulations of shaking-table experimental tests on a benchmark masonry structure, base-isolated by a novel concept of LRM foundation. The foundation is an assembly of steel cells of circular shape, vertically stacked and mutually connected by circular steel laminated elastomeric bearings. The steel cell consists of a hollow steel cylinder, connected by a rubber ring to an annular steel ring, which acts as resonant unit oscillating in the horizontal plane. Numerical simulations are carried out on detailed 3D finite element models of foundation and masonry structure, implemented in ABAQUS.

The results show that the proposed LRM foundation can attenuate the effects of earthquake ground motion by inducing relevant bandgaps with strong attenuation in the low frequency response of the structure. Notably, the effectiveness of the foundation is not limited to the low frequency range but also extends to higher frequencies, which are of interest for mitigating the effects of other ground-borne vibration sources as railway or road traffic, whose typical frequency content attains several tens of Hz up to 100 Hz (Saikia 2014; Huang et al. 2017; Pu et al. 2018; He et al. 2023; Ouakka et al. 2023). Another interesting feature of the proposed LRM foundation is that the resonant units are external, readily accessible and not embedded as in alternative LRM foundations; this allows the resonant units to be easily re-tuned by changing the rubber ring and/or the annular steel ring. Re-tuning is important to make the proposed LRM foundation adaptable to potential variations in the structural properties of the masonry structure to be protected. Indeed, non-negligible variations in mechanical parameters such as stiffness and strength may be envisaged in masonry structures as a result of climate-change related phenomena: e.g., chemical degradation due to the increase of CO2 in the atmosphere and/or high thermal stresses during increasingly frequent heatwaves, as recommended by the "National Climate Change Adaptation Plan" released by the Italian Ministry of Environment and Energy Security in December 2022 (Italian Ministry of Environment and Energy Security 2022).

2 System Under Study

The system under study is a benchmark one-floor masonry structure, base-isolated by a novel concept of LRM foundation. Both masonry structure and LRM foundations are designed for 2:3 scale experimental tests on shaking table under real earthquake records.

The masonry structure is depicted in Fig. 1 through Fig. 3, with main geometry and material parameters given in Table 1. The plan is rectangular shaped. The foundation beam and roof bond beam are made by reinforced concrete (rc). The foundation slab is made by rc and the roof slab by wood. Six wood beams support the roof slab.

The LRM foundation is illustrated in Fig. 4, the detailed geometry is given in Fig. 5 and material parameters in Table 2. It is conceived as an assembly of steel cells of circular shape. Each cell is made of a steel hollow cylinder welded to top and bottom steel plates and hosts a steel annular ring, connected to the steel hollow cylinder via a rubber annular ring; the steel annular ring acts as mass-spring resonant unit oscillating in the horizontal plane. Teflon annular rings are interposed between the steel annular ring and bottom/top steel plates to avoid friction. The cells are vertically stacked and connected to one another by circular steel-laminated elastomeric bearings. High-strength, double-sided thin adhesive layers ensure the connection between the steel-laminated elastomeric bearings and the top/bottom steel plates of two consecutive cells. Each steel-laminated elastomeric bearing includes four steel layers and five rubber layers. Overall, the LRM foundation is composed of 5 cells. In the system under study, four LRM foundations are placed at the four corners of the masonry structure.

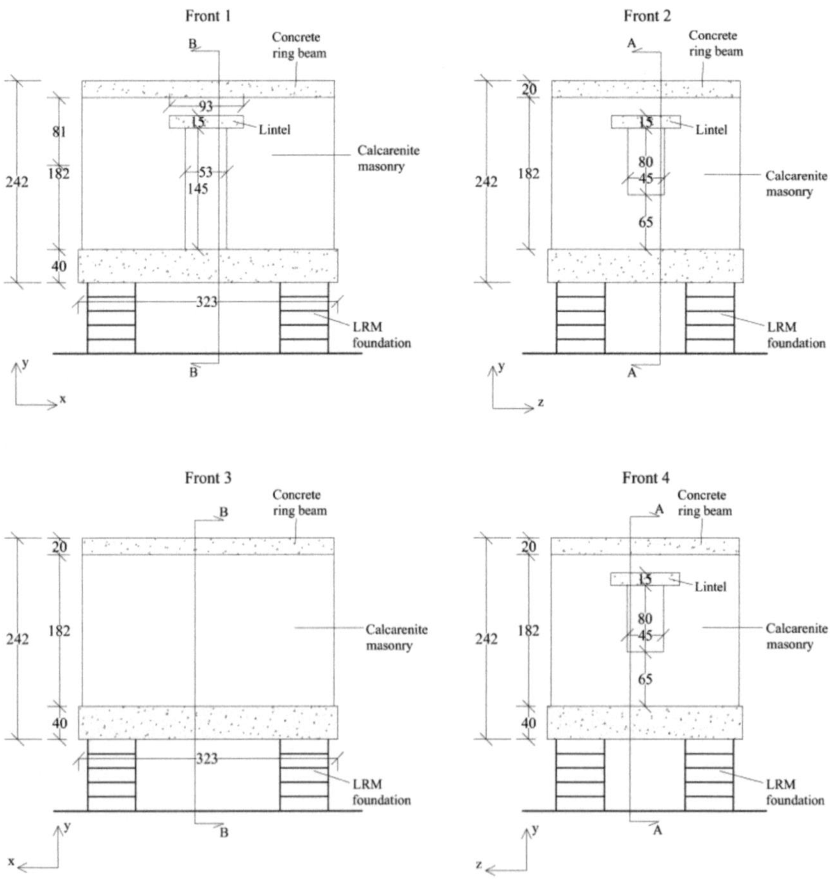

Fig. 1. Side views of benchmark one-floor masonry structure.

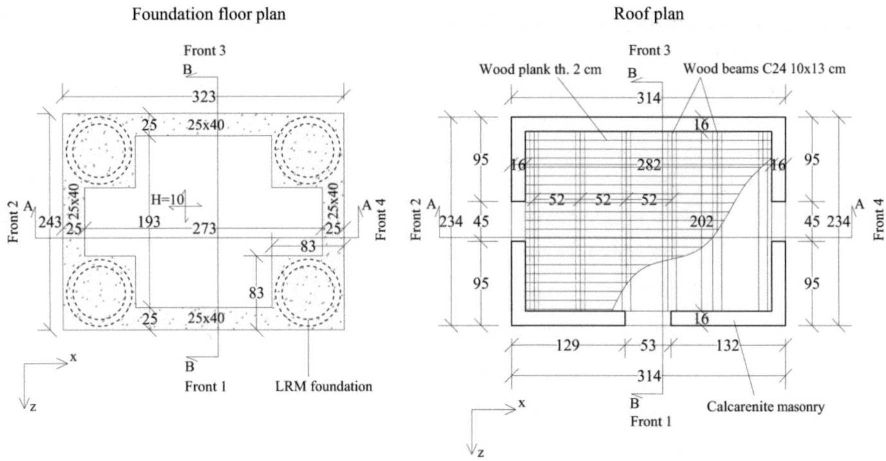

Fig. 2. Roof plan and foundation plan of benchmark one-floor masonry structure.

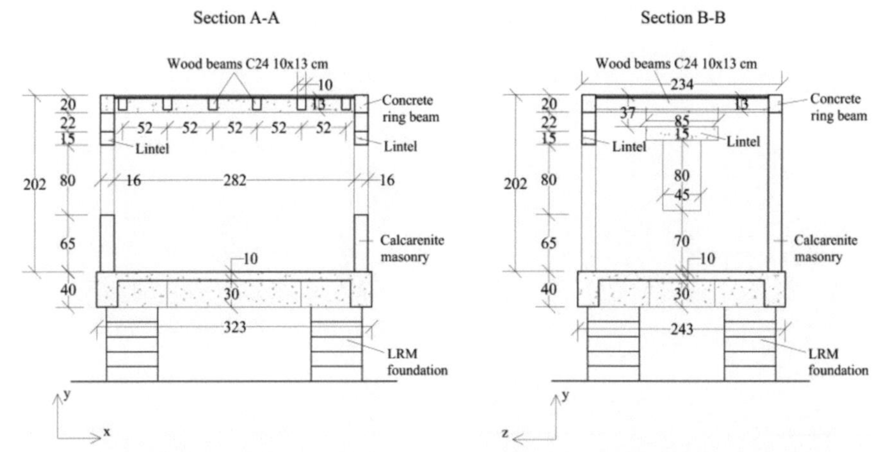

Fig. 3. Vertical plane cross sections of benchmark one-floor masonry structure.

Table 2. Material parameters of LRM foundation

	Young modulus (GPa)	Poisson coefficient	Mass density (kg/m^3)
Steel	210	0.3	7800
Rubber	9×10^{-4}	0.46	1350

Table 1. Geometry and material parameters of benchmark one-floor masonry structure

Wall sides length	2.34 × 3.23 m
Wall height	2.02 m
Wall thickness	16 cm
Roof ring beam cross section	16 × 20 cm
Roof slab thickness	2 cm
Roof beams cross section	10 × 13 cm
Foundation beam cross section	25 × 40 cm
Foundation slab thickness	10 cm
Masonry Young modulus	1410 MPa
Masonry mass density	1800 kg/m^3
Reinforced concrete Young modulus	30 GPa
Reinforced concrete mass density	2500 kg/m^3
Wood Young modulus	11 GPa
Wood mass density	420 kg/m^3

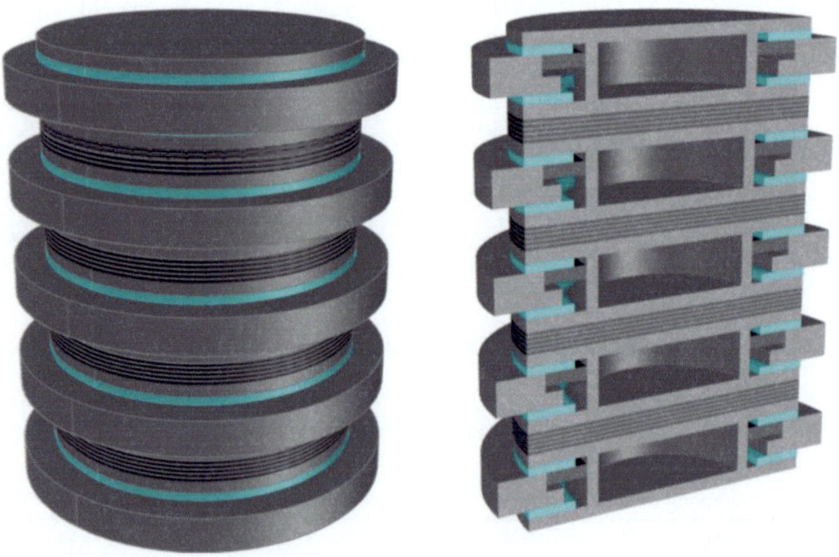

Fig. 4. Proposed LRM foundation.

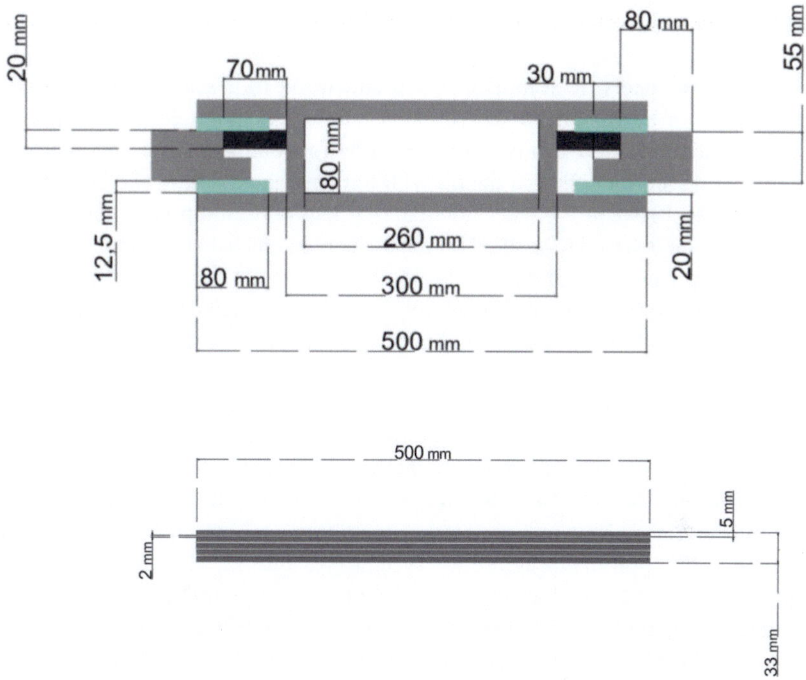

Fig. 5. Unit cell and steel-laminated elastomeric bearings of proposed LRM foundation.

3 Numerical Models

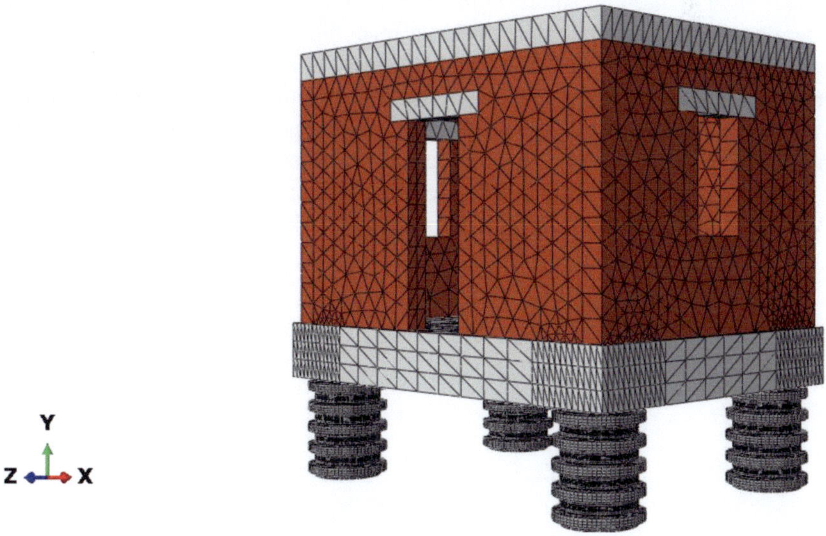

Fig. 6. Finite element model of benchmark one-floor masonry structure on four LRM foundations.

A finite-element model of the system under study is implemented in ABAQUS (see Fig. 6).

The masonry panels, rc foundation slab and wood roof slab, rc bond/foundation beams, wood roof beams are modelled using tetrahedral finite elements, while hexahedral finite elements are employed for the LRM foundations. All components of the LRM foundation are modelled in detail, with the only exception of the Teflon annular rings, interposed between the steel annular ring and bottom/top steel plates to avoid friction, which are reverted to vertical kinematic constraints between the steel plates and the steel annular rings, for simplicity. Meshes of the steel cell and the steel-laminated elastomeric bearings are generated separately and then assembled by the tie-mesh option of ABAQUS. This choice is made to obtain accurate and non-distorted meshes in the thin steel and rubber layers composing the steel-laminated elastomeric bearings. Likewise, meshes of the masonry structure and LRM foundations are created separately and coupled, again, by the tie-mesh option. Overall, 78512 finite elements model each LRM foundation and 34937 finite elements model the masonry structure.

Numerical analyses are implemented using linear modal superposition to calculate the transmittance of the masonry structure under a unit harmonic ground displacement in the z-direction of the reference system (see Figs. 1, 2, 3 and Fig. 6) and its dynamic response under real accelerograms with one horizontal component acting in the z-direction. Damping is included in the model using the "Composite Modal Damping" option in ABAQUS, which allows pertinent damping factors to be assigned to each material in the model. These damping factors are combined into a modal damping ratio for each mode, obtained as weighted averages of the mass matrix associated with each material. Attributing an appropriate damping factor to each material is important in the problem under study, because damping effects play an important role in LRM structural components: in particular, it is known that high level of damping in the resonant units may be detrimental and reduce the attenuation within the bandgaps. A 5% damping factor is attributed to masonry and rc, while a higher damping factor is attributed to rubber (rubber rings of the resonant units, rubber layers of the steel-laminated elastomeric bearings), in agreement with current literature (Pianese et al. 2024): namely, two different values are considered, 10% or 15%. Damping in steel and wood is neglected, for conservativeness. On the other hand, using linear modal superposition to calculate the dynamic response of the masonry structure may be considered an acceptable approximation, as very low stress values are expected due to the isolation effects ensured by the array of LRM foundations (Das et al. 2016; Calabrese et al. 2019).

4 Numerical Results

Figures 7, 8 show the transmittance of the masonry structure under a unit harmonic ground z-displacement, calculated as the ratio of the absolute z-displacement amplitude measured at two reference points, (a) the center of the roof top side and (b) the center of the foundation slab bottom side, to the ground displacement unit amplitude. The transmittance exhibits a strong attenuation zone starting in the vicinity of 10 Hz, which is approximately equal to the tuning frequency of the resonant units within the LRM foundations calculated as

$$f = \frac{1}{2\pi}\sqrt{\frac{k}{m}} = 10.1 \text{ Hz} \tag{1}$$

with m the mass of the steel annular ring and k the stiffness provided by the rubber annular ring provided by the following formula (Xiao et al. 2013):

$$k = \frac{\pi\left(5 + 3.29S^2\right)Gl}{\ln(r_2/r_1)} \tag{2}$$

where $S = l/[(r_1 + r_2)\, ln\, (r_2/r_1)]$ with l, r_1 and r_2 the thickness, inner and outer radii of the rubber ring, G the shear modulus of the rubber. This is an important result for design, demonstrating that the starting frequency of the attenuation zone can be predicted based on mass/stiffness properties of the resonant units within the LRM foundations. Remarkably, the attenuation zone extends over a wide frequency range, which is of interest not only for seismic wave attenuation but also for mitigating the effects of other ground-vibration sources such as railway or road traffic, whose relevant frequency content may attain tens of Hz, up to 100 Hz (Saikia 2014; Huang et al. 2017; Pu et al. 2018; He et al. 2023; Ouakka et al. 2023). In addition, an attenuation zone is also found at lower frequencies below the local resonance bandgap, right above a second resonance peak of the system at approximately 5.5 Hz that, however, is significantly damped. Based on the results in Figs. 7, 8, it can be concluded that the LRM foundations may ensure effective protection of the masonry structure not only in the low frequency range above 5.5 Hz, which is typical of several real earthquake ground motions, but also in the higher frequency range typical of railway or road traffic. This is an important conclusion of this study, demonstrating the effectiveness of the LRM foundation for a wide range of applications. Notice that the observations above hold for both modal damping factors, 10% and 15%, assumed for rubber, with no appreciable differences between the respective transmittances. Therefore, it is apparent that the high damping values of the rubber are not detrimental to the attenuation starting in the vicinity of 10 Hz and have a positive effect on the two resonance peaks below 10 Hz.

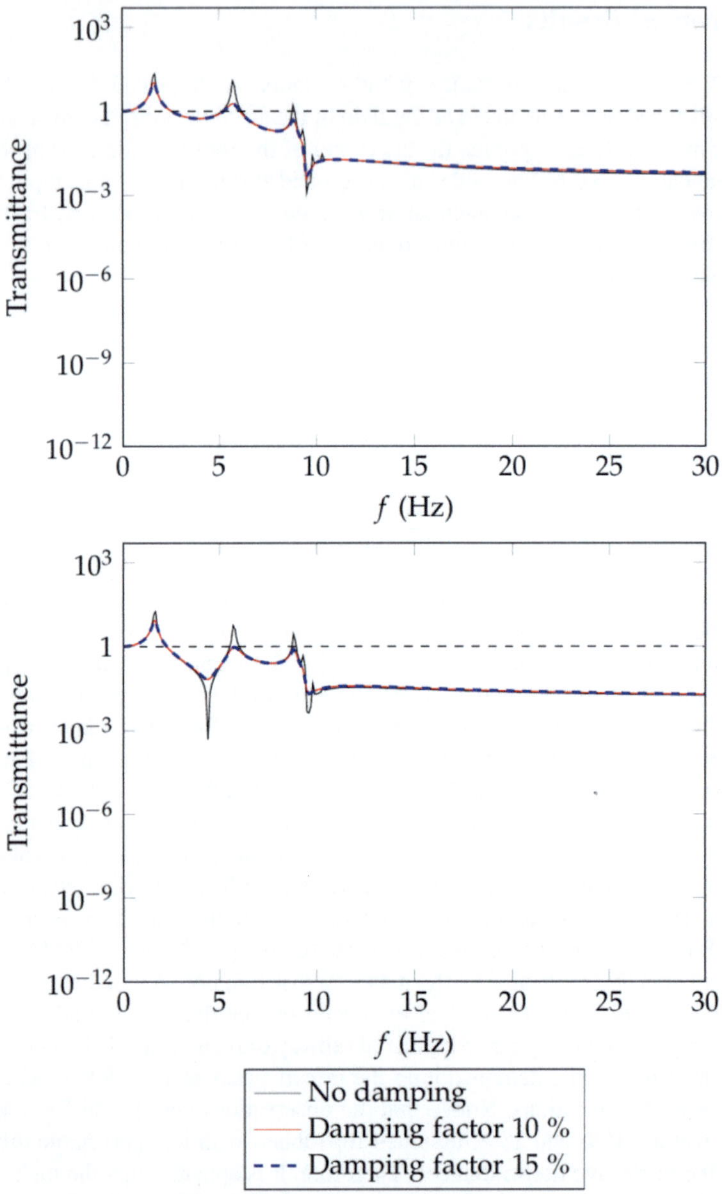

Fig. 7. Transmittance of benchmark one-floor masonry structure in $0 - 30$ Hz frequency range: (a) center of roof top side (top); (b) center of foundation slab bottom side (bottom).

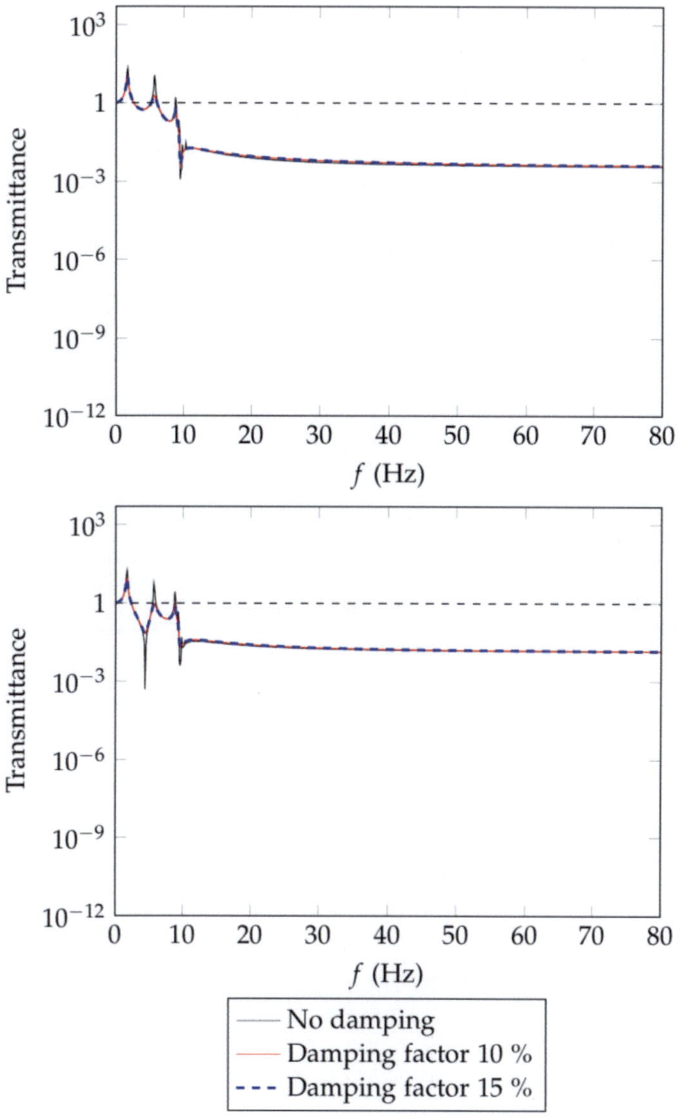

Fig. 8. Transmittance of benchmark one-floor masonry structure in 0 – 80 Hz frequency range: (a) center of roof top side (top); (b) center of foundation slab bottom side (bottom).

To corroborate the results in Figs. 7, 8, Figs. 9, 10 show acceleration and displacement time histories at relevant points of the masonry structure, under two real earthquake records; they are taken from PEER Ground Motion Database (2013) and properly scaled to simulate 2:3 scale experimental testing. Specifically, Figs. 9–10 report the absolute z-acceleration of the center of the roof top side with and without LRM foundations and the z-displacement relative to ground of the center of the foundation slab bottom side; being taken at the foundation slab bottom side, this displacement measures the displacement demand on the LRM foundations. Fig. 9 shows that the LRM foundations reduce significantly the acceleration, with reductions of the peak value up to 58% (see Table 3, where the reduction is calculated as $\left(a_{max}^{(w/o)} - a_{max}^{w}\right)/a_{max}^{(w/o)}$, with a_{max} denoting the maximum absolute value of the acceleration, (w) and (w/0) with or without the LRM foundation). Furthermore, Fig. 10 shows that the maximum displacement relative to ground slightly exceeds 10 cm, which is fully compatible with maxima displacements found for other seismic isolators, see ref. (Habieb et al. 2019).

Table 3. Reduction of peak acceleration in Fig. 9.

Damping factor	Tabas 1978 RSN139	Cape Mendocino 1992 RSN828
10%	48.97	58.67
15%	56.16	55.30

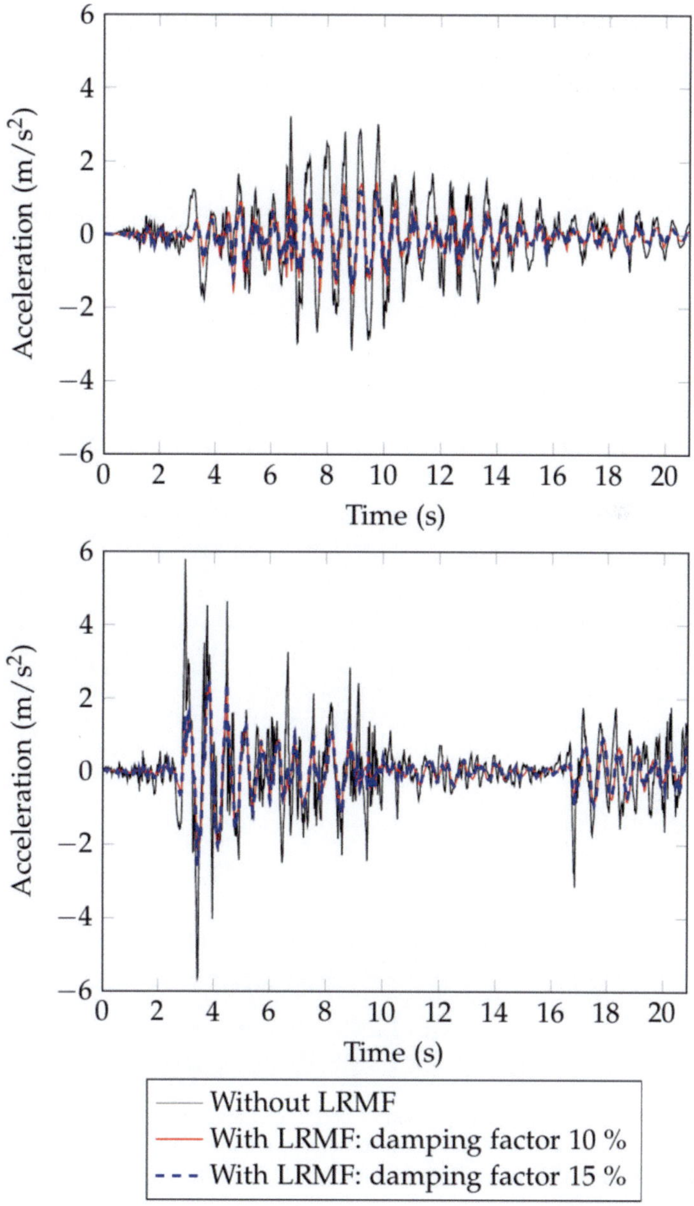

Fig. 9. Absolute z-acceleration of center of roof top side of benchmark one-floor masonry structure under real earthquake records: Tabas 1978 RSN139 (top); Cape Mendocino 1992 RSN828 (bottom). Acronym "LRMF" stands for LRM foundation.

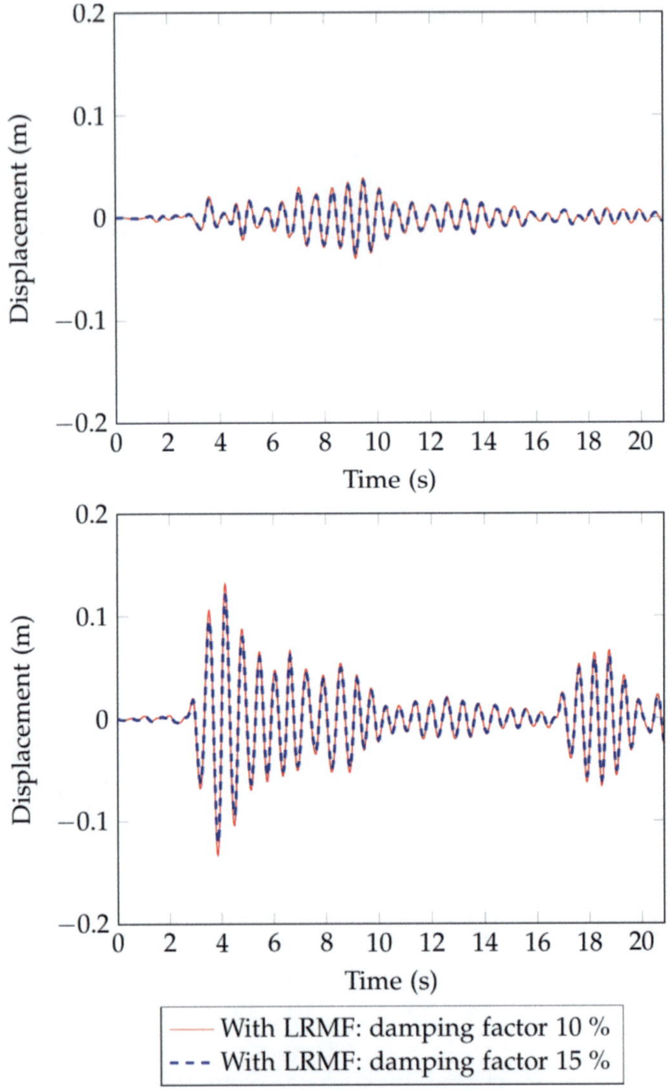

Fig. 10. Z-displacement relative to ground of center of foundation slab bottom side of benchmark one-floor masonry structure under real earthquake records (PEER Ground Motion Database (2013): Tabas 1978 RSN139 (top); Cape Mendocino 1992 RSN828 (bottom). Acronym "LRMF" stands for LRM foundation.

Finally, Figs. 11 and 12 report the maximum normal stress and shear stress measured in the steel-laminated elastomeric bearings of one LRM foundation under Cape Mendocino 1992 RSN828, considering 15% damping factor in rubber. The stress level is low and compatible with the maximum stress level allowable for steel and rubber as well as for the bond. Results do not differ significantly for the 10% damping factor and are omitted for brevity. Order of magnitude and pattern of the maximum stresses is the same under Tabas 1978 RSN139 and pertinent results are not reported for conciseness.

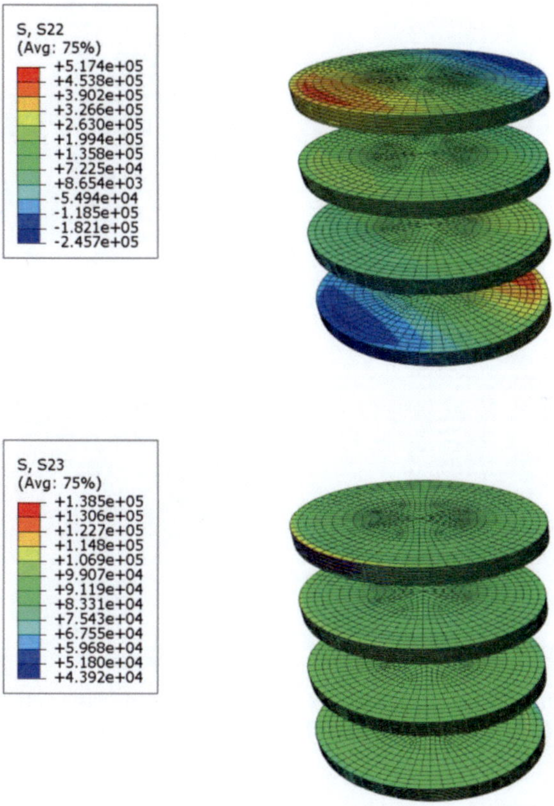

Fig. 11. Maximum stresses in rubber layers of steel-laminated elastomeric bearings in LRM foundation under Cape Mendocino 1992 RSN828: σ_z normal stress (top); τ_{yz} shear stress (bottom).

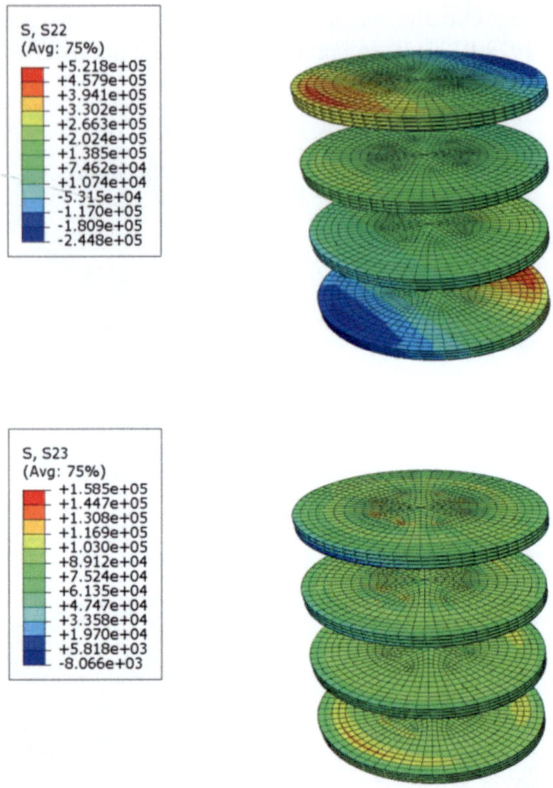

Fig. 12. Maximum stresses in steel layers of steel-laminated elastomeric bearings in LRM foundation under Cape Mendocino 1992 RSN828: σ_z normal stress (top); (b) τ_{yz} shear stress (bottom).

5 Concluding Remarks

This study has discussed the results obtained from numerical simulations of shaking-table experimental tests on a benchmark masonry structure, base-isolated by a novel concept of LRM foundation. The foundation is an assembly of steel cells of circular shape, vertically stacked and mutually connected by circular steel laminated elastomeric bearings. The steel cell consists of a hollow steel cylinder, connected by a rubber ring to an annular steel ring, which acts as resonant unit oscillating in the horizontal plane. Detailed 3D finite element analyses of foundation and masonry structure, implemented in ABAQUS, demonstrate the effectiveness of the LRM foundation not only in the low frequency range typical of earthquake ground motion but also in the higher frequency range typical of other ground-borne vibration sources as railway or road traffic, opening interesting perspectives for engineering applications.

References

Al Ba'ba'a, H., Nouh, M., Singh, T: Formation of local resonance band gaps in finite acoustic metamaterials: a closed-form transfer function model. J. Sound Vib. **410**, 429–446 (2017). https://doi.org/10.1016/j.jsv.2017.08.009

Baravelli, E., Ruzzene, M.: Internally resonating lattices for bandgap generation and low-frequency vibration control. J. Sound Vib. **332**, 6562–6579 (2013). https://doi.org/10.1016/j.jsv.2013.08.014

Burlon, A., Failla, G.: Flexural wave propagation in locally-resonant beams with uncoupled/coupled bending-torsion beam-like resonators. Int. J. Mech. Sci. **215**, 106925 (2022). https://doi.org/10.1016/j.ijmecsci.2021.106925

Burlon, A., Failla, G.: On the band gap formation in locally-resonant metamaterial thin-walled beams. Eur. J. Mech. A/Solids. **97**, 104798 (2023). https://doi.org/10.1016/j.euromechsol.2022.104798

Das, A., Deb, S.K., Dutta, A.: Comparison of numerical and experimental seismic responses of FREI-supported un-reinforced brick masonry model building. J. Earthq. Eng. **20**(8), 1239–1262 (2016). https://doi.org/10.1080/13632469.2016.1140098

Calabrese, A., Losanno, D., Spizzuoco, M., Strano, S., Terzo, M.: Recycled rubber fiber reinforced bearings (RR-FRBs) as base isolators for residential buildings in developing countries: the demonstration building of Pasir Badak, Indonesia. Eng. Struct. **192**, 126–144 (2019). https://doi.org/10.1016/j.engstruct.2019.04.076

Casablanca, O., Ventura, G., Garescì, F., Azzerboni, B., Chiaia, B., Chiappini, M., Finocchio, G.: Seismic isolation of buildings using composite foundations based on metamaterials. J. Appl. Phys. **123**(17), 174903 (2018). https://doi.org/10.1063/1.5018005

Chen, J.S., Sharma, B., Sun, C.T.: Dynamic behaviour of sandwich structure containing spring-mass resonators. Compos. Struct. **93**, 2120–2125 (2011). https://doi.org/10.1016/j.compstruct.2011.02.007

Cheng, Z., Shi, Z.: Novel composite periodic structures with attenuation zones. Eng. Struct. **56**, 1271–1282 (2013). https://doi.org/10.1016/j.engstruct.2013.07.003

Cheng, Z.B., Shi, Z.F.: Composite periodic foundation and its application for seismic isolation. Earthq. Eng. Struct. Dyn. **47**(4), 925–944 (2018). https://doi.org/10.1002/eqe.2999

Clemente, P., Martelli, A.: Seismically isolated buildings in Italy: state-of-the-art review and applications. Soil Dyn. Earthq. Eng. **119**, 471–487 (2019). https://doi.org/10.1016/j.soildyn.2017.12.029

Colombi, A., Zaccherini, R., Aguzzi, G., Palermo, A., Chatzi, E.: Mitigation of seismic waves: metabarriers and metafoundations bench tested. J. Sound Vib. **485**, 115537 (2020). https://doi.org/10.1016/j.jsv.2020.115537

Failla, G., Santoro, R., Burlon, A., Russillo, A.F.: An exact approach to the dynamics of locally-resonant beams. Mech. Res. Commun. **103**, 103460 (2020). https://doi.org/10.1016/j.mechrescom.2019.103460

Failla, G., Burlon, A., Russillo, A.F.: A novel metamaterial multiple beam structure with internal local resonance. Acta Mech. **235**, 5885–5903 (2024a). https://doi.org/10.1007/s00707-024-04006-w

Failla, G., Marzani, A., Palermo, A., Russillo, A.F., Colquitt, D.: Current developments in elastic and acoustic metamaterials science. Philos. Trans. R. Soc. A. **382**, 20230369 (2024b). https://doi.org/10.1098/rsta.2023.0369

Failla, G., Marzani, A., Palermo, A., Russillo, A.F., Colquitt, D.: Current developments in elastic and acoustic metamaterials science. Philos. Trans. R Soc. A. **382**, 20240038 (2024c). https://doi.org/10.1098/rsta.2024.0038

Habieb, A.B., Valente, M., Milani, G.: Base seismic isolation of a historical masonry church using fiber reinforced elastomeric isolators. Soil Dyn. Earthq. Eng. **120**, 127–145 (2019). https://doi.org/10.1016/j.soildyn.2019.01.022

He, C., Zhou, S., Li, X., Di, H., Zhang, X.: Forest trees as a natural metamaterial for surface wave attenuation in stratified soils. Constr. Build. Mater. **363**, 129769 (2023). https://doi.org/10.1016/j.conbuildmat.2022.129769

Huang, J., Liu, W., Shi, Z.: Surface-wave attenuation zone of layered periodic structures and feasible application in ground vibration reduction. Constr. Build. Mater. **141**, 1–11 (2017). https://doi.org/10.1016/j.conbuildmat.2017.02.153

Italian Ministry of Environment and Energy Security: Piano Nazionale di Adattamento ai Cambiamenti Climatici. https://www.mase.gov.it/sites/default/files/archivio/allegati/clima/PNACC_versione_dicembre2022.pdf (2022). Accessed 14 June 2025

Jia, G., Shi, Z.: A new seismic isolation system and its feasibility study. Earthq. Eng. Eng. Vib. **9**, 75–82 (2010). https://doi.org/10.1007/s11803-010-8159-8

Li, Y., He, H., Chen, H., Liu, Y., Wang, W., Dong, J., Jin, F., Fan, H.: Composite wave attenuation mechanism of periodic layered metastructure with embedded rubber-mass resonators. Constr. Build. Mater. **384**, 131322 (2023). https://doi.org/10.1016/j.conbuildmat.2023.131322

Li, Y., He, H., Feng, J., Chen, H., Jin, F., Fan, H.: Layered metastructure containing freely-designed local resonators for wave attenuation. Def. Technol. **32**, 23–31 (2024). https://doi.org/10.1016/j.dt.2023.03.026

Miranda Jr., E., Nobrega, E., Ferreira, A., Dos Santos, J.: Flexural wave band gaps in a multi-resonator elastic metamaterial plate using Kirchhoff-Love theory. Mech. Syst. Signal Process. **116**, 480–504 (2019). https://doi.org/10.1016/j.ymssp.2018.06.059

Naeim, F., Kelly, J.M.: Design of Seismic Isolated Structures: From Theory to Practice. Wiley, New York (1999)

Nauman Masoom, M., Q, u.A.K., Badar, I., Khushnood, R.A., Najam, F.A., Naseer, A.: Development of a new base isolation system using the concept of metamaterials. Eng. Struct. **286**, 116151 (2023). https://doi.org/10.1016/j.engstruct.2023.116151

Ouakka, S., Gueddida, A., Pennec, Y., Djafari-Rouhani, B., Kouroussis, G., Verlinden, O.: Efficient mitigation of railway induced vibrations using seismic metamaterials. Eng. Struct. **284**, 115767 (2023). https://doi.org/10.1016/j.engstruct.2023.115767

PEER Ground Motion database: University of California, Berkeley. https://ngawest2.berkeley.edu/ (2013). Accessed 10 Jun 2025

Peng, H., Pai, P.F.: Acoustic metamaterial plates for elastic wave absorption and structural vibration suppression. Int. J. Mech. Sci. **89**, 350–361 (2014). https://doi.org/10.1016/j.ijmecsci.2014.09.018

Pianese, G., Van Engelen, N., Toopchi-Nezhad, H., Milani, G.: An experimental and numerical insight into the unbonded and partially bonded high-damping fiber-reinforced elastomeric isolators. Soil Dyn. Earthq. Eng. **187**, 109016 (2024). https://doi.org/10.1016/j.soildyn.2024.109016

Pu, X., Shi, Z., Xiang, H.: Feasibility of ambient vibration screening by periodic geofoamfilled trenches. Soil Dyn. Earthq. Eng. **104**, 228–235 (2018). https://doi.org/10.1016/j.soildyn.2017.10.022

Qian, D., Shi, Z.: Bandgap properties in locally resonant phononic crystal double panel structures with periodically attached spring-mass resonators. Phys. Lett. A. **380**(41), 3319–3325 (2016). https://doi.org/10.1016/j.physleta.2016.07.068

Russillo, A.F., Failla, G.: On the free vibrations of locally-resonant structures. Comput. Struct. **241**, 106356 (2020). https://doi.org/10.1016/j.compstruc.2020.106356

Russillo, A.F., Failla, G., Fraternali, F.: Free and forced vibrations of damped locally-resonant sandwich beams. Eur. J. Mech. A/Solids. **86**, 104188 (2021). https://doi.org/10.1016/j.euromechsol.2020.104188

Russillo, A.F., Failla, G.: A novel reduced-order dynamic-stiffness formulation for locally resonant metamaterial plates. Compos. Struct. **280**, 114811 (2022). https://doi.org/10.1016/j.compstruct.2021.114811

Russillo, A.F., Failla, G., Alotta, G.: Ultra-wide low-frequency band gap in locally-resonant plates with tunable inerter-based resonators. Appl Math Model. **106**, 682–695 (2022). https://doi.org/10.1016/j.apm.2022.02.015

Russillo, A.F., Arena, F., Failla, G.: Water-tank metabarrier for seismic Rayleigh wave attenuation. Philos. Trans. R. Soc. A. **382**, 20230363 (2024a). https://doi.org/10.1098/rsta.2023.0363

Russillo, A.F., Kouznetsova, V.G., Failla, G., Geers, M.G.D.: A reduced-order computational homogenization framework for locally resonant metamaterial structures. Comput. Mech. **74**, 743–762 (2024b). https://doi.org/10.1007/s00466-024-02453-9

Russillo, A.F., Failla, G.: Seismic metamaterials for Rayleigh wave attenuation: a novel concept of soil-embedded water-tank metabarrier. Int. J. Solids Struct. **324**, 113656 (2026). https://doi.org/10.1016/j.ijsolstr.2025.113656

Saikia, A.: Numerical study on screening of surface waves using a pair of softer backfilled trenches. Soil Dyn. Earthq. Eng. **65**, 206–213 (2014). https://doi.org/10.1016/j.soildyn.2014.05.012

Santoro, R., Mazzeo, M., Failla, G.: A computational framework for uncertain locally resonant metamaterial structures. Mech. Syst. Signal Process. **190**, 110094 (2023). https://doi.org/10.1016/j.ymssp.2023.110094

Sciutteri, S., Russillo, A.F., Santoro, R., Ricciardi, G., Failla, G.: An inerter-based concept of locally resonant fluid-conveying pipe. Eur. J. Mech. A/Solids. **106**, 105316 (2024). https://doi.org/10.1016/j.euromechsol.2024.105316

Shi, Z., Huang, J.: Feasibility of reducing three-dimensional wave energy by introducing periodic foundations. Soil Dyn. Earthq. Eng. **50**, 204–212 (2013). https://doi.org/10.1016/j.soildyn.2013.03.009

Wang, K., Zhou, J., Cai, C., Xu, D., Ouyang, H.: Mathematical modeling and analysis of a meta-plate for very low-frequency band gap. Appl Math Model. **73**, 581–597 (2019). https://doi.org/10.1016/j.apm.2019.04.033

Xiang, H.J., Shi, Z.F., Wang, S.J., Mo, Y.L.: Periodic materials-based vibration attenuation in layered foundations: experimental validation. Smart Mater Struct. **21**, 112003 (2012). https://doi.org/10.1088/0964-1726/21/11/112003

Xiao, Y., Wen, J., Wen, X.: Broadband locally resonant beams containing multiple periodic arrays of attached resonators. Phys. Lett. A. **376**(16), 1384–1390 (2012). https://doi.org/10.1016/j.physleta.2012.02.059

Xiao, Y., Wen, J., Yu, D., Wen, X.: Flexural wave propagation in beams with periodically attached vibration absorbers: band-gap behavior and band formation mechanisms. J. Sound Vib. **332**(4), 867–893 (2013). https://doi.org/10.1016/j.jsv.2012.09.035

Yan, Y., Laskar, A., Cheng, Z., Menq, F., Tang, Y., Mo, Y.L., Shi, Z.: Seismic isolation of two dimensional periodic foundations. J. Appl. Phys. **116**(4), 044908 (2014). https://doi.org/10.1063/1.4891837

Zheng, H., Miao, L., Xiao, P., Lei, K., Wang, Q.: Novel metamaterial foundation with multi low-frequency bandgaps for isolating earthquakes and train vibrations. Structures. **61**, 106070 (2024). https://doi.org/10.1016/j.istruc.2024.106070

Zhou, X., Liu, X., Hu, G.: Elastic metamaterials with local resonances: an overview. Theor. Appl. Mech. Lett. **2**, 041001 (2012). https://doi.org/10.1063/2.1204101

Experimental Assessment of Climate-Induced Impacts on Heritage Materials: Application of the Damage Index Methodology in the Case Study of Bova

Francesca Giglio and Francesco Armocida(✉)

Department of Architecture and Desing (dAeD), Mediterranea University of Reggio Calabria, Reggio Calabria, Italy
{francesca.giglio,francesco.armocida}@unirc.it

Abstract. This paper presents an advanced methodology for defining a Damage Index to assess the climate vulnerability of material cultural heritage. The Index is based on three core factors: environmental conditions (F_{env}), material properties (F_{mat}), and external drivers (F_{ext}), each represented by specific indicators, weighted and normalized on a 0–3 scale using the Analytic Hierarchy Process (AHP), allowing for structured risk evaluation. This contribution focuses specifically on the experimental assessment of material factors (F_{mat}), aimed at determining the physical and chemical properties of heritage materials through a replicable site-specific approach. The study is structured in three sections: the first outlines the overall framework of the Index; the second details the experimental and simulation procedures for assessing material properties; the third presents its application to the case study of Palazzo Mesiani in Bova (Reggio Calabria). These analyses were conducted by the MATeRICs Interdepartmental Laboratory at Mediterranea University of Reggio Calabria, using non-destructive techniques such as X-Ray Fluorescence (XRF), Scanning Electron Microscopy with Energy Dispersive X-ray Spectroscopy (SEM-EDX), and X-Ray Diffraction (XRD). The data were processed via a dedicated spreadsheet enabling step-by-step calculation of the Index. The case study confirms the methodology's applicability and potential for integration within Living Labs and digital platforms, offering a scalable tool to support adaptive conservation strategies for heritage exposed to evolving climate-related risks.

Keywords: Damage Index · climate vulnerability · cultural heritage · experimental analysis · simulation methodologies

1 Introduction

Climate change is the fastest-growing global threat to the world's natural and cultural heritage. The conservation of built heritage requires complex studies concerning their Global Vulnerability (Day et al., 2020; Smith, 2013). These studies must consider the

C. Nava et al. (Eds.): *Climatic and Structural Safety in Multi-Hazard Regime of Cultural and Natural Heritage*, GREEN, pp. 295–312, 2026.
https://doi.org/10.1007/978-3-032-11926-1_18

current condition of the building, particularly its degree of degradation, as well as the main factors contributing to deterioration. These factors extend beyond the building's structure and include its location and surrounding environmental conditions. Therefore, the assessment of built heritage vulnerability should encompass not only the building itself but also the broader site and environmental context. (Damas Mollá et al., 2022).

Cultural heritage is increasingly threatened by climate-related risks, including erosion, corrosion, thermal stress, and freeze-thaw cycles, posing significant challenges for preservation (Mavrakou et al., 2025; Sesana et al., 2019). The literature review highlights that climate risks often compound other pressures, such as urbanization, overtourism and lack of adaptation policies and methodologies (Georgiadis, 2019; Boeri et al., 2023).

Possible resolution strategies focus on the built environment, advanced, ecological, and regenerative design (Nava, 2023) capable of strengthening the levels of preparedness of the contexts involved and, at the same time, increasing resilience to climate risk. Intervening in this direction requires the experimentation of integrated tools that consider the multiple fragility conditions inherent to the analyzed heritage contexts. Measurement of multiple risk factors and the calculation of site-specific damage indices represent priority tools for identifying risk conditions in the considered heritage contexts.

As stated in Giglio F., et al., 2024, an all-encompassing approach should be employed when evaluating the vulnerability to climate change of cultural heritage buildings and structures. This approach should consider various aspects, such as the long-term and sudden impacts of climate change, non-climatic factors, the capacity to adapt, and the cultural significance to the local community. The calculation of these indices should rely on comprehensive vulnerability assessments, utilization of extended monitoring methods, and execution of policies to overcome barriers to adaptation.

In this context it's proposed a methodology for assessing the vulnerability of cultural heritage materials and structures to climate change (Damage Index) that offers a systematic approach to understanding how climate change affects cultural heritage materials.

The paper is structured in three parts: the first provides an overview that briefly outlines the general methodology for calculating the Damage Index, identifying its main steps; the second describes in detail the methodologies used to assess one of the three components of the index, specifically focusing on material properties (F_{mat}) through the different experimental analyses applied; the third part presents the application of the experimental methodology to the case study of Palazzo Mesiani in Bova (RC), with a specific focus on the description of the in-situ material analyses carried out.

2 The Damage Index: An Endpoint Index for Assessing the Combined Effect of Environmental, Material and External Factors of Climate Change on Cultural Heritage

A vulnerability assessment includes an analysis of the scope and severity of the potential impacts of climate change. The study of risk reduction requires a perception of the hazards, and a representation of the system's response when subjected to the hazards (Esposito et al., 2017). Over the past decade, efforts to assess vulnerability to climate change triggered a process of theory development and assessment practice, which is also reflected in the recent reports of the IPCC (Füsse and Klein 2006).

In this context, vulnerability plays a crucial role in assessing the level of risk and developing measures to reduce risk and adapt to climate change. Vulnerability can be defined as the extent to which a system, subsystem, or component is susceptible to damage when exposed to a risk, such as a disturbance or stressor (Turner et al. 2003). Vulnerability, within the realm of climate change adaptation research, is described as the propensity or predisposition to experience adverse effects (IPCC, 2018). Within the framework of the IPCC, vulnerability encompasses various concepts and elements, including sensitivity or susceptibility to harm, as well as the lack of capacity to cope and adapt (IPCC, 2023).

In Vulnerability Assessment Framework, experimented by Ravan, M., et al. 2023, the primary variable is "susceptibility", which pertains to the inherent vulnerability of the elements exposed to hazards and is function of "sensitivity" that means "an attribute of the system, existing prior to the perturbation, and separate from exposure". The secondary variable is the "lack of resilience", which is characterized by limited access to and mobilization of resources to effectively respond to and recover from the adverse consequences of hazards, therefore a lack of adaptive capacity, that is "the ability of systems, institutions, humans and other organisms to adjust to potential damage, to take advantage of opportunities, or to respond to consequences" (Ravan et al., 2023), even for slow onset events in a long-term perspective.

Following those definitions, the literature review suggested for this study the development of an endpoint index, that consider the comprehensive and combined effect of various midpoint indicators. The Damage Index has been chosen to this purpose, being one of the most representative endpoint indicators. Therefore, it has been considered as a function of the three primary factors:

$$DAMAGE\ INDEX = f(F_{env}, F_{mat}, F_{ext})$$

About what has been established so far, the identification, definition, and quantification of the Damage Index is closely tied to three fundamental components:

- Environmental conditions (F_{env}),
- Material properties (F_{mat}),
- External factors (F_{ext}).

These components collectively form the Damage Index definition (Fig. 1).

It is important to highlight that Damage Index evaluation is just referred to the degradation processes of building materials, such as physical and chemical damages, excluding static and dynamic alterations of structures.

The Damage Index, previously outlined in Giglio et al. 2024, and in Lucanto et al. 2024, adopts a comprehensive approach to assessing the climate change vulnerability of cultural heritage buildings and structures by integrating various types of data, including environmental data, experimental ex-post findings, and non-climatic factors. The Index is calculated by the normalization of experimental data and their weighting through Analytic Hierarchy Process (AHP), to weigh the different indicators, from 0 (low risk) to 3 (high risk) (Fig. 2). The AHP, initially formulated by Saaty and Kearns in 1985,

	Indicators	Ranking criteria	Susceptibility
F env	**Environmental conditions**		1.59
I1	Extreme temperatures, heat waves, and droughts	the frequency and intensity of thermal extremes and heat waves (and precipitation) are expected to be affected by the increase in global surface temperature, resulting in an increased likelihood of storms or heavy rains	
I2	Extreme rainfall, storms and floods	represent a serious multi-level threat with direct and indirect impacts, periodic reappearance, and complicated socio-economic impacts [21]. Possible overloads of structures represent the risks of extreme precipitation, the accumulation of pollutants, penetration into the urban	
I3	Severe wind	the surfaces of the built CH undergo erosive/abrasive effects because the wind is often accompanied by rain, salt, and sand. A further effect is the chemical changes due to	
I4	Sea level rise and wave action	landscapes and buildings may be subject to erosion caused by prolonged contact with water, along with salt intrusion and physical and mechanical impacts from waves. Therefore, CH sites located in coastal areas are at risk of this phenomenon and subsequent coastal flooding, with the	
F mat	**Material properties**		0.56
I5	Deterioration of surface	Monitoring the relative intensity of diffraction of principal phase /composition/ porosity	
I6	Efflorescence index	Intensity of crystalized salt	
I7	Carbonation index	Identification of carbonation product	
I8	Corrosion index	Determination of corrosion product	
F ext	**External drivers**		1.46
I9	Geological-Related Hazards	Earthquakes, landslides, volcanic eruption	
I10	Biological-Related Hazard	Specific metabolic activity related to living organisms (warm or humid climates)	
I11	Human-Induced Hazards	Excessive land use due to urban growth and pollution	

Fig. 1. Excel table for Environmental, Materials and External factors, indicators and ranking criteria.

is a technique used to assign relative importance to different evaluation criteria in a decision-making procedure.

Fig. 2. Numerical score system referred to specific standard of reference

or these reasons, a procedural methodology has been developed for the quantitative definition of the index, based on the variables acquired from the experiments conducted by Ravan et al. (2023), through the application of appropriate calculation methods developed by the authors[1]. The methodology employs:

- Vulnerability, calculated as a function of susceptibility and adaptive capacity:

$$Vulnerability\ (sub - index) = \sqrt{Susceptibility\ x\ Lack\ ofAdaptability}$$

[1] The experimental development of the Damage Index was carried out at the MATeRICs Interdepartmental Laboratory in collaboration with the MATEES Laboratory, involving Prof. P. Frontera and Eng. RtdB A. Malara, together with the authors, Prof. F. Giglio and Arch. PhD F. Armocida.

- Sensitivity, determined through risk classification and weighting criteria:

$$Susceptibility\ I_X = \frac{\left(Sensitivity_n\ x\ Weight_n + Sensitivity_m\ x\ Weight_m\right)}{\left(Weight_n + Weight_m\right)}$$

- Adaptive capacity, which includes human and financial resources and operational plans.

As regards the weighting procedure, the ranking scales used for the pairwise evaluation of criteria/indicators in AHP are based on a standardized five-point scale of 'intensity of importance': 1, 3, 5, 7, and 9 (Fig. 3). The Analytic Hierarchy Process (AHP), originally formulated by Saaty and Kearns in 1985, is a method designed to assign relative importance to different evaluation criteria within decision-making procedures (Saaty & Kearns, 1985; Ravan et al., 2023).

1	Equal importance	Two activities contribute equally to the objective
3	Moderate importance of one over another	Experience and judgment slightly favor one activity over another
5	Essential or strong	Experience and judgment strongly favor one activity over another
7	Very strong importance	An activity is strongly favored and its dominance is demonstrated in practice
9	Extreme importance	The evidence favoring one activity over another is of the highest possible order of affirmation

Fig. 3. Saaty's scale of relative importance (graphical elaboration by Stanic et al., 2017, based on the methodology of Saaty and Kearns, 1985)

The final index is a weighted average of the "vulnerability" sub-indices. The calculation relies on a spreadsheet that classifies and weighs sensitivity, adaptive capacity, and their relative importance, drawing on experimental data for the materials examined. The outcome is a quantitative framework for assessing climate-related risk to building materials, which in turn informs conservation and adaptation strategies.

Finally, the Damage Index is calculated as the average of the three weighted vulnerability indexes (Fig. 4).

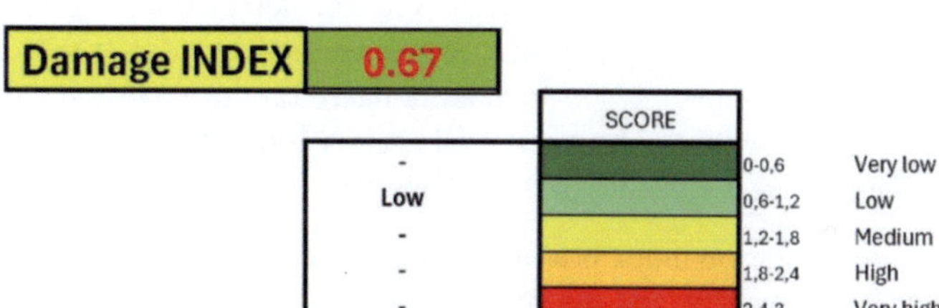

Fig. 4. Excel table indicating an example of the final comprehensive Damage Index

This framework provides a systematic approach to understanding the impacts of climate change on cultural heritage materials by integrating environmental data, material properties, and adaptive capacities into a unified numerical model, it enables accurate vulnerability assessments that reflect the complexity and interdependence of multiple influencing factors.

3 The Methodology of Damage Index for Assessing Material Properties (F_{mat}) Through Experimental Analysis

With reference to the Damage Index calculation methodology outlined in the previous sections, this part delves into the specific procedures related to one of the three components of the index: material properties (F_{mat}). These properties form the basis for both the experimental assessment and the methodological transfer through in-situ analyses conducted on the selected case study.

These experimental analyses (described in paragraph 3) involve the implementation of non-invasive investigations on Palazzo Mesiani, Bova (RC), using advanced instrumentation aimed at determining the chemical-physical properties of the materials entity that constitute the fabric of the case study (XRD, XRF evaluation, Carbonation Test, etc.).

3.1 Material properties (F_{mat})

The F_{mat} factor specifically refers to the degradation processes affecting construction materials, including physical and chemical damage. Drawing on scientific literature, midpoint indices have been identified; these typically represent the average impact of a stressor across different spatial or temporal scales and contribute to calculating the Damage Index. The selected indices (Fig. 5) include:

- Surface deterioration
- Carbonation (relative to the reference substrate)
- Efflorescence
- Corrosion of metallic elements

Each index is linked to specific ranking criteria, which are subsequently used in the calculation of the Damage Index (Table 1).

The methodology for material properties (Fig. 6), starts from the preliminary definition of the midpoint indicators and their respective benchmarks, it was necessary to divide each of them into specific sub-indicators based on the experimental analyses to be carried out in situ or with the support of laboratory tools (see paragraph 2.2.2).

These sub-indicators contribute to their respective definitions:

- The surface deterioration index is a function of surface composition, surface erosion, degree of surface porosity, and the state of surface crystallinity.
- The efflorescence index is a function of extent of efflorescence on the surface, salt deposits, and XRF evaluation.
- The carbonation index is based on carbonation testing.

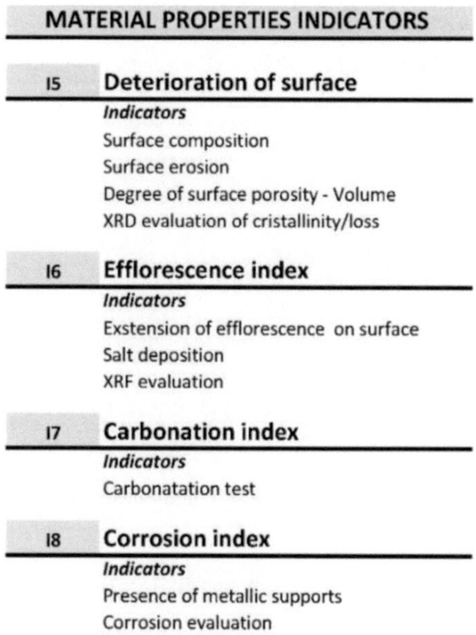

Fig. 5. Sensitivity Factor: Indicators of Material properties

Table 1. Midpoint Indices and Corresponding Ranking Criteria Used for Damage Index Calculation

Indicators (Midpoint Indices)	Ranking criteria
Surface Deterioration	Monitoring the relative intensity of diffraction of the primary phase
Efflorescence Index	Intensity of crystallized salts
Carbonation Index	Identification of carbonation products
Corrosion Index	Determination of corrosion products

- The corrosion index is a function of presence of metallic elements and corrosion evaluation (%).

A third methodological phase involved identifying specific measurement criteria for each sub-indicator, as well as determining the experimental results (Table 2).

The indicators are expressed as percentages instead of absolute units to allow a uniform comparison of different parameters. Using percentages enables:

1. Comparability: It normalizes diverse data like porosity, erosion, and crystallinity, making them directly comparable despite different measurement scales.
2. Relative assessment: Percentages reflect changes or degradation relative to an ideal or original material state, rather than absolute values.

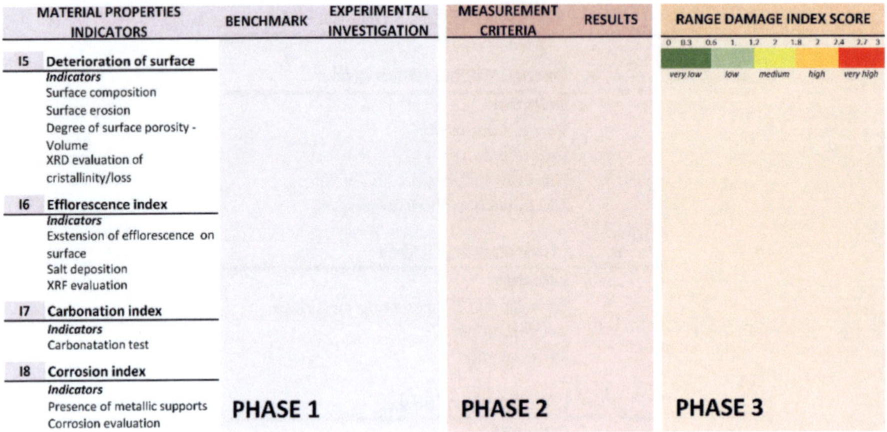

Fig. 6. Schematic representation of methodological phases for the normalization of F_{mat}

Table 2. Correlation between Indicators, Sub-indicators, and Measurement Criteria

Indicators	Sub-indicators	Measurement criteria
Surface Deterioration	Surface Composition	Composition (N compounds)
	Surface Erosion	Percentage of erosion
	Degree of Surface Porosity	Percentage of porosity
	Surface Crystallinity	Percentage of crystallinity
Efflorescence	Extent of Efflorescence	Percentage of coverage
	Salt Deposits	Percentage quantification
	XRF Evaluation	Peak area ratio
Corrosion	Carbonation Test	Percentage of surface area extension
Corrosion	Presence of Metallic Elements	Percentage of metallic supports present
	Corrosion Evaluation	Percentage of corrosion

3. Extent of degradation: They quantify how much damage has occurred compared to the original, undamaged material, which is key for conservation.
4. Spatial distribution: Percentages capture how phenomena like crystallinity and carbonation vary across a surface, showing material heterogeneity and localized degradation.

Thus, the use of percentages allows for uniform comparison across different parameters, represents the severity of degradation relative to an original state, and facilitates the communication of results.

The final phase involved normalizing the results by transposing the experimentally obtained values into the Damage Index.

3.1.1 F_{mat}: Procedural Protocol Through Experimental Instrumental Analysis for the Quantitative Definition of the Indices

A procedural protocol has been implemented for the quantitative definition of the sub-indicators constituting the indices, as illustrated in the following diagram (Fig. 7).

PROCEDURAL PROTOCOL		
	STEP	EXPERIMENTAL APPROACH
1	Identify Damage Factors	Identification of the key factors that contribute to the alteration of buildings or structures
2	Assign Weight to Factors	Assignment of weights to each deterioration factor based on its significance and potential impact on the building's condition. The weighting should reflect the relative importance of each factor in determining deterioration. Consult with experts and stakeholders to determine appropriate weights
3	Data Collection and Standardization	Gather data for each degradation factor. Data sources may include building inspections, maintenance records, construction documents, architectural assessments, and historical records. Standardize the data to ensure that all factors are on a common scale, making it possible to compare and combine them
4	Normalize Data	Normalize the data for each deterioration factor to bring it to a common scale (e.g., from 0 to 1). Normalization is essential when factors have different units or scales. Common normalization methods include min-max scaling or z-score normalization
5	Calculate degration Scores	Multiply each factor's normalized value by its assigned weight to calculate a deterioration score for that factor. Repeat this process for all deterioration factors
6	Aggregate Scores	Sum the deterioration scores for all factors to obtain an overall deterioration score for the building or structure. This overall score represents the degree of deterioration
7	Interpretation and Classification	Establish thresholds or categories that help classify the deterioration scores. For example, you could create categories such as "excellent," "good," "fair," "poor," or "critical" based on predefined ranges of scores
8	Monitoring and Periodic Assessments	Implement a system for regular monitoring and periodic assessments of the building's condition. Update the deterioration index as new data becomes available to track changes over time
9	Sensitivity Analysis	Conduct sensitivity analysis to assess how changes in weighting or data inputs affect the deterioration index. This helps evaluate the robustness of the index
10	Validation and Calibration	Validate the deterioration index by comparing it to observed deterioration in the field, if possible. Calibrate the index based on the observed data and feedback from experts and stakeholders
11	Documentation	Document the methodology, data sources, and assumptions used in developing the deterioration index. Transparent documentation is essential for the credibility and replicability of the index

Fig. 7 Schematization of the procedural protocol (Graphic elaboration from Giglio, F.; Frontera, P.; Malara, A.; Armocida, F. Materials and Climate Change: A Set of Indices as the Benchmark for Climate Vulnerability and Risk Assessment for Tangible Cultural Heritage in Europe. Sustainability 2024, 16, 2067.)

To this end, experimental investigation techniques such as X-ray diffraction (XRD) for the identification of material structures and scanning electron microscopy (SEM) combined with energy dispersive X-ray analysis (EDX) for morphological and chemical characterization were used. The protocol includes both in situ and laboratory investigations. Specifically:

- In situ: X-ray fluorescence (XRF) spectrometry was applied to non-destructively determine the chemical composition of masonry materials at selected planar points, followed by sample collection for further analysis.

- Laboratory investigations: SEM-EDX and XRD analyses were performed to investigate the morphological, chemical, and structural properties of the collected samples. The analysis was conducted by the MATeRICs Interdepartmental Laboratory in collaboration with the MATEES Laboratory, both at the Mediterranea University of Reggio Calabria.[2]

4 Application of Experimental Diagnostic Analyses and Simulation Tools in the Case Study of Palazzo Mesiani, Bova (RC)

Macroscopically, it is possible to observe the presence of heterogeneous materials in this building, due to the combined use of different types of original materials, as well as more recent materials likely introduced during reconstruction and/or maintenance work.

Indeed, from an initial visual inspection, the main materials constituting the structure are evident: local stone, identified as limestone or sandstone, already studied in other buildings in nearby areas and typical of many historic buildings in Calabria; bricks made of local clay, which are significantly present in this structure; mortar used to bind bricks and stones; and cement, found in some areas, likely due to restoration or adaptation work. The experimental analyses and simulation methodologies were first carried out in-situ (Fig. 8) and then processed in the laboratory.[3]

Fig. 8 Representative images of Palazzo Mesiani (Bova, Reggio Calabria)

---

[2] MATeRICs Interdepartmental Laboratory (resp. scient. prof. F. Giglio), MATEES Laboratory (resp. scient. prof. P. Frontera)

[3] The experimental analyses on Palazzo Mesiani were conducted by MATeRICs Interdepartmental Laboratory in collaboration with the MATEES Laboratory, involving Prof. P. Frontera and Eng. RtdB A. Malara, alongside the authors, Prof. F. Giglio and arch. PhD's F. Armocida.

4.1 In-Situ Investigations

Experimental analyses were conducted on various points of the masonry by means of X-ray Fluorescence Spectroscopy (XRF) to determine the chemical composition of materials. The microstructural and chemical characteristics were directly identified on-site using XRF analysis (X-ray fluorescence). Given the building's large dimensions, the most representative points of the entire structure were analyzed, and the results obtained were appropriately averaged (Fig. 9).

Fig. 9. Application of spectroscopy and analysis sampling, Palazzo Mesiani, Bova (RC)

The most representative XRF spectra are reported below (Fig. 10), from which relevant information regarding chemical composition, potential crystalline structures, and alterations experienced by the structure over time were extracted.

As evidenced by the XRF spectra, the main elements identified are calcium, silicon, iron, chlorine, and, in smaller percentages, aluminum and potassium. The main crystalline structures were identified as belonging to limestone matrices (tufa), various types of silicates (calcium and iron), quartz, aragonite, and calcite, which are compatible with the structures of mortars. Additionally, the high percentage of iron (iron oxide) was attributed to bricks, which are mainly characterized by high concentrations of iron oxide, calcium oxide, and low levels of silica.

4.2 Laboratory experimental investigations

Regarding the implementation of the Damage Index, the material characterization analyses include various experimental techniques and methodologies:

- ImageJ Software Analysis on Representative Areas: The ImageJ software is used for quantitative evaluations of visual and structural characteristics of surfaces. Specifically, in the described analyses, the software measures parameters such as porosity and surface erosion (Fig. 11).

- Surface deterioration

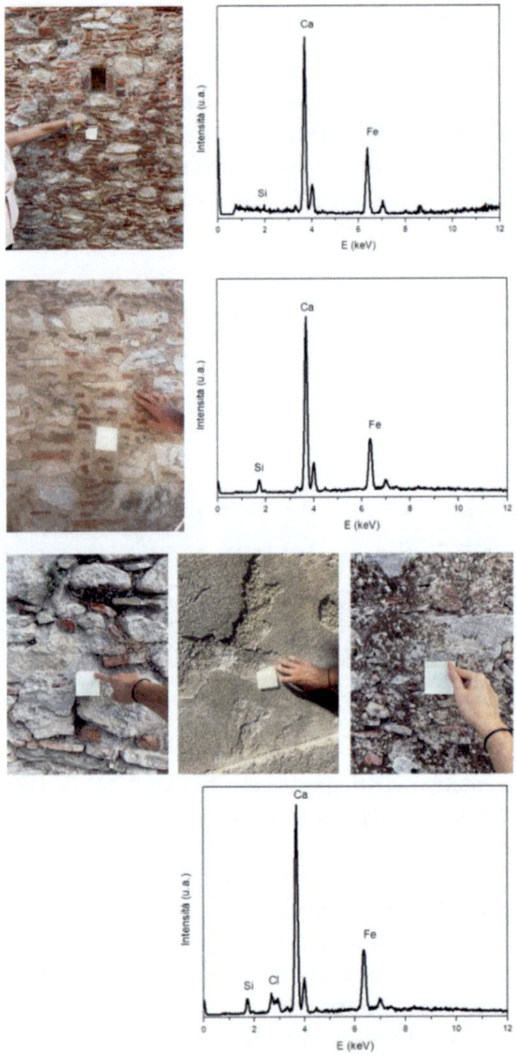

Fig. 10. Representative samples of the structure and corresponding averaged XRF spectra.

- Surface composition: Point-specific XRF analysis identified the chemical composition of surface morphologies, detecting 8 main compounds with a weight percentage >5%. A relative score from 0 to 3 was assigned based on the number of compounds. A low number indicates limited impact, whereas high material heterogeneity can trigger multiple and interconnected degradation processes.
- Degree of surface porosity: Evaluated using ImageJ software by analyzing selected image areas to distinguish voids through contrast and depth calculations. The average porosity measured was 10.7%, significantly influenced by the construction technique involving alternating bricks and stone elements.

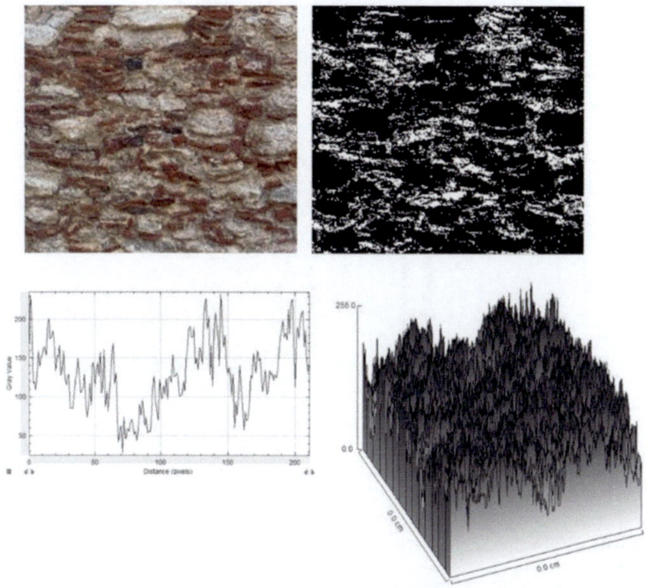

Fig. 11. Analyses obtained using ImageJ software

- Surface erosion: Also analyzed with ImageJ, erosion was quantified as the average depth value across a fixed-width section (50 cm). The surface is generally irregular, and some cavities are due to construction technique rather than erosion. The average erosion percentage was approximately 8.7% (≈4.3 cm per 50 cm at 100%).
- XRD evaluation of crystallinity: Although direct crystallinity evaluation was not possible, the presence of efflorescence and organic deposits (from XRF data) suggests an estimated 10% decrease in surface crystallinity, corresponding to the observed efflorescence extent.

• Efflorescence index

- Extension of efflorescence on surface: Visual inspections of selected areas identified efflorescence and organic deposits. The surface coverage was estimated at 1.1% of the total analyzed area (in m^2).
- XRF evaluation: Targeted XRF spectroscopy detected chlorides typically associated with efflorescence. The ratio between fluorescence peak areas related to efflorescence compounds and those from building materials averaged 0.07 (Fig. 12).
- Salts deposition: XRF analysis confirmed the presence of chloride-based salts indicative of saline efflorescence. These are attributed to moisture-driven salt migration from the subsoil to the wall surface. Although weight % quantification is complex, it is unnecessary for the index calculation given the limited extension observed.

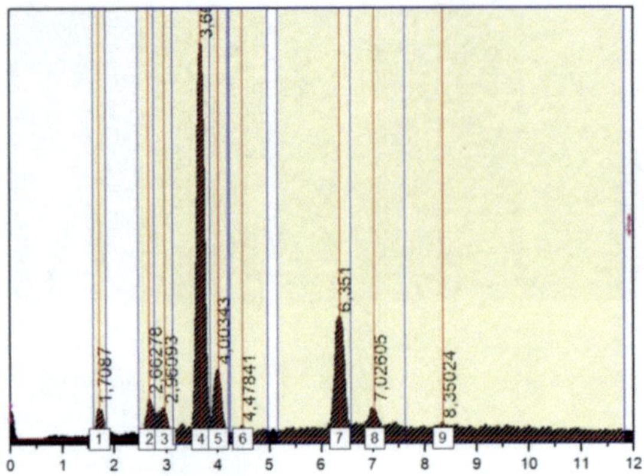

Fig. 12. XRF evaluation elaboration results

- Carbonatation index
  - Carbonatation test: The carbonation test evaluates CO_2 penetration into construction materials and its reaction with calcium hydroxide, forming calcium carbonate. This process reduces pH and may trigger corrosion if metal reinforcements are present. In their absence, carbonation improves durability by increasing hardness and reducing porosity. Although direct measurements of carbonation and surface carbon content were not possible, the presence of calcium compounds indicates that carbonation likely occurred. Given the heterogeneous surface and the estimated distribution of calcium-rich binder mortar (about 50% of the analyzed area, including restored sections), an approximate carbonation score of 50% was assigned.

- Corrosion index
  - Presence of metallic support: No visible metallic elements were found in the analyzed area, and due to the historical period of construction, the presence of internal metal components is unlikely. The score is 0, indicating negligible corrosion risk.
  - Corrosion evaluation: Consistent with the above, no corrosion was observed. The corrosion index is 0, with no measurable impact on degradation.

The data obtained are summarized below (Table 3):

Table 3. Summary of data obtained from experimental analyses

Indicators	Sub-indicators	Measurement	
Surface Deterioration	Surface Composition	Composition (N compounds)	8(>5% by weight)
	Surface Erosion	Percentage of erosion	8.7%
	Degree of Surface Porosity	Percentage of porosity	10.7%
	Surface Crystallinity	Percentage of crystallinity	Not measurable; hypothesized 10% decrease
Efflorescence	Extent of Efflorescence	Percentage of coverage	1.1% of the surface covered
	Salt Deposits	Percentage quantification	Minimal; irrelevant for calculation
	XRF Evaluation	Peak area ratio	XRF peak area ratio: 0.07
Corrosion	Carbonation Test	Percentage of surface area extension	50% of the surface
	Presence of Metallic Elements	Percentage of metallic supports present	0
	Corrosion Evaluation	Percentage of corrosion	0

5 Conclusions

The study, developed within the framework of Pilot Project 4.7.1 T4Y Spoke 4, highlights how climate change poses an increasing challenge to the preservation of tangible cultural heritage, emphasizing the importance of systematic and multidisciplinary approaches for risk assessment and mitigation. The proposed Damage Index, integrating environmental conditions (Fenv), material properties (Fmat), and external drivers (Fext), provides a comprehensive framework for evaluating vulnerabilities and supporting evidence-based conservation strategies. Through quantitative and experimental methodology, this contribution provides practical tools to identify degradation processes and support conservation strategies.

In particular, the application of these tools, tested in heritage contexts characterized by evident formal and structural fragilities, can foster the adoption of climate adaptation strategies calibrated to the criticalities and risk conditions experimentally identified, through a site-specific approach. This makes it possible to outline an integrated and replicable model, capable of supporting decision-makers and the key stakeholders involved in heritage management in defining policies, methodologies, and targeted adaptation actions, with particular attention to historic contexts that are poorly resilient and vulnerable to physical, chemical, and biological degradation processes (e.g., surface erosion, cracking, thermal shocks) induced by climate change.

The case study of Palazzo Mesiani in Bova (RC) demonstrates the methodology's applicability, underscoring the importance of combining on-site analyses, laboratory experimental data, and simulation tools to obtain precise results.

The methodology also shows potential for integration into digital tools and Living Labs for heritage, enabling dynamic and participatory monitoring frameworks. In this perspective, the Damage Index may play a key role in operationalizing climate adaptation at local scale, aligning European and international frameworks such as the IPCC, and promoting knowledge-based, context-specific conservation practices.

Future research will focus on incorporating the remaining factors (Fenv and Fext) trough site-specific climate and socio-imfrastructural data, enabling a full evaluation of the Damage Index and validating its applicability across diverse heritage sites and climatic scenarios.

Ultimately, this work contributes to the ongoing effort of bridging the gap between climate science and conservation practices, offering a scientifically robust, experimentally grounded, and policy-relevant methodology for addressing the increasing vulnerability of built heritage in the face of climate change.

Acknowledgments. The development of the indices was conducted within the Interdepartmental laboratory section MATeRICs, Material Analysis Material Analysis and Technologies for Restoration, Innovation and Climate Exploration (resp. prof Francesca Giglio, dAeD Department), Mediterranea University of Reggio Calabria. The work was carried out in collaboration with Prof. Patrizia Frontera, as well as with RtdB Eng. Angela Malara, Department of Civil, Energy, Environmental and Materials Engineering (DICEAM), Mediterranea University of Reggio Calabria, and PhD's Arch. Francesco Armocida (dAeD Department).

Author Contributions The study was conducted collaboratively, with both authors contributing to all stages of research and writing. Francesca Giglio led conceptualization, theoretical framing, literature review, (Introduction, Sects. 2 and 3). Francesco Armocida performed experimental analyses, data collection, processing, and application to the case study, including in-situ and laboratory work, simulations, and drafting of methodological and results sections (Sects. 3.1.1, Sect. 4 and Conclusions).

References

Day, J.C., Heron, S.F., Markham, A.: Assessing the climate vulnerability of the world's natural and cultural heritage. Parks Stewardship Forum. **36**(1), 144–153 (2020) https://escholarship.org/uc/psf

Smith, P.F.: Climate Change and Cultural Heritage: A Race Against Time. Routledge, London (2013) ISBN 9781138189423

Damas Mollá, L., Sagarna, M., Zabaleta, A., Aranburu, A., Antiguedad, I., Uriarte, J.A.: Methodology for assessing the vulnerability of built cultural heritage. Sci. Total Environ. **845**, 157314 (2022). https://doi.org/10.1016/j.scitotenv.2022.157314

Mavrakou, T., Polydoros, A., Cartalis, C., Philippopoulos, K.: Assessing the climate change related risk of cultural heritage sites. Sci. Cult. **11**(1) (2025). https://doi.org/10.5281/zenodo.14274602

Sesana, E., Bertolin, C., Loli, A., Gagnon, A.S., Hughes, J., Leissner, J.: Increasing the resilience of cultural heritage to climate change through the application of a learning strategy. In: Moropoulou, A., Korres, M., Georgopoulos, A., Spyrakos, C., Mouzakis, C. (eds.) Transdisciplinary Multispectral Modeling and Cooperation for the Preservation of Cultural Heritage.

TMM_CH 2018, vol. 961, pp. 447–455. Springer (2019) https://doi.org/10.1007/978-3-030-12957-6_29
Georgiadis, T.: Role of climate and city pattern. In: Fabbri, K. (ed.) Urban Fuel Poverty, pp. 41–62. Elsevier, Amsterdam, Netherlands (2019)
Boeri, A., Longo, D., Fabbri, K., Roversi, R., Boulanger, S.: The relation between outdoor microclimate and people flow in historic city context: the case study of Bologna within the ROCK project. Sustainability. **15**(7527), 1–27 (2023). https://doi.org/10.3390/su15097527
Nava, C.: Tecnologie emergenti per il progetto rigenerativo. 5 questioni teoriche su innovazione e sostenibilità dalla prassi della Ricerca di frontiera. ARACNE, Roma (2023) ISBN 979-12-218-0826-1
Giglio, F., Frontera, P., Malara, A., Armocida, F.: Materials and climate change: a set of indices as the benchmark for climate vulnerability and risk assessment for tangible cultural heritage in Europe. Sustainability. **16**(5), 2067 (2024). https://doi.org/10.3390/su16052067
Esposito, F., Russo, M., Sargolini, M., Sartori, L., Virgili, V.: Building back better: idee e percorsi per la costruzione di comunità resilienti. Carocci, Roma (2017) ISBN 9788843090600
Füssel, H.M., Klein, R.J.T.: Climate change vulnerability assessments: an evolution of conceptual thinking. Clim. Change. **75**, 301–329 (2006). https://doi.org/10.1007/s10584-006-0329-3
Turner II, B.L., Kasperson, R.E., Matson, P.A., McCarthy, J.J., Corell, R.W., Christensen, L., Eckley, N., Kasperson, J.X., Luers, A., Martello, M.L., Polsky, C., Pulsipher, A., Schiller, A.: A framework for vulnerability analysis in sustainability science. Proc. Natl. Acad. Sci. USA. **100**(14), 8074–8079 (2003). https://doi.org/10.1073/pnas.1231335100
IPCC: Summary for Policymakers. In: Global Warming of 1.5 °C. An IPCC Special Report on the impacts of global warming of 1.5 °C above pre-industrial levels and related global greenhouse gas emission pathways, in the context of strengthening the global response to the threat of climate change, sustainable development, and efforts to eradicate poverty. IPCC. Available at: https://www.ipcc.ch/sr15/chapter/spm/ (2018)
IPCC: Summary for policymakers. In: Lee, H., Romero, J. (eds.) Climate Change 2023: Synthesis Report. Contribution of Working Groups I, II and III to the Sixth Assessment Report of the Intergovernmental Panel on Climate Change, pp. 1–34. IPCC, Geneva, Switzerland (2023) https://doi.org/10.59327/IPCC/AR6-9789291691647.001
Ravan, M., Kabdaslı, S., Kaya, A., Henriques, A., Ferreira, B.: A vulnerability assessment framework for cultural heritage sites: the case of the Roman ruins of Tróia. Int. J. Disaster Risk Sci. **14**(1), 26–41 (2023). https://doi.org/10.1007/s13753-023-00449-y
Lucanto, D., Nava, C., Mangano, G.: Digital prototyping and regenerative design toward carbon-neutrality and a climate resilient built environment: a multi-scale assessment of environmental multi-risks. Buildings. **14**(12), 3934 (2024). https://doi.org/10.3390/buildings14123934
Saaty, T.L., Kearns, K.P.: The analytic hierarchy process. In: Saaty, T.L., Kearns, K.P. (eds.) Analytical Planning, pp. 19–62. Elsevier, Amsterdam (1985) https://doi.org/10.1016/B978-0-08-032599-6.50008-8
Stanic, V., Fafandjel, N., Matulja, T.: A methodology for improving productivity of the existing shipbuilding process using modern production concepts and the AHP method. Brodogradnja. **68**, 37–56 (2017) https://doi.org/10.21278/brod68303

The Digital Atlas: Structure, Information, Interoperability of Physical and Digital Models

Domenico Lucanto(✉) and Asmae Hanida

ABITAlab, dArTe, Mediterranea University of Reggio Calabria, Reggio Calabria, Italy
domenico.lucanto@unirc.it, asmae.hanida@emadu.ueuromed.org

Abstract. This paper presents the "Digital Atlas", an open, three-layer platform developed within the TECH4YOU PP 4.7.1 framework to connect climate-change scenarios with the practical implementation of regenerative design strategies. It integrates advanced digital prototyping, a "parametric pipeline", which refers to computational workflow linking climate data to design rules, and additive manufacturing to produce and verify site-specific Nature-Based Solutions, based on IPCC RCP 4.5 and 8.5 scenarios. Tested in the Grecanica area (Palazzo Mesiani, Bova; Palizzi Marina), the Digital Atlas structure, informations and interoperability addresses multi-scale climate challenges, from localized architectural conservation under thermal stress to urban-scale risks such as flooding, coastal erosion, and sea level rise. Eco-Reef modules made of mussel-shell-enriched biocement demonstrated rapid biofilm formation after three weeks, validating the "phygital approach", the iterative integration of physical and digital models. The integrated Decision Support System (DSS) provides actionable information for zero-carbon transitions and resilient development. The Atlas offers a replicable methodology to Mediterranean coastal communities, promoting ecological regeneration, preservation of cultural and natural heritage, and active/passive decarbonization processes while illustrating how digital-physical interoperability an accelerate climate adaptation.

Keywords: Digital Atlas · interoperability · physical prototyping · digital models · climate adaptation · regenerative design · parametric modeling · decision support systems

1 Introduction

Climate change poses increasing risks to cultural and natural heritage. This study responds with the Digital Atlas, an innovative decision-support framework that operationalizes regenerative design principles through visualization, interpretation and development. The latest IPCC assessments underscore the urgent need for scientifically based adaptation strategies and mitigation interventions capable of addressing both medium- and long-term impacts. In response, the PP 4.7.1 pilot project, with its actions 8 and 9, collects data, information, and resources to contribute to the development of a Digital Atlas, an innovative decision support framework designed to operationalize regenerative

C. Nava et al. (Eds.): *Climatic and Structural Safety in Multi-Hazard Regime of Cultural and Natural Heritage*, GREEN, pp. 313–327, 2026.
https://doi.org/10.1007/978-3-032-11926-1_19

design principles through visualization, interpretation, and development to respond to the effects of climate change on built and natural heritage. The Atlas integrates three interconnected levels, digital prototyping, physical prototyping, and an interactive user-profile DSS dashboard, to help small municipalities and coastal communities adapt to climate risks. 'Phygital' in this paper refers to the iterative loop between physical and digital prototypes, while 'parametric pipeline' denotes the automated link between climate data and design parameters. These levels facilitate the development of robust, evidence-based climate change mitigation strategies that can be adapted to both small municipalities in Grecanical Area of Calabria inland areas and coastal communities in the Mediterranean. Focusing on two distinct case studies (Fig. 1), Palazzo Mesiani in Bova and the coastal settlement of Palizzi Marina, the Atlas demonstrates its ability to propose zero-carbon technological solutions capable of addressing multi-scale climate risks, ranging from localized conservation of architectural heritage to broader landscape management. This experimental integration provides a replicable framework for interdisciplinary collaboration and nature-based, climate-resilient, and carbon-neutral interventions.

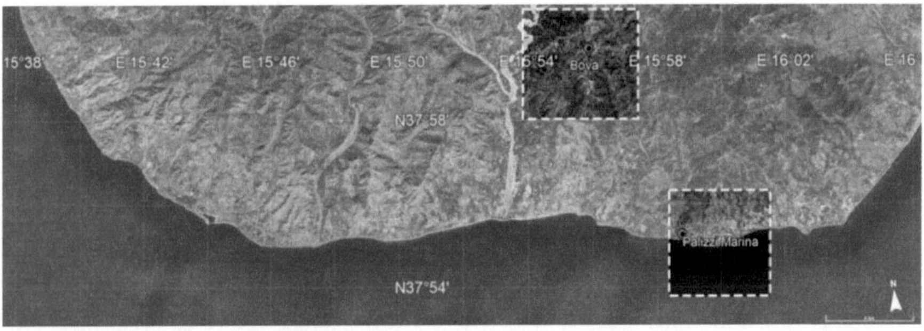

Fig. 1. Map of the Grecanic Area

1.1 Theoretical Background

The theoretical foundation of this research is deeply rooted in the intersection of regenerative design, emerging technologies serving climate mitigation and adaptation strategies, and cultural and natural heritage conservation—concepts that are increasingly critical in contemporary debates on architectural and environmental design. Emerging paradigms emphasize how regenerative design and digital technologies are essential to effectively address climate challenges (Nava, 2019). Cultural and natural heritage assets, in the context of climate adaptation strategies, require specifically designed vulnerability assessments that incorporate both climate data and heritage-specific risk indicators (Gallina et al., 2016; Conti et al., 2021). Advanced parametric environments—Rhino+Grasshopper, enhanced by climate engine plugins (Ladybug, Butterfly, ENVI-met) and fed with multiple forcing datasets (IPCC RCP4.5/8.5)—now integrate climate intelligence directly into the generative rules that model the geometry. Because each design variable (slope, porosity, density, façade depth, etc.) is linked to future performance metrics, DSS-type tools can support the testing of technological solutions for carbon neutrality such as Nature Based Solutions or passive design measures whose effectiveness remains stable over time, replacing point forecasts with quantified risk bands and, in turn, dramatically reducing the epistemic uncertainty that once plagued early-stage decision-making[1] (Nava, Filice, Laganà, 2024). Furthermore, the evolving resilience framework emphasizes the need to integrate ecological, technical, and psychological dimensions, which are crucial for addressing immediate impacts and comprehensively planning future adaptation measures (Hoseinzadeh et al., 2023; Mangano and Leuzzo, 2022). Carbon neutrality, deeply rooted in contemporary architectural[2] and urban development practices, is systematically supported by regenerative and circular principles in the design, construction, disassembly, and recovery of raw materials processes, striving for zero environmental impact, improving ecosystem health, and optimizing the technological functioning of built environment transformations (He et al., 2023; Ye et al., 2024). Finally, advances in digital fabrication, particularly additive manufacturing, play a key role in enabling precise and detailed physical prototyping. This technological capability not only facilitates better understanding and response to site-specific climatic and morphological conditions, but also strengthens the practical implementation and iterative refinement of regenerative and adaptive design solutions (Kim, 2019; Tabassum & Park, 2024).

2 Methodology

The proposed methodological framework is specifically designed to overcome the traditional limitations inherent in architectural and environmental design approaches by integrating advanced digital survey techniques, parametric modeling, climate simulations, and physical prototyping, aimed at developing a DSS integrated into the digital atlas. Key to this methodology is seamless interoperability between detailed digital modeling and iterative physical prototyping, which includes parametric simulations, IPCC scenario-based climate modeling, and additive manufacturing techniques. The Atlas also

[1] Similar suggestions for the transition from developing a climate risk atlas are provided in C. Nava, F. Filice, D. Laganà; in ReKap, ABITAlab). This operational proposal for Reggio Calabria combines advanced parametric workflows with policy-ready protocols, demonstrating how scenario-based modeling directly influences urban regeneration strategies and strengthens the coherence of green and blue infrastructure across municipal sectors.

[2] Several peer-reviewed studies and European policy papers converge on four pillars underpinning a climate-neutral built environment by 2050. The EU-level directives that mandate zero-emission performance (European Commission, 2021) promote the use of alternative materials such as bio-based or recycled materials, can cut embodied CO_2 by 20–30% relative to conventional aggregates (Churkina et al., 2020; Pomponi & Moncaster, 2017). Other research demonstrating that electrified plant, off-site and innovative digital fabrication and process can halve construction-phase emissions when powered by renewable electricity (D'Amico et al., 2022) and support digital process by enabling Early Design carbon calculators, smart-building management systems and the emerging Digital Building Logbook—provide continuous feedback for reducing both operational and embodied carbon across the asset life cycle (Paiho et al., 2022; BPIE, 2023).

promotes multi-stakeholder engagement through a Decision Support System (DSS) integrated into an interactive, scalable, and accessible user-profile platform. This effectively communicates complex climate scenarios and informs targeted adaptation and mitigation measures, thus paving a resilient and carbon-neutral path for the conservation of cultural and natural heritage (Lehmann, 2019; Blanco et al., 2021). The experimental setup for the Digital Atlas is designed to be populated with digital prototypes capable of converting raw spatial data into decision-ready artifacts. Figure 2 (attached) explains how hazard, vulnerability and exposure interact in the IPCC framework to underpin the Atlas's risk analysis, summarizes the workflow and describes in detail how each phase produces and validates specific research outputs. The synthesis required for schematization allows us to identify specific intervention points where different technological solutions, whether NBS, SUDS, or RES devices, could effectively mitigate environmental risks.

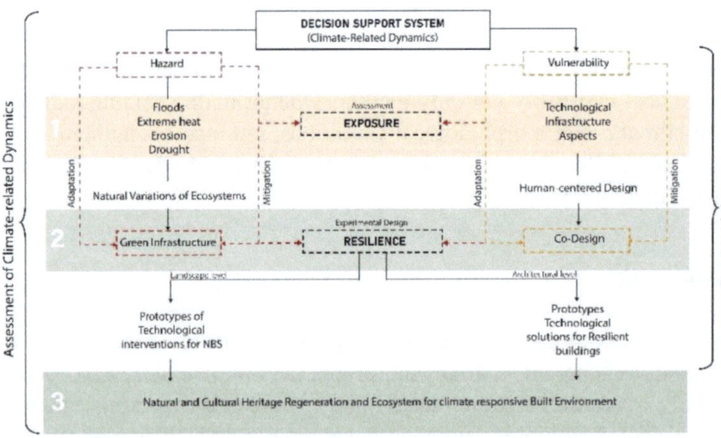

Fig. 2. Resilience as a result of complex interactions among the determinants of risk (hazard, vulnerability and exposure), based on the IPCC framework.

The resilience implementation phase enabled the technological and infrastructural aspects of vulnerability through a regenerative design approach that not only minimizes climate damage but often also produces co-benefits, from energy savings and cost efficiency to improved occupant comfort and resilience, aligning the architecture with carbon neutrality. Technological innovation in building materials is a key factor in reducing the carbon footprint of construction. This phase used advanced parametric modeling to facilitate robust and automated data interoperability, ensuring seamless integration with the DSS platform. The final results included practical technological interventions, in this case NBS prototypes, ensuring regeneration and climate response strategies for the built environment.

The methodological approach is based on a mixed methods strategy, emphasizing iterative, integrative and multiscalar analyses. Its operational core is parametric modelling, using specialized software such as Grasshopper, combined with environmental plugins such as Ladybug Tools and ENVI-met. Figure 2 distils the workflow into three steps that run from raw data to actionable design choices. By chaining these stages in an event-driven architecture, the Digital Atlas turns global climate scenarios into **site-specific, verifiable and continuously updatable design intelligence**—the essence of experimental rigour for climate-adaptive heritage and coastal planning. We begin with *top-down* climate drivers (IPCC scenarios RCP 4.5 and 8.5 for 2030-2090) and marry them with *bottom-up* GIS layers: drone-derived DTMs, land-use maps and building surveys. Inside a parametric engine (Grasshopper + Ladybug + ENVI-met) the data are translated into site-specific hazards. For Palizzi Marina the focus is sea-level rise, coastal erosion and runoff; for Bova it is solar stress, cooling-degree days and material decay. The final column links each hazard to a strategy bundle.

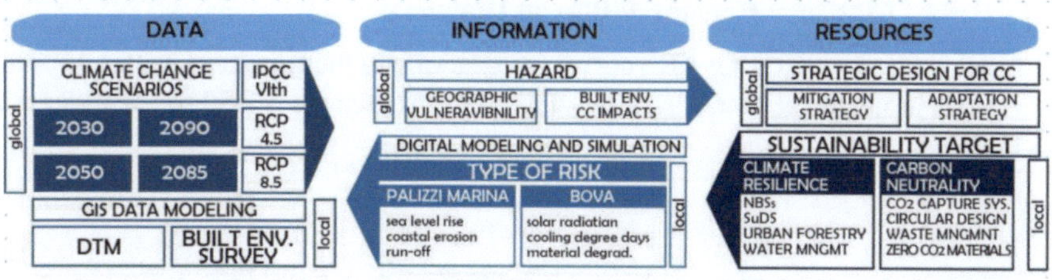

Fig. 3. Methodological Framework for the Climate Change oriented design workflow

3 Digital Atlas——Structure, Information Flow and Phygital Interoperability

The Digital Atlas, as mentioned, informs and structures the dashboard of the three-layer cyber-physical platform. This innovative platform offers a coherent and iterative approach, transforming global climate trajectories into accurate and actionable data.

The Digital Modeling Layer serves as the foundation and provides environmental data, projecting different climate effects on the built environment based on IPCC climate change scenarios.

The Physical Prototyping Layer informs technological solutions with simulation data and proposes advanced technological solutions. Here, digital geometries are translated into tangible prototypes through large-format 3D printing.

Finally, the Decision Support System (DSS) Layer serves as an interoperability hub, seamlessly integrating data flows where citizens, technicians, projects and administrators can access targeted information, ranging from simplified visualizations and narrative maps to detailed data and technical assessments.

3.1 Digital Prototyping Layer

The digital prototyping layer assimilates SfM meshes at the building scale and GIS rasters at the cluster and urban scale, then passes them through a Rhino/Grasshopper pipeline: Ladybug, Honeybee and ENVI-met quantifying solar load, runoff and microclimate, while custom Python nodes post-process the results. The meteorological forcing comes from Meteonorm EPW files generated according to two IPCC pathways—RCP4.5 for 2030 and 2090, RCP8.5 for 2050 and 2085—selected for complementary reasons: (i) RCP4.5 (2030 and 2090) represents a future of moderate CO2 emissions mitigation. Using this pathway at the beginning (2030) and end (2090) of the analysis window allows us to capture both short-term design stresses and long-term benefits (Abbass etal., 2022); (ii) RCP8.5 (2050 and 2085) depicts a "business-as-usual" trajectory with continued emissions growth. These projections reveal how the built environment would behave if decarbonization stalled, providing a high-stakes benchmark for adaptive capacity testing (Roberts etal., 2019).

3.1.1 Palizzi Marina—Coastal & Pluvial Diagnostics

Sea-Level Trajectories Using AR6 curves integrated into the Climate-Central API, still-water rise is projected at **+0.60 m (2050-4.5, medium luck) to +1.50 m (2100-8.5, bad luck)**.

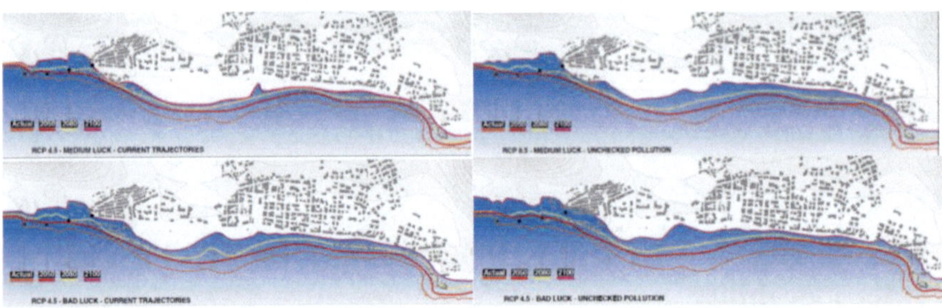

Fig. 4. High SX—SLR Projections Current Trajectory (RCP 4.5, medium luck): 2050, 2080, 2100. High DX—SLR Projections Unchecked Pollution (RCP 8.5, medium luck): 2050, 2080, 2100. Bottom SX—SLR Projections Current Trajectory (RCP 4.5, bad luck): 2050, 2080, 2100. Bottom DX—SLR Projections Unchecked Pollution (RCP 8.5, bad luck): 2050, 2080, 2100.

Run-Off Analytics Cell-by-cell Q=C·I·AQ=C·I·AQ=C·I·A calculations reveal peak discharges of **1 232 m^3 h^{-1}** in the sealed urban basin "B" and **502 m^3 h^{-1}** in the coastal basin "C" under 2085-8.5 forcing. Bio-retention SuDS sized on these hydrographs would reduce the 30-year event by ~35%[2].

Micro-Climate Clusters This ENVImet simulation illustrates the Micro-climate clusters chosen in Palizzi Marina, clearly identifies **heat-stressed urban clusters** and the **regenerative effect of green and coastal zones** in Palizzi Marina. The model validates a strong need for **nature-based interventions**, improved **morphological ventilation**, and **landscape-integrated cooling strategies**, essential in a Mediterranean coastal town facing climate-related heat challenges. According to these advanced analysis, we notice a considerable **heat accumulation around dense buildings**, especially in the **north-central zone** (compact urban fabric), and South-central block clusters.

These areas exhibit **UTCI above 31.20 °C**, suggesting **thermal discomfort** due to poor airflow, high radiation absorption, and limited vegetation. These data feed directly

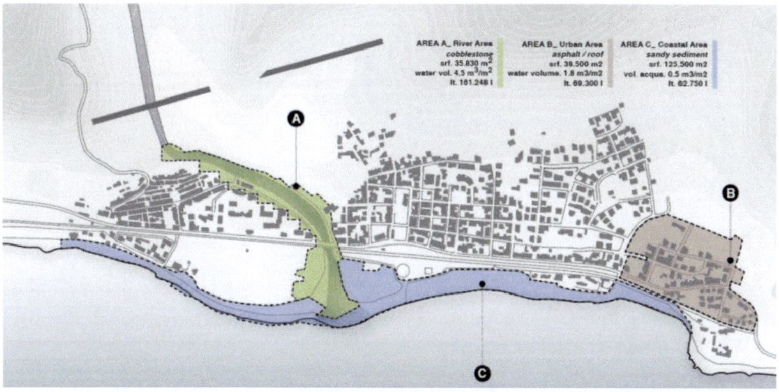

Fig. 5. Simulation results identifying key flood-prone areas in Palizzi Marina during intense rainfall events.

into Atlas dashboards, ranking adaptation priorities (vegetative shading, permeable paving) without yet involving material prototyping.

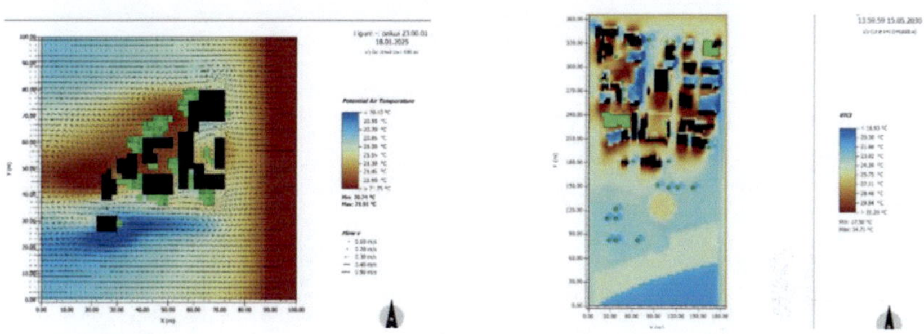

Fig. 6. Environmental simulation of two clusters in Palizzi Marina, illustrating the Potential air Temperature and UTCI

3.1.2 Bova—Palazzo Mesiani Solar-Thermal Stress

SfM-derived meshes coupled with Ladybug/Dragonfly indicate an **annual irradiance increase of +10.7%** by 2085-8.5. Summer peaks hit 7.5 kWh m^{-2} on south- west façades, well above CCDI class-4 thresholds (Fig. 7 and Fig. 8). Radiation deltas range from +4% (spring 2030) to +12% (summer 2085).

3.2 Physical Prototyping Layer

The physical prototyping level translates the studies conducted at the digital prototyping level into physical models, rigorously validating their ecological and structural effectiveness. The first experiment following this process, guided by the atlas structure,

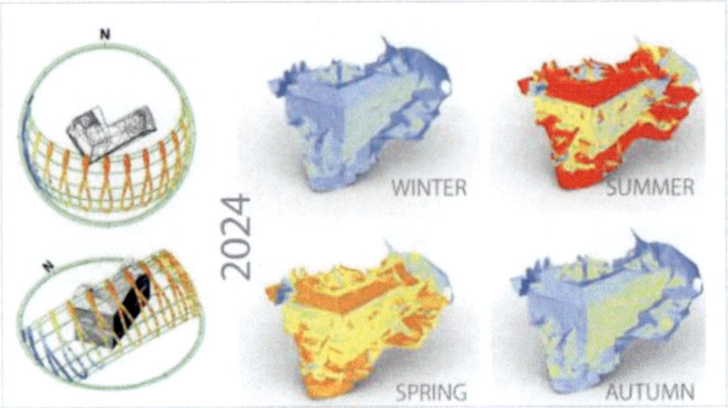

Fig. 7. Seasonal solar irradiance under current (2024) mean daily irradiance by season (winter, spring, summer, autumn) on the Palazzo Mesiani mesh.

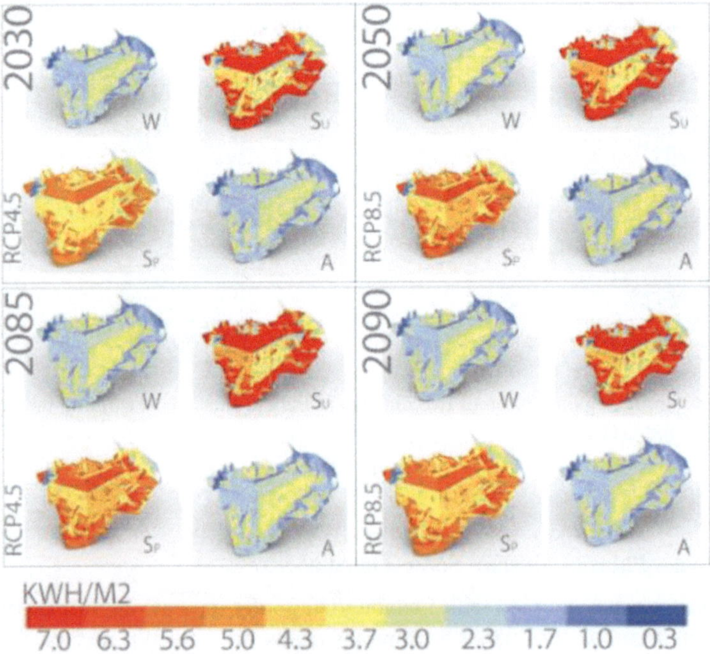

Fig. 8. Seasonal solar irradiance. Top left: 2030 seasonal irradiance RCP4.5.Top right: 2050 seasonal irradiance RCP 4.5. Bottom left: 2085 seasonal irradiance RCP 8.5. Bottom right: 2090 seasonal irradiance RCP 8.5.

information flow, and phygital interoperability framework and the methodological workflow (see Par. 2), involves the Eco-Reef modules, biocementitious structures specifically designed to mimic the natural functions of coral reefs and therefore suitable as NBS prototypes. These modules use a composite material formulated for marine applications, composed of 65% high-performance two-component cement mortar, 20% water, and 15% finely ground Mediterranean mussel (Mytilus galloprovincialis) shell powder. The integration of mussel shells provides essential surface microroughness and increased calcium carbonate ($CaCO_3$) availability, significantly promoting early colonization by marine organisms[3] (Stel'makh et al., 2023; El Biriane et al., 2020).

The design process for the Eco-Reef modules, as shows (Fig. 9), utilized sophisticated computational growth generate coral shapes defined based on biological and morphological parameters derived from natural coral reef ecosystems, accurately reproducing their structural complexity. The finalized digital models were then subjected to a slicing process, translating the complex three-dimensional geometries into sequential 4 mm-thick layers.

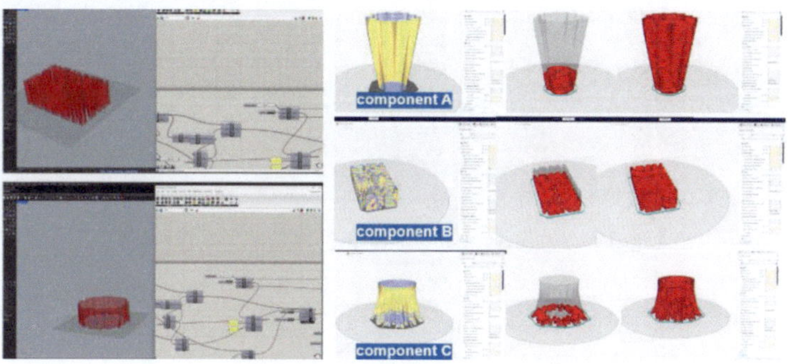

Fig. 9. Computational design example and slicing process for 3D-Printing

3.2.1 Additive Manufacturing and Geometric Complexity

Eco-Reef prototypes were fabricated at the ABITAlab Prototyping Laboratory, with advanced additive manufacturing technology, specifically Liquid LDM via a WASP 40100 3D printer. The laboratory equipment employed in this process was funded through. The LDM method allowed for the precise realization of highly in-tricate geometries directly inspired by natural coral structures. Following the print-ing phase, each module under a room-temperature curing and drying process to ensure robustness and maximize bioreceptivity for marine colonization.

[3] This approach is based on studies that have shown that the addition of biogenic materials (e.g., mollusc shells) to cementitious matrices can significantly increase the surface availability of calcium carbonate ($CaCO_3$) and microroughness, essential parameters for the adhesion of microbial biofilms and subsequent colonization by marine organisms such as filamentous algae, bacteria, and filter-feeding larvae (Stel'makh et al., 2023; El Biriane et al., 2020). Stel'makh et al. (2023) found that the addition of shell powder improves the microstructure of concrete by increasing porosity and roughness, characteristics directly correlated with the increase in the ecological capacity of artificial marine surfaces. El Biriane et al. (2020) also confirmed that mussel shell powder allows for the production of cementitious composites with adequate mechanical properties, while simultaneously ensuring a higher degree of ecological compatibility compared to traditional cements.

The Eco-Reef prototypes were fabricated at the ABITAlab Prototyping Laboratory, with advanced additive manufacturing technology, specifically Liquid LDM via a WASP 40100 3D printer. The laboratory equipment employed in this process was funded through. The LDM method allowed for the precise realization of highly intricate geometries directly inspired by natural coral structures. Following the printing phase, each module under a room-temperature curing and drying process to ensure robustness and maximize bioreceptivity for marine colonization.

Fig. 10. Additive manufacturing laboratorial process

3.2.2 Prototype Development and Ecological Aspects

Three distinct module designs were developed, each optimized for specific ecological functionalities. The figure below, (Fig. 11), that shows the composition of different component of the module,

Filtering Cone (Module A): Cylindrical structure (Ø 16 cm × 20 cm height) featuring dense perforations to create turbulence and foster colonization by filter-feeding organisms (e.g., sponges, ascidians); **Reef Block (Module B)**: Rectangular module (25 × 18 × 6 cm) with parametrically designed coral-like ridges providing protected microhabitats for juvenile marine species and crustaceans; **Hybrid Cylinder (Module C)**: Cylindrical shape (Ø 24 cm × 9 cm height) combining the porosity of Module A with structural niches from Module B, facilitating comparative biological colonization studies.

These modules, assembled vertically, formed cohesive structures approximately 40 cm in height, facilitating easy installation, monitoring, and potential removal without permanent alteration of the seabed (Fig. 12).

3.2.3 Field Deployment and Initial Observations

Following installation on 24 June 2025 at Palizzi Marina (37.918486 N, 15.979145 E), the modules were inspected underwater at two key intervals one week and three weeks post-deployment. Figure X captures the same vantage points in both campaigns, revealing clear signs of early ecological conditioning:

Week 1—The bare composite appears pale-beige, with crisp, well-defined bead lines and no visible organic cover. Fine sediment has begun to settle in horizontal ledges, but the lattice remains mechanically pristine and free of biofilm.

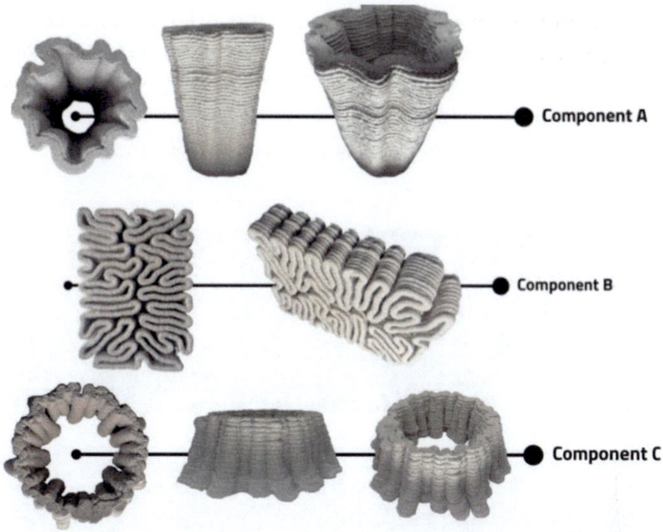

Fig. 11 Eco-Reef module types

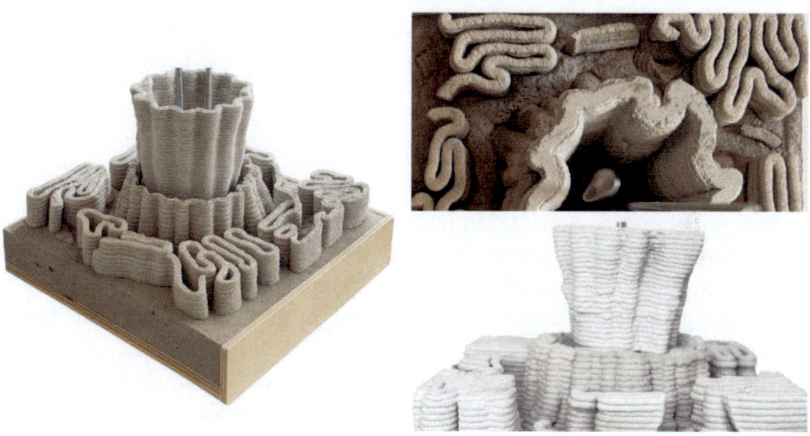

Fig. 12. Final eco-reef NBS prototype

Week 3—A uniform ochre-brown film coats virtually every surface, softening ridge edges and infilling valley channels. This film—confirmed the development of bacterial-diatom biofilm (Ruhl & Dixson, 2020). Sparse strands of filamentous algae and trapped sand grains occupy crevices, indicating the structure's growing role as a sediment trap and biological substrate.

Across both campaigns the chain-anchored bases remained secure, and no structural damage or displacement was observed (Fig. 13). The accelerated shift from a sterile print to a microbially conditioned surface within three weeks demonstrates that the mussel-shell-enriched cementitious mix effectively jump-starts reef-like colonization under real-world hydrodynamic conditions (Temmink et al., 2022). Continuous photographic and video monitoring confirm that these early-succession stages provide the foundational biofilm needed for later settlement of invertebrates and larger reef assemblages, validating the design's ecological performance in situ.

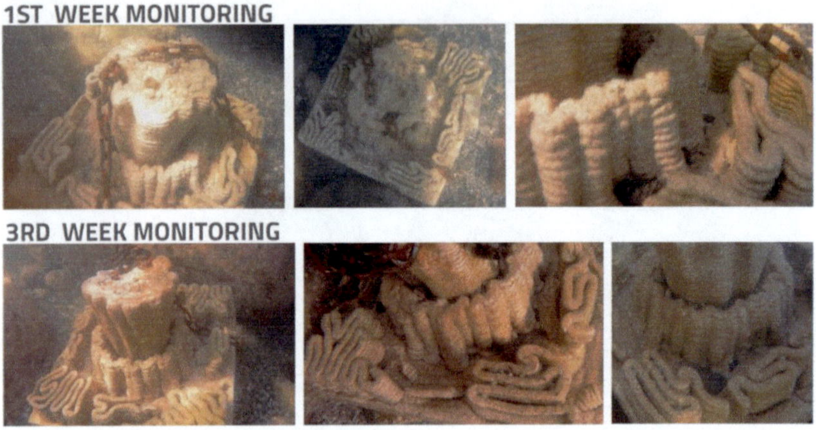

Fig. 13. Eco-Reef module after 1 week (top row) versus 3 weeks (bottom row) underwater: progressive development of a uniform biofilm, softened ridge edges, and initial sediment trapping

3.3 Decision-Support System Layer

The Decision Support System (DSS) layer ensures seamless interoperability between digital and physical models, providing customized and contextualized outputs accessible through an intuitive platform. This layer supports real-time synchronization and dynamic management of different data streams, promoting scalable and replicable decision-making processes.

By uniting high-fidelity digital prototypes, eco-reef prototypes and an open, standards-based DSS, the Digital Atlas establishes a replicable blueprint for evidence-driven climate adaptation. Its novelty rests on three pillars: (i) a parametric pipeline that translates IPCC macro-scenarios into site-specific digital protype; (ii) a phygital verification loop where additive-manufactured artefacts empirically ground the simulations; and (iii) an interoperability stack that democratises complex data while safeguarding traceability. Collectively, these factors position the Atlas as an operational tool to apply regenerative design strategies seeking to combine, ecosystem regeneration and carbon-neutral transition within a unified decision framework.

4 Discussions

The findings presented underline the importance of a tightly integrated digital-to-physical workflow in enhancing climate-resilient heritage conservation and ecosystem regeneration. The use of parametric models allowed dynamic adaptation to diverse climate scenarios, offering significant flexibility and precision in designing interventions. The deployment of Eco-Reef modules illustrated the practical benefits of incorporating bio-based sustainable materials into adaptive strategies. However, the effectiveness of these interventions in broader contexts requires careful consideration of local ecological conditions, long-term maintenance requirements, and community engagement in "Living Lab" activities. Additionally, while the Digital Atlas successfully demonstrated interoperability and adaptability, ongoing refinement of data integration processes remains critical. In addition to technical performance, effective governance is essential: local authorities, heritage agencies and community groups must agree on data stewardship, maintenance responsibilities and decision-rights to ensure long-term success. The Atlas's open standards support horizontal scalability, cloning the schema and injecting new aerial surveys is sufficient to initialize the Atlas for another coastal community.

5 Conclusions—Significance of the *Digital Atlas* framework

The study confirms that **the structure, information, interoperability of physical and digital models inside a Digital Atlas** provides a replicable methodology to address global climate trajectories into local, regenerative design-based proposal. The platform's innovation aspects underlies three interlocking advances.

First, a **multi-scalar parametric risk assessment method** that converts heterogeneous inputs—IPCC projections, digital prototype into hazard and performance indicators without manual file-handling. The automatic recomputation of full chains when a single EPW is updated transforms the Atlas from a static survey into a self-updating decision engine, critical for the non-linear nature of climate science.

Second, the **phygital verification loop** bridges the epistemic gap between prediction and implementation. Sensor-rich prototypes feed empirical coefficients back into the digital layer, progressively tightening model fidelity. This closed-loop structure underwrites design confidence: whether selecting pore diameters for coastal defences or louvre depths for heritage façades, stakeholders act on metrics that have been both simulated and measured.

Third, an **interoperability stack**, REST/JSON APIs, ISO 19115 metadata, extended CityGML/IFC semantics, ensures that every actor, from local citizens to regional planners, accesses information at the appropriate granularity. The same architecture supports horizontal scaling: cloning the schema and injecting new aerial surveys is sufficient to initialise the Atlas for another coastal community.

Taken together, these pillars demonstrate that a rigorously engineered coupling of digital twins, additive manufacturing and open data can reconcile ecological regeneration, cultural continuity and carbon-neutral targets. Upcoming work will integrate GNSS-derived subsidence fields and Monte-Carlo ensemble runs, yet the current results already position the Digital Atlas as a practical template for Mediterranean coastlines—and beyond—striving for resilient, low-carbon futures.

Acknowledgment. This study was carried out within the Innovation Ecosystem Tech4You—"Technologies for climate-change adaptation and quality-of-life improvement," Mission 4, Component 2, Investment 1.4 of Italy's National Recovery and Resilience Plan (PNRR), funded by the

European Union—NextGenerationEU (project code ECS_00000009, CUP C33C22000290006). The authors gratefully acknowledge Actions 8 and 9 of Pilot Project 4.7.1 ("Open platform phygital space for advanced, dynamic co-design of built-environment interventions"), led by Prof. Consuelo Nava, which provided the coordination and organization of the team, the research theoretical framework and the capital investment for the prototype equipment installed in the ABITAlab Prototyping Laboratory.

References

Abbass, K., Qasim, M.Z., Song, H., Murshed, M., Mahmood, H., Younis, I.: A review of the global climate change impacts, adaptation, and sustainable mitigation measures. Environ. Sci. Pollution Res. **29**, 42539–42559 (2022)

ABITAlab Prototyping Laboratory. Prototipi e Dimostratori:. ABITAlab – Università Mediterranea di Reggio Calabria. Disponibile online: https://www.abitalab-unirc.com/prototipi-dimostratori. Accessed 22 July 2025

Blanco, E., Pedersen Zari, M., Raskin, K., Clergeau, P.: Urban ecosystem-level biomimicry and regenerative design: linking ecosystem functioning and urban built environments. Sustainability. **13**, 404 (2021)

BPIE – Buildings Performance Institute Europe: Digital Building Logbooks: A Key Tool for Decarbonising the EU Building Stock. BPIE, Brussels (2023)

Camba, C., Mier, J.L., Carral, L., Lamas, M.I., Álvarez, J.C.: Erosive degradation study of concrete augmented by mussel shells for marine construction. J. Marine Sci. Eng. **9**(10), 1087 (2021)

Churkina, G., Reyer, C.P.O., Organschi, A., et al.: Buildings as a global carbon sink. Nature Sustain. **3**, 269–276 (2020)

Conti, M.E., Battaglia, M., Calabrese, M., Simone, C.: Fostering sustainable cities through resilience thinking: the role of nature-based solutions (NBSs). *Sustainability.* **13**, 12875 (2021)

D'Amico, B., Pomponi, F., Hart, J.: Electrification of construction equipment for low-carbon building sites. J. Clean. Prod. **357**, 131901 (2022)

El Biriane, M., Barbachi, M., Bakkali, S.E., Mahi, M.E.: Properties of sustainable concrete with mussel shell waste powder. Open Civil Eng. J. **14**, 350–362 (2020)

European Commission: Proposal for Recast of the Energy Performance of Buildings Directive (EPBD), COM (2021) 802 Final. European Commission, Brussels (2021)

Gallina, V., Torresan, S., Critto, A., Sperotto, A., Glade, T., Marcomini, A.: A review of multi-risk methodologies for natural hazards: consequences and challenges for climate-change impact assessment. J Environ. Manag. **168**, 123–132 (2016)

He, B., Yuan, X., Qian, S., Li, B.: Carbon neutrality: a review. J. Comput. Inf. Sci. Eng. **23**, 060809 (2023)

Hoseinzadeh, H., Mehraein, M., Tavakoli, M., Heidarnejad, P.: Investigation of zero-energy and zero-carbon buildings for the hot-dry climate of Iran. Case Stud. Thermal Eng. **45**, 102739 (2023)

Kim, D.Y.: A design methodology using prototyping based on digital–physical models in the architectural design process. Sustainability. **11**, 4416 (2019)

Korniejenko, K., Oliwa, K., Gądek, S.: Additive manufacturing techniques in artificial reef construction: materials, processes and ecological impact. Appl. Sci. **15**(8), 4216 (2025)

Lehmann, S.: Low Carbon Cities: Transforming Urban Systems. Routledge, London (2019)

Mangano, G., Leuzzo, A.: Co-design e tecnologie abilitanti: 15 esperienze progettuali per la ricerca di frontiera e le comunità in transizione. Aracne, Rome (2022)

Nava, C.: Tecnologie emergenti per il progetto rigenerativo. Cinque questioni teoriche su innovazione e sostenibilità dalla prassi della ricerca di frontiera. Aracne, Pomezia (2019)

Nava, C., Filice, F., Laganà, D.: Proposta operativa tra protocolli e indirizzi per la città di Reggio Calabria. In: Nava, C., Filice, F., Laganà, D. (eds.) ReKap: Il progetto di adattamento climatico come dispositivo per la rigenerazione urbana, pp. 94–107. Aracne, Rome (2024) ISBN 979-12-218-1200-8

Paiho, S., Ala-Juusela, M., Caputo, P., Ferigo, F.: BIM-based lifecycle carbon assessment and optimisation of smart buildings. Energy Build. **259**, 111912 (2022)

Pomponi, F., Moncaster, A.: Circular economy for the built environment: a research framework. J. Clean. Prod. **143**, 710–718 (2017)

Roberts, S.H., Axon, C.J., Goddard, N.H., Foran, B.D., Warr, B.S.: Modelling socio-economic and energy data to generate business-as-usual scenarios for carbon emissions. J. Clean. Prod. **207**, 980–997 (2019)

Ruhl, S., Dixson, D.L.: 3-D-printed coral reefs as a restoration tool: influence of design on coral larval settlement. PLoS ONE. **15**(4), e0231153 (2020)

Stel'makh, S.A., Shcherban', E.M., Beskopylny, A.N.: Composition, technological and microstructural aspects of concrete modified with finely ground mussel shell powder. Materials. **16**, 82 (2023)

Tabassum, R.R., Park, J.: Development of a building evaluation framework for biophilic design in architecture. Buildings. **14**, 3254 (2024)

Temmink, J.H.M., van der Heide, T., Oteman, B., et al.: Initiating and upscaling mussel-reef establishment with life-cycle-informed restoration: successes and future challenges. Ecol. Eng. **175**, 106496 (2022)

Valenzuela Matus, I., Góis, J., Vaz-Pires, P., Alves, J.L.: Artificial reefs through additive manufacturing: design, purposes and fabrication processes. Rapid Prototyping J. **30**(3), 505–519 (2024)

Ye, J., Yang, F., Wang, J., Meng, S., Tang, D.: A bibliometric review of green-building policies. Buildings. **14**, 2607 (2024)

Solar Energy-Driven Design for a Building Demonstrator through the Restoration of Ruins in Bova (RC)

Giuseppe Mangano(✉) and Eliana Catalano

ABITAlab, dAeD, Mediterranea University of Reggio Calabria, Reggio Calabria, Italy
{giuseppe.mangano,eliana.catalano}@unirc.it

Abstract. The energy transition pathway requires integrated decarbonization strategies based on renewable energy production and storage, oriented towards "zero" emission models in line with European directives. These strategies can be implemented across multiple scales through an energy-driven urban design approach that combines adaptive design principles with the organizational framework of Renewable Energy Communities, aiming to meet energy demand within adaptable, decentralized, and resilient spatial configuration. This study applies a Pre-Design workflow to the restoration of disused buildings and ruins in the historic center of Bova (RC), reconfigured as a technological demonstrator and productive node within a local REC. The novelty lies in the integrating heritage renovation with parametric and microclimatic modeling to transform abandoned cultural assets into active energy infrastructure, bridging the gap between conservation and climate-neutral innovation. The workflow integrates advanced digital tools for contextual analysis, enabling optimized energy design scenarios that reduce carbon emissions and maximize renewable energy generation, while adapting solutions to the specific microclimatic and morphological conditions of the urban cluster. The proposed approach incorporates passive and active technologies into a replicable model which, beyond architectural rehabilitation, includes experimental implementation within Positive Energy Districts at a supra-local scale. Adoption of this model can play a decisive role in accelerating climate-neutral pathways, while preserving cultural and identity capital, and fostering innovation that delivers substantial gains in efficiency, environmental quality, and functional integration within hybrid "historic-energy" clusters. By positioning heritage buildings as energy-positive demonstrators, the study provides a transferable methodology that extends the role of cultural heritage from passive conservation to proactive climate action.

Keywords: Decarbonization · Cultural Heritage · Regenerative Pre-Design · Energy optimization workflow · Energy-driven urban design · Integrated interventions

C. Nava et al. (Eds.): *Climatic and Structural Safety in Multi-Hazard Regime of Cultural and Natural Heritage*, GREEN, pp. 328–337, 2026.
https://doi.org/10.1007/978-3-032-11926-1_20

1 Introduction

The transition towards low-carbon urban and territorial models demands a synergistic integration of decarbonization strategies, cultural heritage valorization, and technological innovation (European Commission, 2019). Within this context, adaptive design approaches have emerged as effective frameworks for addressing environmental, energy, and social challenges simultaneously, while enhancing users' psycho-physical well-being and the overall quality of urban environments (Lucchi, 2022). When applied to the regeneration of historic settings, these approaches successfully couple heritage preservation with the integration of advanced energy systems, fostering virtuous cycles of climate change adaptation and mitigation (Mangano & Catalano, 2024).

Recent advances in the concept of energy-driven urban design have reinforced an integrated understanding of urban form as a decisive factor in determining energy and environmental performance (Fonseca et al., 2023). Simulation-based workflows—supported by multi-objective optimization and parametric modelling—allow the identification of configurations that minimize energy demand while maximizing the potential for renewable self-production (Mangano, 2024). In this perspective, the combined use of morphological, climatic, and environmental analyses becomes a prerequisite for the development of design scenarios capable of meeting the demands of the energy transition (IEA, 2021).

Insights from the most recent WEF 2023 Report (WEF, 2023) highlight the urgency of addressing biodiversity-related challenges, particularly relevant in the inland areas of the Grecanica region—the focus of the present study. These areas also hold potential in mitigating migration-related emergencies, owing to their capacity to be repopulated and revitalized. Such conditions can act as catalysts for sustainable development and innovation. Renewable Energy Communities (RECs), with their ability to proactively navigate the ongoing transition—especially under climate change pressures—offer a platform for co-design processes, multiscale analyses, network-building, and bottom-up innovation. By acting as demonstrators of novel governance and technological models, RECs can enable the replication and scaling of these processes to neighboring territories, forming interconnected networks of energy communities (Mangano, 2024).

At the European level, RECs were formally introduced through Directive (EU) 2018/2001 and transposed into Italian legislation via Legislative Decree 199/2021 (European Commission, 2019). These communities represent cooperative governance models capable of delivering environmental, economic, and social benefits through the localized sharing of renewable energy. When implemented in historic contexts and remote inland areas, RECs can facilitate the adaptive reuse of ruins and underutilized buildings, transforming them into active hubs for renewable energy generation and community cohesion.

The present case study is situated within a high-value cultural landscape undergoing progressive depopulation and structural decay. This work applies an integrated solar energy-driven design methodology within the framework of a local-scale REC, assessing its energy, environmental, and social performance. By combining adaptive reuse principles with renewable energy strategies, the study aims to provide a replicable model for sustainable regeneration in heritage-rich but vulnerable territories.

2 Methodological Approach

The research adopts an integrated and adaptive solar-energy-driven methodology (Fig. 1) aimed at reducing annual energy demand, increasing on-site renewable energy production, and enhancing environmental quality, while ensuring full compatibility with the morphological, historical, and landscape constraints of the intervention context. The methodological framework is applied within the Pre-Design phase and supports strategies for the establishment of Renewable Energy Communities (RECs) and Positive Energy Districts (PEDs) through the implementation of Regenerative Digital Design and advanced predictive models. Within this approach, the urban energy system is conceptualised as the set of processes, infrastructures, and technologies that enable the acquisition, transformation, distribution, and final use of energy within an urban area. This system integrates three main components: the agents, such as users and managers; the objects, namely the infrastructures and technologies that ensure the operation of energy flows; and the environment, which provides the necessary resources. In this framework, the so-called "layer of energy systems" refers to the spatial dimension occupied by infrastructural elements responsible for the transport, storage, conversion, distribution, and on-site generation of energy. Although often imperceptible to inhabitants, this layer constitutes the operational backbone of the urban energy network.

2.1 Operational Phases

The first phase operates predominantly at the urban cluster scale [CL] and begins with the collection and definition of the design context. It integrates climatic datasets, derived from EPW files, with future projections based on IPCC climate scenarios (AR6, SSP5-8.5), in order to assess potential impacts of climate change on the intervention site. Detailed morphological surveys, carried out through photogrammetric techniques and three-dimensional modelling, are combined with the mapping of existing energy infrastructures and in-depth historical-architectural documentation, allowing for the identification of typological, constructive, regulatory, and environmental constraints. Based on these analyses, Key Performance Indicators (KPIs) are defined to establish a direct correlation between the regulatory framework, building energy performance, efficiency standards, and environmental protocols, in full compliance with European and Italian directives. These indicators guide the definition of the REC/PED organisational model, setting targets for reducing annual energy demand, maximising renewable energy production, and integrating with the historical and environmental context. The second phase takes place at the building scale [BL] and employs Regenerative Digital Design methods and advanced predictive models to optimise both passive and active components of the building form. The process focuses on refining building geometry to improve the form factor and reduce heat-dissipating surfaces, configuring the architectural and technological integration of photovoltaic systems—with particular attention to Building Integrated Photovoltaics (BIPV)—while ensuring morphological adaptation in accordance with heritage preservation requirements. The use of parametric tools enables the rapid generation and iteration of alternative design scenarios, while environmental simulations assess microclimatic performance. Simulation outputs are incorporated into a multi-objective optimisation process aimed at identifying configurations that achieve the best balance between energy performance, mitigation of the Urban Heat Island (UHI) effect, and increased permeable surfaces, leading to the selection of the most effective design solution.

The final phase concerns the feasibility verification and testing of the selected hybrid building model against climatic scenarios, ensuring compliance with the KPIs defined in the first phase and adherence to European and Italian energy directives. Performance evaluation is carried out across multiple time horizons (2024, 2030, 2050) and climate projections (IPCC SSP5-8.5) to confirm the project's resilience under future conditions. The methodology aligns with the concept of Zero-Heat Building (nZHB) within the broader NZEB–PEB strategy (He, 2019), with the aim of optimising the building envelope, increasing renewable energy production, and achieving greenhouse gas emissions neutrality (Qui et al., 2024). In this context, the optimisation process also addresses the reduction of UHI effects by combining energy conservation strategies with measures to limit anthropogenic heat emissions, promoting replicable, low-carbon design solutions for historic and climate-sensitive urban contexts.

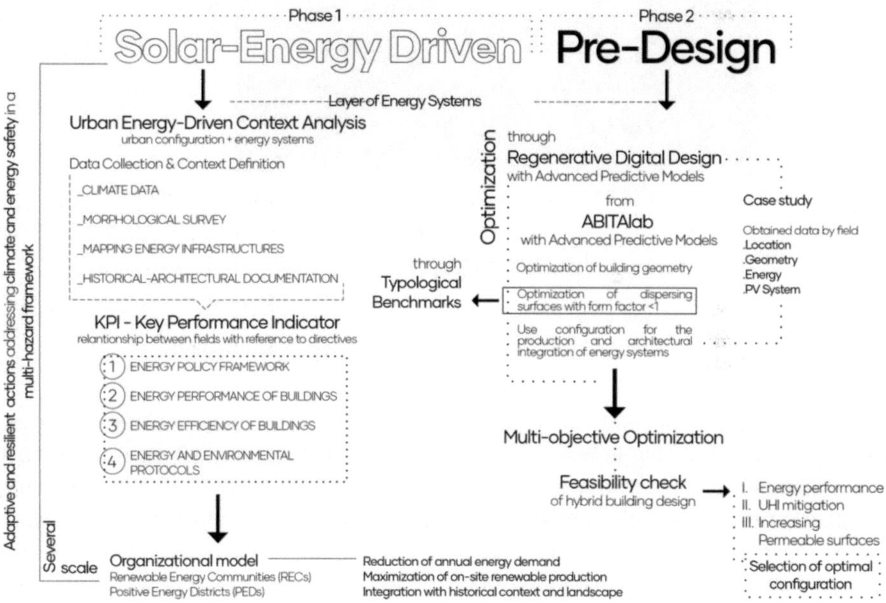

Fig. 1. Pre-design workflow for solar-energy driven optimization through regenerative digital design. Source: Elaboration by PhDs E. Catalano, ABITAlab dAeD UniRC

3 Application

The application focused on a set of disused buildings and ruins located in the historic centre of Bova, selected within a cluster for the establishment of a RECs. As illustrated in Fig. 2, the Living Lab scheme clarifies the relationship between ruins, public buildings, and the wider cluster, showing their potential as active energy nodes within the REC. The main objective of the solar energy-driven approach is to transform existing buildings, even those in a state of decay, into structures capable of producing, storing, and distributing solar energy, thereby contributing to the creation of an autonomous energy system integrated within the community context (Nava et Mangano, 2023). The design process began with an in-depth contextual analysis, employing advanced photogrammetric surveying and parametric modeling technologies to reconstruct the buildings and their surrounding environment digitally.

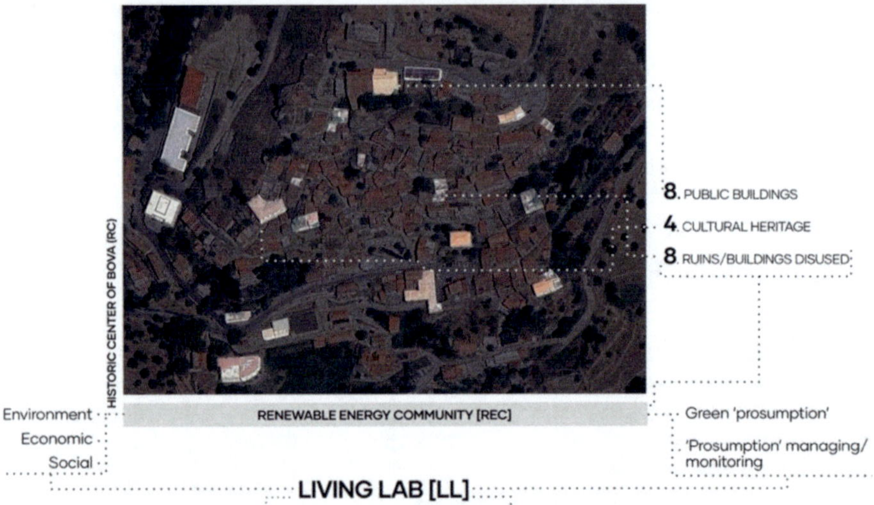

Fig. 2 Scheme of the Living Lab for CERs with application to the case of the historic centre of Bova (RC). Source: Elaboration by PhDs E. Catalano, ABITAlab dAeD UniRC

The high-precision three-dimensional modeling enabled the assessment of critical factors for solar integration—solar radiation availability, seasonal shadow patterns, and orientation relative to the solar path. These predictive analyses directly informed design decisions: photovoltaic modules were prioritized on south-facing roofs, and semi-transparent panels were adopted on facades where simulations indicated both high solar exposure and daylight compatibility, maximizing photovoltaic system efficiency. This approach is applied at every stage of the design process, considering the solar behavior of buildings in relation to their orientation and the integration of photovoltaic systems. Photovoltaic modules were designed for installation on roofs, façades, and other architectural elements without compromising the historical integrity of the buildings. Innovative solutions were also considered, such as semi-transparent photovoltaic panels and green roofs integrated with photovoltaic modules, which not only generate energy but also enhance environmental comfort, with a positive effect on the surrounding microclimate. Simulations explored various building configurations and their associated energy storage potential, making use of advanced storage technologies to ensure a continuous energy supply even during hours without solar irradiation. In particular, the feasibility of implementing smart grids—intelligent networks enabling efficient and decentralized management of energy production, distribution, and storage—was examined to reduce dependence on external energy sources. A key element is its ability to integrate solar energy with other sustainable solutions, such as natural ventilation and passive lighting. The buildings were designed with layouts that promote natural summer cooling and passive winter heating, thereby further reducing energy demand and improving indoor comfort. In addition, photovoltaic windows and solar walls were employed to increase the solar energy collection surface, improving overall energy efficiency without compromising the aesthetic and historical quality of the buildings. The project also included an iterative optimization phase, enabling progressive refinement of the adopted energy solutions through advanced simulations to test energy performance under current and future climate scenarios. The integration of projected climate scenarios for 2030 and 2050, in line with IPCC simulations, allowed for the design of buildings resilient to climate change, anticipating variations in solar radiation and meteorological patterns. The project generated an integrated model that harmonizes technological innovation with the enhancement of cultural heritage, thereby contributing to the creation of a decentralized and sustainable local energy system with a positive impact at both the community and environmental scales.

4 Results

The mapping of buildings suitable for the establishment of the Renewable Energy Community (REC) of Bova identified eight public buildings, four with historical value, and eight ruins or abandoned structures (Fig. 2). The optimised configuration—developed using the Rose Energy Community platform and validated through simulations with City Energy Analyst (Figs. 3, 4) —assigns distinct functional roles: historic buildings as consumers, public buildings as prosumers, and rehabilitated ruins as photovoltaic producers, equipped with monocrystalline silicon modules rated at 410 Wp (1 kWp $\approx$ 7 m^2). Fig. 3 visualizes the solar potential across the cluster, highlighting rooftops with the highest exposure and supporting the decision to prioritize PV installations on these surfaces. Fig. 4 complements this by mapping the spatial distribution of demand and the temporal peaks, which guided storage sizing and confirmed the allocation of consumers, prosumers, and producers. The renewable potential assessment revealed high solar intensity on specific rooftops, while the demand analysis provided the spatial and temporal distribution of energy loads, with differentiated hourly consumption profiles by building use type. As shown in Fig. 4, the peaks in temporal demand guided the sizing of storage systems, while the spatial distribution confirmed the functional roles of consumers, prosumers, and producers. Results estimate an annual production of 467.70 MWh, of which 26.56 MWh is self-consumed and 7.80 MWh is fed into the grid, with an estimated reduction of 227.77 tCO_2. From an economic perspective, the system ensures an energy valorisation of €21,620, and a shared energy contribution of €167.60. Microclimatic analyses show a decrease in average summer indoor temperatures and an increase in green surface areas, contributing to the mitigation of the urban heat island effect.

The economic estimate, obtained through the Rose Energy Community software, is presented as a preliminary assessment supporting the overall evaluation of the REC model. As detailed economic modelling does not fully fall within the disciplinary scope

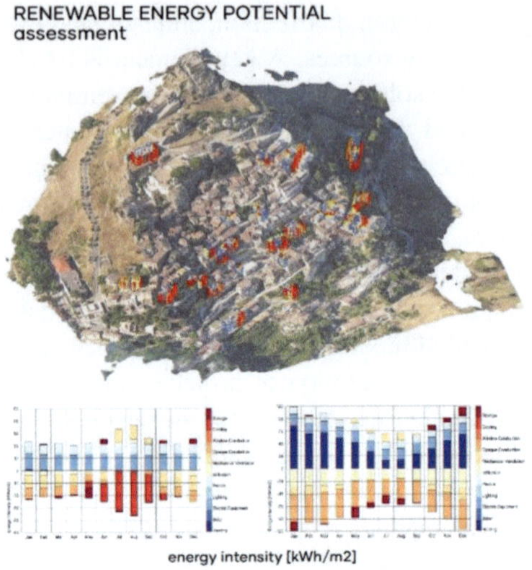

Fig. 3. Simulation of Solar Energy Potential using solar collectors, photovoltaic panels and solar-thermal panels. Source: Elaboration by PhDs E. Catalano, ABITAlab dAeD UniRC

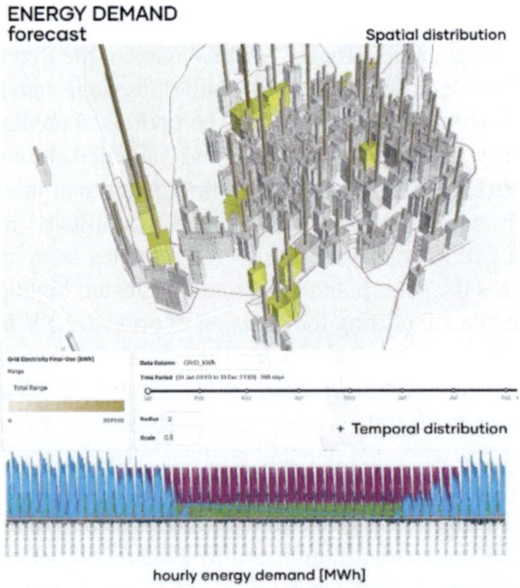

Fig. 4. Energy Demand forecast according to spatial distribution in the cluster of Bova and Temporal distribution of peak energy demand. Source: Elaboration by PhDs E. Catalano, ABITAlab dAeD UniRC

of this research, the analysis is limited to indicative values of energy valorisation and shared contributions. The energy balance shows that the ruins and abandoned structures, once renovated, offer the highest potential for photovoltaic production. Consequently,

Table 1 Energy Potential, Economic Estimate and Environmental Impact for the simulated REC configuration in Bova. Source: elaboration by authors

Energy Potential	
Energy Produced	467,70 MWh
Energy Input	465,95 MWh
Energy Consumed	26,56 MWh
Energy Withdrawn	18,76 MWh
Direct Self-Consumption	7,80 MWh
Economic Estimate	
Shared Energy Contribution	167,60 €
Community Revenue	1.138,09 €
Energy Input Valorisation	21.620 €
Self-Consumption Rate	1,67%
Shared Energy Rate	3,40%
Shared Energy Contribution	167,60 €
Environmental Impact	
CO_2 avoided	227,77 t

they were designated as primary energy producers within the REC configuration, while historic buildings, with lower production capacity, were assigned a complementary role.

5 Discussions

The results confirm the strategic role of energy optimisation strategies applied to historic building stock in advancing decarbonisation and the transition towards highly sustainable energy models. In the Bova (RC) case study, the integration of high-efficiency photovoltaic systems, combined with the optimisation of building morphology and the use of advanced microclimatic simulations, enabled the achievement of a positive energy balance, generating a surplus that can be fed into the local Renewable Energy Community (REC) network. Multi-objective simulations demonstrated that compact morphological configurations, integrated with passive solutions such as natural ventilation, shading systems, green roofs, and green façades, can significantly reduce primary energy demand and improve indoor thermal comfort, while simultaneously producing beneficial effects on urban microclimatic quality. Based on these findings, the design adopted compact layouts integrated with passive solutions ensuring that each selected measure corresponded to a specific performance improvement revealed by the simulations The adoption of BIPV (Building Integrated Photovoltaics) technologies, compatible with historical and landscape constraints, proved the feasibility of combining cultural heritage enhancement with technological innovation, while preserving the formal integrity of the buildings. The adopted methodological framework—based on parametric digital tools and participatory processes inspired by the Living Lab model—shows high potential for replicability and scalability in similar contexts characterised by high climate vulnerability, depopulation phenomena, and degradation of the built heritage.

6 Conclusions

The study highlighted how the integration of regenerative digital workflows, geometric optimisation, and both active and passive solar strategies constitutes an effective approach for transforming abandoned buildings into productive nodes within local energy networks. The analysis, conducted at both the urban cluster scale and the individual building scale, demonstrated the capacity of the proposed solutions to reduce CO_2 emissions, increase energy self-sufficiency, and enhance environmental comfort, thereby contributing concretely to achieving the climate neutrality targets set for 2050. The combination of predictive climate data, advanced environmental simulations, and historical-landscape compatibility criteria enabled the development of a replicable intervention model capable of integrating energy efficiency, climate resilience, and cultural heritage preservation. Looking ahead, the application of this methodology to other vulnerable historic contexts could foster the creation of positive-energy districts and networks of Renewable Energy Communities, thereby amplifying the environmental, economic, and social impact of urban regeneration strategies.

Acknowledgment. This study is co-authored by G. Mangano and E. Catalano. G. Mangano and E. Catalano are the authors of paragraph 4, 5 and 6. G. Mangano is the author of paragraphs 1. E. Catalano is the author of paragraphs 2 and 3. This study was carried out within the Innovation Ecosystem Tech4You– "Technologies for climate-change adaptation and quality-of-life improvement," Mission 4, Component 2, Investment 1.4 of Italy's National Recovery and Resilience Plan (PNRR), funded by the European Union—Next Generation EU (project code ECS_00000009, CUP C33C22000290006). The paper illustrates the results of ABITAlab's research for the NRRP 'Tech4You. Pilot Project 4.7.1 Open 'phygital space' user profiling platform for advanced and dynamic codesign of built and ex-novo interventions' (P.I. Prof. C. Nava): Action 8 'Atlas of predictive models and adaptive technologies on climate scenarios and reliability of the artifacts of historic heritage buildings, with studies and elaborations of advanced parametric design [resources-data-information]' and Action 9 'Enabling Technologies and Additive Manufacturing [3D printing] of prototypes of systems and components on testing cases [at settlement and building scale]'. For further information, visit www.abitalab.unirc.it.

References

European Commission: Clean Energy for all Europeans. Publications Office, Act 2019 (2019). https://doi.org/10.2833/9937

He, B.J.: Towards the next generation of green building for urban heat island mitigation: zero UHI impact building. Sustain. Cities Soc. **50**, 101647 (2019). https://doi.org/10.1016/j.scs.2019.101647

International Energy Agency (IEA): CO_2 Emissions in 2022 (2023a). https://www.iea.org/reports/co2-emissions-in-2022

International Energy Agency (IEA): World Energy Outlook 2023 (2023b) https://www.iea.org/reports/world-energy-outlook-2023
IPCC: Synthesis report. Contribution of working groups I, II and III to the sixth assessment report of the intergovernmental panel on climate change. In: Core Writing Team, Lee, H., Romero, J. (eds.) Climate Change 2023. IPCC, Geneva, Switzerland, pp. 35–115 (2023). https://doi.org/10.59327/IPCC/AR6-9789291691647
Lucchi, E.: Energy efficiency of historic buildings. Buildings. **12**, 200 (2022). https://doi.org/10.3390/buildings12020200
Mangano, G.: Tecnologie avanzate per le Comunità Energetiche. Innovazione Green nella Ricerca Applicata, Aracne Editrice (2024)
Mangano, G., Catalano, E.: Energy-Environmental Pre-Design for Decarbonization. Application of a Regenerative Digital Design Workflow in the Cultural Heritage Cluster of Bova (RC), CrossMed2024 (2024)
Nava, C., Mangano, G.: Living Lab for the design and activation of energy communities in the inner areas. J. Technol. Architect. Environ. ***26***, 120–130 (2023). https://doi.org/10.36253/techne-14495
Qiu, Z., Yong, Q., Wang, J., Liao, L., Yu, B.: A multi-objective optimization framework for performance-based building design considering the interplay between buildings and urban environments. Energy Conv. Manag. **315**, 118793 (2024). https://doi.org/10.1016/j.enconman.2024.118793
Shi, Z., Fonseca, J.A., Schlueter, A.: A review for simulation-based urban form generation and optimization modeling for energy-driven urban design at the district scale. Build. Environ. (2017). https://doi.org/10.1016/j.buildenv.2017.05.006
World Economic Forum: Global Risks Report (2023)

Correction to: Climatic and Structural Safety in Multi-Hazard Regime of Cultural and Natural Heritage

Consuelo Nava, Aurora Angela Pisano, Giuseppe Mangano, and Francesca Giglio

Correction to:
Chapters 5 and 6 in: C. Nava et al. (Eds.): *Climatic and Structural Safety in Multi-Hazard Regime of Cultural and Natural Heritage*, GREEN, https://doi.org/10.1007/978-3-032-11926-1

In the original version of the book, the following belated corrections are to be incorporated:

In chapter 5 "Interval modelling and propagation of uncertainties in masonry structures, Figure processed incorrectly due to alt text injection ("q" added).

The erratum book has been updated with the changes.

In the original version of the book, the following belated corrections are to be incorporated:

In chapter 6 "Exploring the impact of climate change on the progressive collapse resistance of an existing reinforced concrete building, Author corrections provided in annotated pdf were missed during production and not carried forward in subsequent stages.

The erratum book has been updated with the changes.

The updated version of these chapters can be found at
https://doi.org/10.1007/978-3-032-11926-1_5
https://doi.org/10.1007/978-3-032-11926-1_6

C. Nava et al. (Eds.): *Climatic and Structural Safety in Multi-Hazard Regime of Cultural and Natural Heritage*, GREEN, pp. C1–C2, 2026.
https://doi.org/10.1007/978-3-032-11926-1_21

Author Index

C. Nava et al. (Eds.): *Climatic and Structural Safety in Multi-Hazard Regime of Cultural and Natural Heritage*, GREEN, p. 339, 2026.
https://doi.org/10.1007/978-3-032-11926-1

MIX
Papier aus verantwortungsvollen Quellen
Paper from responsible sources
FSC® C105338
FSC
www.fsc.org

If you have any concerns about our products,
you can contact us on
ProductSafety@springernature.com

In case Publisher is established outside the EU,
the EU authorized representative is:
Springer Nature Customer Service Center GmbH
Europaplatz 3, 69115 Heidelberg, Germany

Printed by Libri Plureos GmbH
in Hamburg, Germany